Agricultural Policies in a New Decade

Agricultural Policies in a New Decade was written in preparation for the 1990 Farm Bill in the United States in collaboration with the Food and Agriculture Committee of the National Planning Association. Originally published in the same year, this study aims to inform on the key issues affecting the new bill including budget austerity, the effects of agriculture on environmental quality, the safety and security of the U.S. food supply and international markets and trade policy. This title will be of interest to students of Environmental Studies and policy makers.

Agricultural Policies in a New Decade

Edited By
Kristen Allen

RFF PRESS
RESOURCES FOR THE FUTURE

First published in 1990
by Resources for the Future, Inc.

This edition first published in 2016 by Routledge
2 Park Square, Milton Park, Abingdon, Oxon, OX14 4RN
and by Routledge
711 Third Avenue, New York, NY 10017

Routledge is an imprint of the Taylor & Francis Group, an informa business

Publisher's Note
The publisher has gone to great lengths to ensure the quality of this reprint but points out that some imperfections in the original copies may be apparent.

Disclaimer
The publisher has made every effort to trace copyright holders and welcomes correspondence from those they have been unable to contact.

A Library of Congress record exists under LC control number: 89043667

ISBN 13: 978-1-138-94907-2 (hbk)
ISBN 13: 978-1-315-66907-6 (ebk)

AGRICULTURAL POLICIES IN A NEW DECADE

NATIONAL CENTER
FOR FOOD AND
AGRICULTURAL
POLICY
Resources for the Future

and

FOOD AND AGRICULTURE
COMMITTEE
National Planning Association

ANNUAL POLICY REVIEW
1990

AGRICULTURAL POLICIES IN A NEW DECADE

KRISTEN ALLEN
Editor

Resources for the Future
and
National Planning Association
Washington, D.C.

Published by

Resources for the Future	National Planning Association
1616 P Street, N.W.	1424 16th Street, N.W.
Washington, D.C. 20036	Washington, D.C. 20036

Library of Congress Catalog Card Number 89-43667
ISBN 0-915707-54-3

ABOUT THE NATIONAL CENTER FOR FOOD AND AGRICULTURAL POLICY

The National Center for Food and Agricultural Policy was established in 1984 by Resources for the Future, an independent, nonprofit research and educational organization. The National Center's mission is to conduct research, policy analysis, and outreach on emerging policy issues, with particular emphasis on the linkages of U.S. agriculture with environmental quality and natural resources, food safety and quality, and health, with consideration of the international dimensions. Its purpose is to contribute to the public debate on policy issues affecting the food and agricultural system.

ADVISORY COMMITTEE

RESOURCES FOR THE FUTURE

Robert W. Fri, *President*
Paul R. Portney, *Vice President*
Edward F. Hand, *Secretary-Treasurer*

RESOURCES FOR THE FUTURE (RFF) is an independent nonprofit organization that advances research and public education in the development, conservation, and use of natural resources and in the quality of the environment. Established in 1952 with the cooperation of the Ford Foundation, it is supported by an endowment and by grants from foundations, government agencies, and corporations. Grants are accepted on the condition that RFF is solely responsible for the conduct of its research and the dissemination of its work to the public. The organization does not perform proprietary research.

RFF research is primarily social scientific, especially economic. It is concerned with the relationship of people to the natural environmental resources of land, water, and air; with the products and services derived from these basic resources; and with the effects of production and consumption on environmental quality and on human health and well-being. Grouped into four units—the Energy and Natural Resources Division, the Quality of the Environment Division, the National Center for Food and Agricultural Policy, and the Center for Risk Management—staff members pursue a wide variety of interests, including forest economics, natural gas policy, multiple use of public lands, mineral economics, air and water pollution, energy and national security, hazardous wastes, the economics of outer space, climate resources, and quantitative risk assessment. Resident staff members conduct most of the organization's work; a few others carry out research elsewhere under grants from RFF.

Resources for the Future takes responsibility for the selection of subjects for study and for the appointment of fellows, as well as for their freedom of inquiry. The views of RFF staff members and the interpretation and conclusions of RFF publications should not be attributed to Resources for the Future, its directors, or its officers. As an organization, RFF does not take positions on laws, policies, or events, nor does it lobby.

ABOUT THE FOOD AND AGRICULTURE COMMITTEE

The Food and Agriculture Committee (FAC) of the National Planning Association has been active in public policy analysis over a period of nearly five decades. Its membership includes representatives from all segments of the agriculture and agribusiness sector. [A full list of the membership of the FAC appears as the appendix to this volume.] By symbolizing the tremendous scope and complexity of this important sector, the Committee is in a unique position to make important contributions to food and agricultural policy research and analysis.

EXECUTIVE COMMITTEE

Lynn Daft, FAC Chairman; Able, Daft and Earley

R. J. Hildreth, FAC Executive Committee Chairman; Farm Foundation

Anne Chaffin Bostick, Nabisco Brands, Inc.

Mark R. Drabenstott, Federal Reserve Bank of Kansas City

Robert M. Frederick, National Grange

Dale M. Hoover, North Carolina State University

Dean Kleckner, American Farm Bureau Federation

Charles K. Mann, Harvard Institute for International Development

Edward E. Masters, National Planning Association

Vernon R. McMinimy, Rolling Meadows, Ill.

John Miller, Sandy Spring, Md.

George E. Rossmiller, National Center for Food and Agricultural Policy

John Stencel, Rocky Mountain Farmers Union

NATIONAL PLANNING ASSOCIATION

THE NATIONAL PLANNING ASSOCIATION (NPA) is an independent, private, nonprofit, nonpolitical organization that carries on research and policy formulation in the public interest. NPA was founded during the Great Depression of the 1930s when conflicts among the major economic groups—business, labor, agriculture—threatened to paralyze national decisionmaking on the critical issues confronting American society. It was dedicated to the task of getting these diverse groups to work together to narrow areas of controversy and broaden areas of agreement and to provide on specific problems concrete programs for action planned in the best traditions of a functioning democracy. Such democratic planning, NPA believes, involves the development of effective governmental and private policies and programs not only by official agencies but also through the independent initiative and cooperation of the main private sector groups concerned. And to preserve and strengthen American political and economic democracy, the necessary government actions have to be consistent with, and stimulate the support of, a dynamic private sector.

NPA brings together influential and knowledgeable leaders from business, labor, agriculture, and the applied and academic professions to serve on policy committees. These committees identify emerging problems confronting the nation at home and abroad and seek to develop and agree upon policies and programs for coping with them. The research and writing for these committees are provided by NPA's professional staff and, as required, by outside experts.

In addition, NPA's professional staff undertakes research designed to provide data and ideas for policymakers and planners in government and the private sector. These activities include research on national goals and priorities, productivity and economic growth, welfare and dependency problems, employment and manpower needs, and technological change; analyses and forecasts of changing international realities and their implications for U.S. policies; and analyses of important new economic, social and political realities confronting American society. In developing its staff capabilities, NPA has increasingly emphasized two related qualifications. First is the interdisciplinary knowledge required to understand the complex nature of many real-life problems. Second is the ability to bridge the gap between theoretical or highly technical research and the practical needs of policymakers and planners in government and the private sector.

All NPA reports are authorized for publication in accordance with procedures laid down by the Board of Trustees. Such action does not imply agreement by NPA board or committee members with all that is contained therein unless such endorsement is specifically stated.

CONTENTS

PART 3. THE DYNAMIC FORCES SHAPING THE 1990 U.S. AGRICULTURAL LEGISLATION

PART 4. PRACTICAL OPTIONS FOR A 1990 FARM BILL

PART 5. EPILOGUE

FOREWORD

Prior to the passage of each major farm bill over the past several decades, a wide-ranging debate has taken place on the issues and the options for the new legislation. The administration analyzes alternatives. Congress holds hearings. The academic community writes papers and holds conferences. And the farm organizations, the commodity groups, and an ever-expanding array of interest groups speak out on their particular views. Preparation for the 1990 farm bill will be no exception.

In any negotiation—and a farm bill debate is a negotiation—the more comprehensive the knowledge base used as a point of departure, the more fruitful the discussion. Even more than with past farm bills, the 1990 bill will have its own set of specific issues that need to be addressed. The most obvious and important concerns for the 1990 farm bill debate include the need for budget austerity; the effects of agriculture on environmental quality, especially the problem of groundwater and surface water pollution from agricultural production and processing practices; the integrity, safety, and security of the U.S. food supply; the importance of international markets for a significant share of U.S. agricultural production, and hence the vital role of U.S. trade policy and the multilateral trade negotiations in the General Agreement on Tariffs and Trade; and the need for U.S. agriculture to remain competitive in both domestic and international markets. This volume provides a comprehensive discussion of the concepts, constraints, and concerns relevant to these issues, which will, most certainly, be a part of the 1990 farm bill debate.

The volume is the culmination of a unique eighteen-month effort by the National Center for Food and Agricultural Policy (NCFAP) at Resources for the Future and the Food and Agriculture Committee (FAC) of the National Planning Association. This cooperative endeavor has joined the academic community, government, and the private sector in a comprehensive analysis and discussion of 1990 farm bill issues.

The project drew on the respective strengths of the two cosponsoring organizations. The FAC's almost half-century-old tradition of off-the-record dialogue among key representatives from virtually all aspects of the U.S. food and agricultural system was paired with NCFAP's analytical capabilities and proven record for conducting nonpartisan studies and then broadly disseminating the results.

The review process for the chapters was extensive. Each was presented for review and critique at an FAC meeting; revised by its author(s) and circulated for further comment to 300 top policymakers and leaders in the agricultural community throughout the country; and, finally, scrutinized at a conference for all authors, held in November 1989. The work represents a wide range of views, yet remains impartial as to the relative merit of any particular policy direction. This last characteristic in particular should make the findings valuable to those who propose and to those who dispose agricultural policy legislation.

While the fourteen chapters cannot be represented as consensus documents of the agricultural policy community (that would be a tall order indeed), they certainly include the deliberative thought of a large number of representatives of that community and have already provoked a good deal of thoughtful discussion. The chapters in this volume add considerably to the knowledge base that will help us toward the informed development of the 1990 farm legislation. We commend this book to all those interested in this important legislation.

GEORGE E. ROSSMILLER
Director
National Center for Food and Agricultural Policy

MICHAEL F. BROWN
Research Director
Food and Agriculture Committee

Washington, D.C.
December 1989

PREFACE AND ACKNOWLEDGMENTS

In 1990 the United States must reauthorize or replace the current agricultural legislation—the Food Security Act of 1985—or face a financially crippling reversion to permanent agricultural legislation, the Agricultural Act of 1949. The debate that preceded the 1985 act was protracted, emotional, heated, and, at times, acrimonious. Few of the actors still on the agricultural policy stage have the stomach for another debate of that intensity. Also, many of them are quite satisfied with the general plot and may look in 1990 for only minor changes in the script. So why, with apparently so little interest in writing a new and different agricultural bill in 1990, should the National Center for Food and Agricultural Policy and the Food and Agriculture Committee have embarked upon a major project to examine options for that bill? And why should it be the subject of the National Center's 1990 Annual Policy Review?

U.S. agriculture is resting in a relatively comfortable position at present; but, as Sir Isaac Newton pointed out, anything at rest will stay there only as long as no forces act upon it. There are now several forces in action that will necessitate some changes in U.S. agricultural policy and that, consequently, may push the sector out of its comfortable position. The most important of these forces are the U.S. federal budget deficit; the heightened public awareness of and concern about deteriorating environmental quality, agriculture's role in that deterioration, and relatedly, the safety of the food supply; and the ongoing trade negotiations under the General Agreement on Tariffs and Trade, where agriculture has been made a centerpiece of the current round. As the 1990 legislation is formulated, responses to these new forces will have to be balanced against the traditional concerns of agricultural policy, including farm income, consumer food prices, agribusiness interests, rural community vitality, and trade and foreign policies.

This book looks at the economic, social, and political environment in which the next agricultural legislation will be written. The prologue describes the diverse mix of goals and interests that now seem to pervade what is commonly called a "farm bill."

Part 1 provides a detailed discussion of the current legislation—particularly for those unfamiliar with the provisions—including a description of the main commodity, environmental, and trade provisions and how they work; a discussion of the market setting in which the next bill will be written, which gives an idea of the budget, price, production, income, and trade implications of continuing current policy for another four or five years; and suggestions for relatively minor modifications that could be made to the current legislation to address some immediate concerns.

Part 2 looks at alternatives to the current policies. Each chapter draws on concepts that have surfaced in agricultural policy discussions in recent years and that address certain problems embodied in the current policies. The alternative policy approaches discussed are decoupling, targeting, and supply control.

Part 3 focuses on some of the new and old forces most likely to influence the writing of the 1990 legislation, including stabilization, environmental quality, agricultural trade issues, and the competitiveness of U.S. agricultural products on world markets.

Part 4 lays out some realistic policy options for 1990, given the pressures and constraints, and discusses their implications.

Finally, the epilogue examines the broader environment in which the U.S. agricultural sector exists and points out the need to move our thinking beyond the traditional commodity program approach if U.S. agriculture is to move into the twenty-first century in tandem with the rest of the U.S. economy.

The purpose of this book is to raise the level of debate and to contribute to informed discussion on the next agricultural bill, not to advocate any policy, program, or position. The authors were chosen for their expertise and interest in the topics of their chapters, not because they hold any particular point of view on agricultural policy. Each paper was reviewed by many agricultural policy practitioners, but the authors are ultimately responsible for the final version of their work. The editor has attempted to ensure a certain degree of consistency in terminology, but made no attempt to impose a single point of view throughout the volume.

This project would not have been undertaken without generous financial support from a number of organizations. I gratefully acknowledge contributions from the following:

The Farm Credit Council of America

The American Farm Bureau Federation

U.S. Department of Agriculture, Economic Research Service, and Office of the Assistant Secretary for Science and Education

The Northwest Area Foundation

The International Bank for Reconstruction and Development

Pioneer Hi-Bred International, Inc.

W. K. Kellogg Foundation

Resources for the Future

National Planning Association

Many, many people helped make this project, and hence this book, a reality. It is not possible to thank each and every one of them individually, but some do deserve special recognition. George E. Rossmiller, Michael Brown, and Linda Pierce, program associate with NCFAP, were part of the project from the beginning and contributed immeasurably in terms of intellectual stimulation, financial and budgetary management, and logistical support and coordination. The tireless efforts of Sue Higby and Debra Montanino, of the Institutional Relations unit of Resources for the Future, are also gratefully acknowledged.

Throughout the project we were fortunate to be able to draw on a number of experts and leaders in the agricultural community for support and assistance. John Lee and many of his capable staff at the Economic Research Service of the U.S. Department of Agriculture contributed in numerous ways to the success of the project. Thanks go to the members of the Food and Agriculture Committee of the National Planning Association, who were the sounding board for each of the papers that became chapters in this final product (a full list of FAC members is given as the appendix to this book); to the authors who contributed chapters; and to the task force that worked on the policy options chapter.

Then when the going *really* got tough toward the end of the project and the papers had to be pulled together into a coherent set of chapters, a number of people made special contributions. Both Jon Visser and Courtney Harold, NCFAP intern and policy assistant, respectively, proved to be invaluable assistants and have my heartfelt thanks for the many and varied tasks they performed, usually at short notice and on a tight deadline. Thanks also to Anita Washington, program assistant, who was responsible for producing a clean and readable manuscript and much more; to Nancy A. Winchester, who was the very able and obliging project editor; and to Dorothy Sawicki, who oversaw the whole book production process as managing editor.

Finally, my sincere thanks go to my family: David and Colin Trechter, who provided great moral support and who were most understanding about my late nights and weekends at work, and to Jeffrey Allen Trechter, who was considerate enough to delay his arrival until the manuscript was handed over for typesetting.

Washington, D.C.
December 1989

KRISTEN ALLEN
Policy Associate
National Center for Food
and Agricultural Policy

PROLOGUE

REFLECTIONS ON THE PAST, CHALLENGES FOR THE FUTURE
An Examination of U.S. Agricultural Policy Goals

KRISTEN ALLEN

One main reason for an examination of the goals and objectives of U.S. agricultural policies is that policymakers will face some hard choices in the early 1990s. The Food Security Act of 1985 (FSA) legislated the commodity, trade, and general provisions for the period from crop year 1986 through crop year 1990. Its stated purpose was "to extend and revise agricultural price support and related programs, to provide for agricultural export, resource conservation, farm credit, and agricultural research and related programs, to continue food assistance to low-income persons, to ensure consumers an abundance of food and fiber at reasonable prices, and for other purposes" (U.S. House of Representatives, 1985). Its more obvious objectives, judging from the provisions, were to maintain farm income while expanding agricultural exports and drawing down government-held stocks of agricultural commodities, particularly grains and cotton. A broader goal, not limited to agriculture, was to bring down the massive budget deficit.

The FSA set a slightly new course for the commodity programs. It lowered the nonrecourse loan rates for grains and cotton and set them as a percentage of a moving average of market prices, making those U.S. commodities competitive in world markets and effectively ending the United States' role as the sole supporter of world markets. (In the short run, however, the U.S. loan rates still set the price floor, except for rice and cotton, where marketing loans are in effect.) By freezing the program yields for price-supported crops and by freezing wheat and feed grain target prices for the first two years and then lowering them gradually, the act made a small step toward reducing the program-induced incentives for producers to try to maximize production on their planted acres. It allowed the Secretary of Agriculture a considerable amount of discretion in implementing the program provisions. To address some of the concerns—particularly soil erosion—of environmental interests, and simultaneously to reduce surplus commodities, the act provided for a total of 40 million to 45 million acres of erodible land to be taken out of production for ten years and enrolled in the Conservation Reserve Program.

Other environmental concerns were addressed by the conservation compliance and sodbuster and swampbuster provisions.

Over the years the programs that provide support and protection for agriculture have come under fire from economists, politicians, nonfarm interest groups, and even farmers themselves. The most common complaints in recent years include that these programs—

- Entail high budget costs, regressive in nature and with the benefits going to a very small percentage of the population;
- Involve high costs to the economy in general, over and above the federal budget outlays, in terms of high consumer costs, resource misallocation, and efficiency losses;
- Are inequitable;
- Contain inefficiencies and design inconsistencies;
- Are unable to cure rural economic ills;
- Contribute to externalities, especially insofar as they encourage practices that damage or deplete soil, water, and air resources;
- Cause the United States to be a residual supplier of many agricultural commodities and thus the residual adjuster in the international market;
- Fail to adapt to the changing realities of agriculture and agricultural trade.

Thus, in the first year of the act there were many calls for changes, ranging from relatively marginal (although potentially very costly) changes that would have extended some of the newer provisions (such as marketing loans) to more commodities, to dramatic changes in direction. Among the suggestions for more dramatic changes were proposals for mandatory controls on supply and, at the opposite end of the policy spectrum, for severing the link between current production and direct payments. There was also talk of trying to target payments to particular categories of farms, based on size, need, or sales of agricultural products.

More recently there has been much greater acceptance of the FSA, and many observers believe that the programs are working. U.S. food prices are low compared to prices in other industrial countries. The total cost to the federal government for the agricultural programs is coming down, although it is still at high levels. U.S. agricultural exports have increased, although not solely, and maybe not even primarily, as a result of the act. Farm income has been maintained at a relatively high level, which is not unrelated to the high level of government outlays.

Nevertheless, proposals containing elements similar to those mentioned above—controlling supplies, breaking the link between payments and production, and targeting payments—can be expected to surface again when the new legislation is discussed, and each of them gives different weighting to the many goals and objectives that have been assigned to agricultural policies.

This chapter looks at the programs typically covered by agricultural acts, recognizing that they are just some of many factors influencing U.S. agriculture, and at the sometimes strange mix of goals of the programs.

Governments subsidize and protect agriculture for a number of reasons—economic, social, political, and philosophical—and in a number of ways. The reasons include these: the special characteristics of the agricultural sector that have, in the past, set it clearly apart from other economic activities; the importance of the entire food and fiber complex to the economy and the nation's social fabric; the nation's historical ties to an agrarian culture and consequently the popular views of agriculture and farming; and political expediency. The agricultural support programs were conceived and born of economic necessity, grew and prospered by drawing on social and philosophical arguments that trace their lineages back to Thomas Jefferson, and continue largely because of astute political bargaining and coalition building on the part of all groups interested in agricultural policy.

THE WIDE SPECTRUM OF FACTORS AFFECTING AGRICULTURE

One of the most important developments to shape U.S. agriculture over the past decade has been the integration of agriculture with other segments of the U.S. economy and with the economies of other countries. The prosperity of U.S. farmers is now determined as much by fiscal and monetary policies in the United States and in other industrial countries as it is by the annual rainfall in the Farm Belt. Changes in policies designed to protect the environment may have as great an influence on net farm income as do changes in target prices. A change in transportation rates or in the availability of transportation services may be more important for farm families than a change in nonrecourse loan rates. The traditional agricultural programs, so long perceived as the dominant factor shaping U.S. agriculture's well-being, are now just one of many factors that influence the sector (figure 1).

Even the term *agricultural policy* is somewhat misleading. It suggests a single program or a coherent set of programs; yet in its entirety, U.S. agricultural policy includes a number of different programs, each designed to address different problems or aspects of a problem. They range from the agricultural commodity price and income support programs to the following: export promotion programs, subsidized loans from the Farmers Home Administration, food stamps and domestic food donation programs, transportation subsidies, food aid under Public Law (P.L.) 480 (the Food for Peace program), and federally supported research and extension. In addition, the U.S. agricultural sector benefits from—and sometimes loses as a result of—provisions in tax policies and general macroeconomic policies. Apart from a general orientation toward improving conditions in the U.S. agricultural sector, the goal or set of goals for the programs that come under the rubric of agricultural policy are difficult to define in specific, achievable terms.

The importance of the many factors that influence agriculture is widely recognized by agricultural practitioners, analysts, and observers, yet much of the effort that goes into looking for ways to maintain or improve the economic health of the sector still focuses on the commodity programs. They

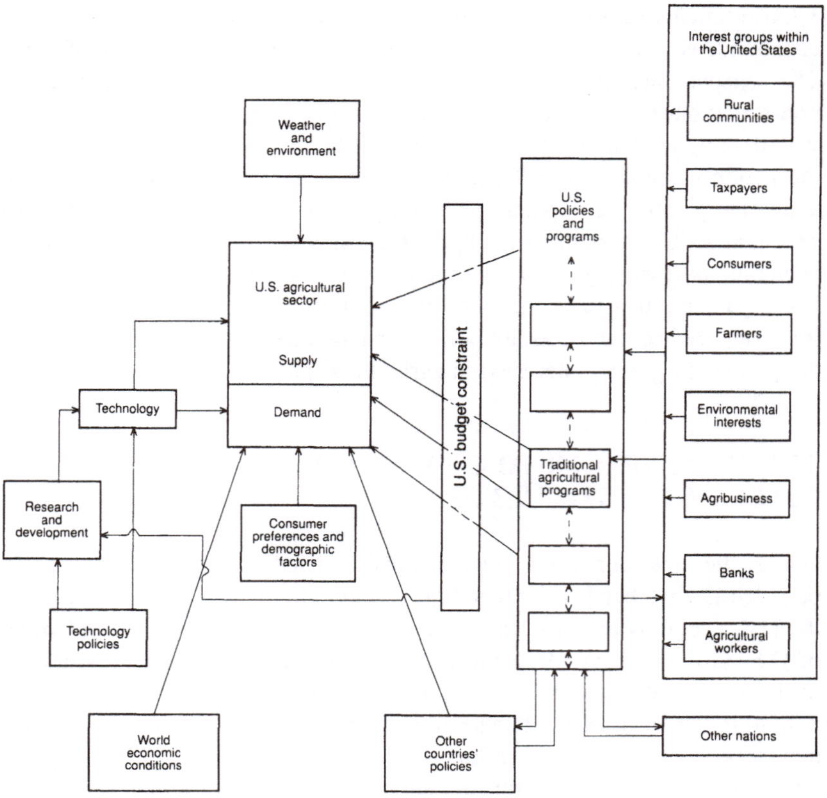

FIGURE 1. Partial schematic of factors affecting the U.S. agricultural sector.

are the most visible agricultural programs, and they are legislated on a regular and frequent basis. The current legislation, the Food Security Act of 1985, will expire in 1990, and if it is not extended or replaced by that time, the commodity program provisions will revert to the Agricultural Act of 1949, which fixes price supports at levels between 75 and 90 percent of parity.[1] This volume focuses mainly on options for the major agricultural programs, but the relative importance of these programs for the agricultural economy should not be overestimated. They cannot and should not be expected to cure all the economic ills of the agricultural sector, and programs directed at agriculture should not be expected to make rural America prosperous.

[1]Parity prices would restore agricultural commodity purchasing power to the level that prevailed during the period 1910–1914. Parity payments are no longer used in the commodity programs. In the Agriculture and Consumer Protection Act of 1973 the parity formula was dropped and target prices were set using a formula that took account of the previous target price and an index of production costs. In the Agriculture and Food Act of 1981, the target prices for each year of the act were set by legislation (USDA, 1984). In the Food Security Act of 1985, Congress set the minimum target prices.

THE ECONOMIC FOUNDATIONS

Two overarching goals of the agricultural programs have been to ensure prosperity in the agricultural sector and to ensure a safe and reliable food supply.[2] The economic arguments for intervention rest on society's real need for a reliable food source and the perceived and real vulnerability of farmers to forces that usurp some of their control over the production and marketing of their products. A number of special characteristics that in combination set agriculture apart from other enterprises have traditionally been cited as reasons for government intervention. Among them, the following characteristics are key and seem to be the basis for the design of many of the commodity support programs:

- The biological risks—weather, disease, and pests—which are often unpredictable and over which farmers (historically) have had little control;
- The competitive structure of the industry, which causes farmers to be price takers on both the input and output sides;
- An inelastic demand for agricultural commodities with respect to price (meaning that a decrease in the price of a product decreases producers' total revenue) and with respect to income (meaning that, in an era of slow domestic population growth, demand for a product or group of products can be increased significantly only by increasing exports);
- The rapid rate of technological change in agriculture, putting farming on what Willard Cochrane termed the "technological treadmill,"[3] meaning that the gains from adopting new technologies go to early adopters; when a large number of farmers adopt the technology the commodity price drops and all other producers must adopt the technology or be squeezed out;
- Relative immobility of resources, meaning that resources (land, labor, and capital) are trapped in agriculture over a wide range of returns because the returns to resources are less than the acquisition price but more than the salvage price.

While none of these characteristics alone would provide sufficient argument that agriculture is a unique sector, taken together they do suggest that it has special traits that distinguish it from most other industries. But, as discussed in a later section, changes in the economic and policy environments and in technologies over the past fifty-five years have in many ways

[2] This section is adapted from Kristen Allen and Barbara J. Elliott, "The Current Debate and Economic Rationale for U.S. Agricultural Policy," in M. Ann Tutwiler, ed., *U.S. Agriculture in a Global Setting: An Agenda for the Future* (Washington, D.C., Resources for the Future, 1988).

[3] For a more detailed discussion of the technological treadmill theory see Willard Cochrane, *Farm Prices: Myth and Reality* (Minneapolis, Minnesota University Press, 1958), chap. 5; or Willard Cochrane, *The Development of American Agriculture* (Minneapolis, Minnesota University Press, 1979), chap. 19.

made these characteristics less important or relevant and thus put agriculture on a more equal footing with other sectors.

A Set of Diverse Programs

The U.S. government has (as have many other national governments) for many years employed a number of federally funded agricultural programs that promoted the development of the sector, helped counteract these characteristics' negative financial consequences for farmers, and allowed agriculture to receive special consideration in other ways.

For example, the Morrill Acts of 1862 and 1890 enabled the establishment of land-grant colleges in each state by providing land or funding. Both acts were part of an effort to apply science to the practical problems of agriculture and to make higher education more available to farm families (at a time when farm families were a large portion of the total population). The state experiment stations were established under the Hatch Act of 1887, and the Smith-Lever Act of 1914 established the extension service. The Federal Farm Loan Act of 1916 established the federal land banks, the cooperatively owned loan agencies that were to provide low-interest loans for land mortgages. Subsequent needs for shorter-term and emergency credit prompted the development of the Farm Credit Administration in the 1930s.

The most visible agricultural programs, the price and income support programs, began in the 1930s, following a period of serious economic hardship for the agricultural sector that began in the early 1920s. The Agricultural Adjustment Acts of 1933 and 1938 laid out the basic design for the major commodity programs that have continued to the present. Building on the characteristic of a price-inelastic demand for agricultural products and the recognition of agriculture's propensity to overproduce, the programs set out to raise farm incomes by raising the price of agricultural commodities through restricting their production. The nonrecourse loan and parity payments were essential parts of the programs for the basic commodities, which included wheat, corn, cotton, and other crops at the discretion of the Secretary of Agriculture. Acreage allotments or marketing quotas, or both, were used for the crops mentioned and for rice, peanuts, and tobacco.

The federal all-risk crop insurance program was initiated in 1938, at first only for wheat but extended to other crops a year later. This program now provides federally subsidized insurance for most major crops to protect against low yields.

The search for new markets to overcome the low income elasticity of demand for most agricultural products has prompted the U.S. government to institute demand enhancement and donation programs for both domestic and foreign recipients. School lunch and breakfast programs, food "give-aways," food stamps, the Food for Peace program, and a variety of export credit and export enhancement programs have been designed to expand the demand for U.S. agricultural commodities.

A Changed Environment

The world and the U.S. agricultural sector have changed considerably since the 1930s, when the forerunners of today's programs were initiated. The portion of the population actively involved in agricultural production has dwindled; the percentage of the total population in farm occupations has dropped from 21 percent in 1930 to less than 3 percent in 1987 (U.S. Department of Commerce and USDA, 1988). A similar trend has occurred in the farm population, which in 1930 was about 25 percent and is now about 2 percent (U.S. Department of Commerce and USDA, 1988). Farm asset values and farm income have become more concentrated, and now a very small proportion of the farm population controls quite a large proportion of the total assets and receives a large portion of the total farm income.

Agricultural production and processing also have become more concentrated; farms have become larger and more specialized in production, and in some food industries vertical integration, from production through processing and distribution, has become common. In production, dependence on purchased inputs—including fertilizers, pesticides, and fuel—and on borrowed funds has increased.

To reduce some price risk, farmers can use futures, options markets, and forward contracting. Most farmers have ready access to a large amount and broad range of information, which helps them make production, marketing, and management decisions. For many farms, income from agricultural product receipts is no longer the sole (often not even the main) source of income for the farm family.

Since the end of World War II, the world economy has become increasingly important to the United States, even as the United States' relative importance in that economy has declined. The expansion of both international trade and international capital markets, the move by the United States in 1973 to a floating exchange rate, the rise of the Organization of Petroleum Exporting Countries (OPEC), and the buildup of international debt, particularly since the 1970s, have all contributed to fundamental changes in the world economy that limit the ability of the United States to control its fortunes in that economy.[4] Because the relevant market for many U.S. agricultural products is not just the United States—exports in the 1980s have accounted for between 18 and 30 percent of agricultural receipts—these changes in the world economy are crucial for U.S. agriculture. U.S. agricultural trade fluctuates not just with changes in commodity prices but also with world economic growth, exchange rate shifts, and policy changes in the many countries that are part of the world trading system.

---

[4]For a summary of the changes in the world economy and the U.S. position in it, see M. Ann Tutwiler and Barbara J. Elliott, "An Interdependent and Fragile Global Economy," in M. Ann Tutwiler, ed., *U.S. Agriculture in a Global Setting: An Agenda for the Future* (Washington, D.C., Resources for the Future, 1988).

In short, many of the arguments used in favor of price and income support programs for farmers—some of which draw on the special characteristics of agriculture outlined earlier—are less valid justifications for public support for agriculture than they were in the 1930s when the programs were initiated. Yet support for the agricultural sector continues, suggesting that economic arguments alone are not the only rationale for the programs. Indeed, as Paul Thompson (1988) notes, the symbolic importance of farming to a population whose recent forebears were farmers, ranchers, or rural merchants cannot be discounted, and neither can the political strength of groups, including farmers, who have enjoyed benefits from the commodity programs.

THE FAMILY FARM AS A SYMBOL

Agriculture and farming have an importance in the public mind that exceeds their importance to the economy alone. Thompson (1988) suggests that there are four key reasons for agriculture being valued both by individual Americans and by the public in general. These reasons are (1) the dependence of society on agriculture—of consumers for food and fiber supplies, and of producers and processors for a source of income; (2) agriculture's intimate association with health and environmental quality; (3) agriculture's importance in international relations; and (4) the tradition of American agriculture that gives family farms and the activity of farming a moral and aesthetic importance. He also points out that there is nothing fundamentally *agricultural* in the first three reasons: they could apply equally well to a number of other economic activities. For many Americans family farms do, however, provide a link with the past—and often the not-so-distant past.

In characterizing farming and family farms, many people draw on the philosophy of the nineteenth-century agrarian populists, who held up farming as a way of life, as a vocation chosen for reasons beyond simply earning a living or return on investment (Thompson, 1988). The terms *healthy* and *natural* to describe life-style are frequently used in reference to farming, although there is now probably little uniquely healthy or natural about modern U.S. farming practices. The view of farming as something more than just a vocation, however, may partly explain why people continue to farm even when returns are relatively low or uncertain.

The potential to partake of even a small part of that healthy life-style also seems to have some value in the public mind. The existence of farmland, particularly that relatively close to cities and metropolitan areas, has what Winters (1988) refers to as "amenity value." He discusses two aspects of the amenity value of farmland, the visual and the access. The visual amenity might also be referred to as the "bucolic vistas" value, which, put simply, means that people derive utility from being able to see what they consider to be rural scenes when they travel outside the metropolitan areas. Several U.S. states have passed "right-to-farm" laws that exempt farmers from local noise and sanitary laws in built-up areas, which appear to cater to this social need.

The access aspect of the amenity value of farmland refers to the idea that the public has more chance of being able to pursue recreational activities on land that is in farming use (and preferably small or family farming), than on land that is under shopping malls, office buildings, or condominiums.

During the 1960s and 1970s many people dreamed of escaping from the urban rat race and getting back to the land. Although some people fulfilled the dream at that time, many others put it on hold while they pursued careers in cities, but they continued to view a small farm as a goal for retirement or as a vacation property. Some of this collective dream may derive from the view of farming as a healthy and simple life-style, but it also may have some roots in what Thompson refers to as "Farming and Populism," as championed by Texas Commissioner of Agriculture Jim Hightower. Hightower lauds the hard work and independence of farm families but also focuses on the farmer's role as an entrepreneur, independent of big business and wage labor. As Thompson puts it, Hightower views farming as "one of the few careers open to the entrepreneur in contemporary society, and one that is open to people with relatively little formal training and specialized skills" and as somewhere that "the little guy can make it" (Thompson, 1988, p. 38). Commissioner Hightower does not appear to be deterred from this view by the large capital expenditures and managerial know-how required to enter farming that often put the dream well beyond most "little guys."

While this public perception of farming and agriculture may represent unrealistic and somewhat romantic views of the U.S. agricultural sector today, it is of considerable importance to understanding the political strength of agricultural and farm groups in the U.S. political process. It does not, however, necessarily provide a sufficient argument for a particular farm or agricultural program or, indeed, for any special farm- or commodity-based program at all.

A BROAD POWER BASE BUILT FROM MANY SPECIALIZED INTERESTS

The Food Security Act of 1985 is a testament to what Browne (1988) calls an explosion of interest in agricultural policymaking.[5] While the general U.S. farm and commodity interests were well represented during the long legislative debate that preceded the bill's passage, other groups also sought to have their interests heard, and some achieved their goal. Among these other groups seeking the attention of Congress were consumer and taxpayer representatives, environmental interests, rural communities, agribusiness firms, banking interests, representatives of agricultural workers, and foreign governments, foreign farm groups, and impoverished foreign consumers.

[5]This section draws extensively on Browne (1988).

This explosion of interests confounds attempts to define a single, comprehensive direction for agricultural policy. Agricultural policymaking has become a process of mutual accommodation, with a large number of narrow provisions being fitted together to become a broad piece of legislation that often contains internal inconsistencies and provides few clues as to the real goals of the policy. At the same time, however, the process has become somewhat more democratic by increasing the number of actors in the policy decision process and potentially opening the process to a wider audience.

Browne describes the changes in the model of agricultural policymaking from the old "iron triangle" to a broader, multifaceted model. The old iron triangle incorporated the U.S. Department of Agriculture (USDA), general farm organizations, and the agricultural subcommittees in Congress. In the 1940s, as some of the general farm organizations became closely identified with one or another of the major political parties and their positions on the form of farm support, individual commodity groups gained prominence in the policymaking process and eventually took the place of the general farm organizations in the iron triangle. The model of agricultural policymaking expanded in the 1970s and 1980s to include, in addition to the three key groups, the bevy of new interest groups, an increasing number of other federal departments and agencies (among them the Office of Management and Budget, the Environmental Protection Agency, the Food and Drug Administration, the Office of the U.S. Trade Representative and the Department of State) and more congressional subcommittees. Agricultural policy has been forced to address many more issues than just farm prices and incomes. Farm, food, fiber, trade, environmental, health, rural, macroeconomic, and foreign policies have all become important to agriculture's welfare, and thus we see, increasingly, attempts to write a farm bill that tries to take account of all of these new pressures.

The main objectives of agricultural policy for the farm groups remain price and income support and, at least for some, commodity market stability, but the overall goals of U.S. agricultural policy have broadened in response to the changed physical, economic, and political environment in which agriculture operates and with the increasing awareness among nonfarm groups that the programs have been at least partly responsible for some of agriculture's deleterious effects on the environment and for sending perverse market signals to producers. In 1985 many nonfarm interest groups traded their support for the commodity programs in return for consideration of their agendas (or some portion thereof) and inclusion of some of their objectives in the final legislation. Farm organizations took a "co-opt to survive" approach to the potential threat to their programs from these other groups and thus broadened the base of support for the programs by building coalitions with other groups. Browne notes that the strength of the farm groups continues to be in their ability to provide useful and timely input during policy deliberations and in the general recognition of agriculture's importance to the U.S. economy and society. At the same time, however, the position of all the groups seeking to influence agricultural policy is precarious. No single organization, or even set of organizations, can be insiders on all agricultural policy matters.

The agricultural policy process has drawn together sometimes dissimilar groups that can unite behind a common objective or support a single program that may meet several objectives. For example, producers and environmental interest groups could both support the Conservation Reserve Program, instituted under the Food Security Act of 1985, because it addressed both producers' concerns about surplus commodities and environmentalists' concerns about damage to fragile lands caused by some farming practices. Environmental groups and consumers can find common ground in calling for closer regulation of the use of chemicals on agricultural crops. More recently, some farmers have joined this movement, as they have come face to face with the detrimental effects on their families' and livestocks' health of pesticide-contaminated wells. It could be said that agricultural policy often makes strange bedfellows, but without the coalition building and support of many different groups, many agricultural programs would not likely survive.

AGRICULTURAL POLICY CHALLENGES FOR THE 1990s

Some of the toughest challenges facing U.S. policymakers in the next few years will be in agricultural policy, particularly in the commodity programs. Historically farmers' programs, designed to raise and stabilize farm incomes by raising and stabilizing commodity prices, commodity programs have, for reasons outlined in the preceding sections, taken on more and more objectives in recent years. Replacing or extending the FSA will require skillful juggling of many interests—farm, consumer, environmental, taxpayer, business, foreign, and rural.

Farm-Level Concerns

While farmers as a group are still and will likely continue to be the main target of the price and income support programs, the group is very heterogeneous. Farmers include people who live and work on a farm as a full-time occupation, those who farm as their principal source of income but who also have off-farm jobs to make ends meet, people who have nonfarm occupations but who enjoy farm life and sell enough agricultural products to justify their property's being defined as a farm, and those who earn enough in their nonfarm occupations or businesses to need a convenient outlet for some of the excess cash at tax time. Because annual incomes from farming range from actual losses to millions of dollars, it is difficult to generalize about the income of farmers and farm families. Farmers or producers are individuals, families, family or nonfamily partnerships, and corporations.[6] The current agricultural programs make little

[6]Most farms raising the major crops are owned or operated, or both, by families or some form of family partnership or corporation. See Congressional Budget Office, *Diversity in Crop Farming: Its Meaning for Income-Support Policy* (Washington, May 1985) for an analysis of farm ownership based on the 1982 Census of Agriculture.

distinction among farms, in terms of farm income or firm organization, in determining who is eligible for program benefits.[7]

A part of this diverse group will continue to look to the agricultural programs to raise, support, or protect farm incomes. A closely related goal is the maintenance of asset values or net worth, and the importance of this goal should not be underestimated. Farmers also, however, have a keen interest in how these two goals are achieved. One stated preference is that income support be through the market and price mechanism. For large producers, raising market prices shifts the distribution of benefits in their favor as they produce larger quantities of the commodity. Some groups do pursue such rent-seeking behavior when voicing their preferences for the design of farm price and income support programs. Receiving price supports through the market also makes the income transfer less visible and therefore less apt to be viewed as a direct payment (or welfare). Other farmers often call for a "fair deal" or "fair market value" in their commerce, vis-à-vis other sectors of the U.S. economy and farmers in other countries; that is, they view a market-oriented agriculture as the preferable case but also believe that some government intervention—through price support loans, marketing orders, or other orderly marketing mechanisms—is necessary to enforce equitable distribution of returns and to counteract agricultural support programs in other countries. Few farm or commodity groups have publicly voiced support for a complete shift to direct income transfer payments. Price and income stability will likely continue to be a goal of producers but probably one of secondary importance compared to the goal of income support.

Consumers and the Food Supply

Consumers' interests will become increasingly important to the largely urban Congress. Consumers are usually presumed to be interested primarily in a safe, assured, and nutritious food supply at prices as low as possible. Traditionally, consumers have also been assumed to prefer stable food prices—particularly stability on the upward side—because highly volatile commodity prices increase uncertainty. Other stability goals—for example, a desire for a stable range of food choices or for a certain level of consistency in food quality—may also be sought.

Demographic changes in the population will continue to bring forth changes in consumers' tastes and preferences and hence their views on the performance of the food system. The increasing proportion of elderly in the population will change overall dietary and nutrition requirements. Rising affluence and changes

[7]A $50,000 payment limitation does apply to a portion of direct payments as an attempt to restrict very large payments being made to what are perceived to be wealthy farmers and landowners. The limit is frequently circumvented, however, by farm reorganization. Legally, a corporation is treated as a "person," even when it is made up of persons who are farmers in their own right. (See Sumner, in this volume, for a more thorough discussion of farmers' reactions to the payment limitation.)

in basic family structure and behavior create demands for convenience foods at the same time that many consumers are indicating a preference for "healthier" foods, those lower in fats, salt, sugar, and total calories (Kramer, 1988). At the other end of the scale are significant numbers of families living in poverty, also frequently associated with nontraditional family structure patterns or life-styles. These families, which often include children and elderly people, will continue to look for lower-priced food and rely on government assistance, including food stamps and food donations, to survive.

Perhaps the most intense focus for consumer interests in 1990, however, will be on the safety of the food supply, particularly with respect to chemical residues and microbial contamination. This issue has become a key concern of consumers; according to the *Washington Post,* a recent survey showed that 75 percent of consumers list pesticide contamination of foods as one of their worries about the food supply (Sugarman, 1988). These consumer preferences and concerns may ultimately influence traditional agricultural programs as well as programs that regulate food quality and safety.

The heightened concern about chemical residues in food has proved to be a coalescing point for consumer, environmental, and agricultural worker interest groups. This coalition has successfully focused attention—of policymakers, the media, and the general public—on agriculture's use of chemicals and the effects of those chemicals on humans. Concern about toxic chemical residues in food and water supplies and about the effects of agricultural chemicals on those who must apply them to crops or drink water contaminated by them has given rise to public calls for stricter surveillance or regulation of their use. In California, the Safe Drinking Water and Toxic Enforcement Act of 1986 (commonly known as Proposition 65) was a reaction to this concern and the perception that neither the federal nor state governments were doing enough about ensuring an environment safe from toxic chemicals.

New Technologies

Another issue on which a similar or perhaps a broader coalition could emerge is the new technologies now being applied to agriculture—biological and information. These technologies potentially offer new opportunities for those involved in farming and the food and fiber system (see Carr, 1988). They will, however, also give rise to questions that will have to be answered to the satisfaction of an ever more critical and skeptical public. Witness, for example, the announcement in August 1989 by five major grocery chains and a manufacturer of premium ice cream that they would not accept milk from cows treated with the hormone bovine somatotropin (Schneider, 1989).[8] Genetically engineered organisms have come to be viewed with suspicion by many environmental interest groups, even though they may offer an alterna-

[8] One of the grocery chains did subsequently modify its stance to say that it would not market the milk until bovine somatotropin was approved for general use.

tive to chemical use in some areas of agricultural production. Some of these technologies will undoubtedly prove to be substitutes for agricultural labor, especially unskilled labor. And the adoption of the technologies may fuel the ongoing trend toward the larger, more centralized, more vertically integrated farming operations that are anathema to those who value highly the cultural heritage of a nation of family farmers.[9]

Environmental Quality

Environmental interest groups have taken an active interest in agriculture and agricultural policy since the 1970s. Their primary interest in the agricultural policy process is to promote or work for policies that minimize agriculture's adverse effects on the environment, particularly on land, water, and air resources, and on wildlife that depend on those resources. In the competition for land (soil), water, and air resources, the food and agricultural system has in the past tended to win. There are now good reasons to suspect that there will be much stronger competition from other uses in the future. With surplus commodities being acquired by the government in most of the past seventeen years, many observers are questioning the wisdom of allowing water to be sold for irrigation at heavily subsidized rates when urban areas are required to pay high water rates. In addition, agricultural processes that contribute to the pollution of water supplies, both surface water and groundwater, will continue to come under close scrutiny, as will processes that contribute to soil loss (especially when the soil then contributes to water pollution). It is quite possible that the FSA environmental provisions—conservation compliance, and sodbuster and swampbuster—will form the model for environmental provisions in future agricultural legislation.

Budget Pressures and Difficult Trade-offs

Perhaps the strongest driving force in the 1990 debate on U.S. agricultural policy will be the need to cut federal spending overall. The Balanced Budget and Emergency Deficit Control Reaffirmation Act of 1987 (popularly known as Gramm-Rudman-Hollings) mandates deficit targets of $64 billion in 1991, dropping to $28 billion in 1992 and to zero in 1993. These required reductions will come at a time when there is apparently a general desire on the part of the U.S. public for increased spending on such social programs as education, drug law enforcement and drug rehabilitation and education programs, and the development and maintenance of infrastructure. Reducing spending on agricultural programs has become a perennial "goal" in agricultural policymaking in recent years and is unlikely to

[9]See U.S. Congress, Office of Technology Assessment, *Technology, Public Policy, and the Changing Structure of American Agriculture* OTA-F-285 (Washington, D.C., U.S. Government Printing Office, March 1986) for a detailed discussion of the potential effects of information and biological technologies on U.S. farms and the structure of the agricultural sector.

change in the foreseeable future. When taken in combination with the desire to enhance or even maintain farm incomes at the fairly high levels of recent years, to increase U.S. agricultural exports, and to keep domestic food prices under control, this goal presents policymakers with a serious dilemma.

Most taxpayers prefer lower tax rates and fewer taxes. The strength of this preference was sharply brought home to 1988 U.S. presidential candidates who tried to inject a sense of fiscal responsibility into their campaigns by pointing out the direct correlation between large government expenditures and the need for large government income.

Higher farm incomes can come only from higher prices (as domestic demand is relatively inelastic with respect to price) or lower costs. Lower costs are difficult to achieve in the current economic and policy climate. Higher prices can come only from consumers (domestic or foreign) or from taxpayers (through the government). Convincing foreign buyers to pay higher prices for U.S. agricultural products in a highly competitive international market has proved difficult, and attempts to persuade them to increase their purchases, through lower market prices and the Export Enhancement Program (EEP), have proved expensive for U.S. taxpayers. Therefore, it seems that the burden for continued price and income support will fall largely on the U.S. citizens, either as consumers, as taxpayers, or both.

Part of the dilemma arises from the conflicting desires of the U.S. public. On the one hand, they seem to want to provide some support to what they perceive as "needy" farmers, but on the other they seem increasingly disenchanted with seeing subsidies, often large ones, go to already wealthy farmers. This particular mix of interests gives rise to perennial interest in targeting or limiting subsidy payments to farmers. But, in addition to being difficult to achieve and requiring policymakers to nominate explicitly a "preferred target group" (see Sumner, in this volume), neither targeting nor payment limitations contribute greatly to reducing overall spending on farm programs.

Concerns of Agribusiness Firms

Agribusiness firms also have a strong interest in agricultural policy. These firms operate at all levels of the food and fiber system. They provide inputs for agricultural production, processing, and marketing; they purchase and sell raw commodities; they process commodities; they market intermediate and end products; and they transport agricultural products. Although it is difficult to generalize when discussing a large, heterogeneous group, one of the primary interests of these firms is to maintain U.S. agricultural capacity and output. Another goal, particularly for those firms that sell durable goods and capital equipment to farmers, is to maintain farm income at levels that enable farmers to purchase tractors, combines, and other vehicles, machinery, and equipment. As many agribusiness firms produce production inputs—chemicals, equipment, seeds, and so forth—they are also interested in policies that affect how commodities are produced, including regulatory policies. Agribusiness firms that buy raw agricultural commodities, either for intermediate use or for resale, have a strong interest in low commodity prices and can be expected to join forces with consumers on this issue and the issue of price volatility, as large and rapid price changes could necessitate costly alterations in production decisions or procedures.

Trade and Foreign Policy Issues

Although other nations have no formal role in making U.S. agricultural policy, their interests are not always ignored in the process and many have had no qualms about articulating their views of recent U.S. agricultural programs, often through sophisticated and highly paid U.S. lobbyists. For example, overt U.S. actions (such as the creation of the EEP) that lower other countries' sales, either in volume or in price, have generated strong negative reactions from many countries—e.g., Australia, Canada, and other Cairns Group countries,[10] and the European Community (EC)—with which the United States has close ties in other areas.

Agricultural policy actions in the international arena cannot be taken in isolation from other foreign policy concerns. Agricultural products are not the only ones traded internationally; gains in agricultural trade must often be weighed against losses in other sectors and vice versa. Some actions to promote sales of U.S. agricultural products can have adverse effects on export opportunities for other U.S. products, just as disputes over trade in manufactured goods can damage existing or potential markets for agricultural products. Nor is trade the only concern. National security, historical ties, and ideological reasons promote or discourage alliances, quite independent of trade interests, that sometimes result in outcomes that are at odds with domestic agricultural policies.

The most pressing concern of international origin for U.S. policymakers as they debate the 1990 legislation will be the multilateral trade negotiations under the General Agreement on Tariffs and Trade (GATT), where agriculture has been made one of the high-priority issues for the current Uruguay Round. Many observers view the domestic agricultural policies of the industrial countries as largely to blame for the worldwide imbalance between the supply of and demand for agricultural commodities that has kept world prices relatively low for most of the 1980s.

The Uruguay Round is scheduled to continue until the end of 1990, so it is unlikely that any further agreements will have been reached on changing national agricultural policies before the U.S. 1990 agricultural bill is written. (In the April 1989 statement from the GATT, countries agreed to freeze their levels of agricultural protection for the rest of 1989 and to reduce them

[10] The other Cairns Group countries are Argentina, Brazil, Chile, Colombia, Fiji, Hungary, Indonesia, Malaysia, New Zealand, the Philippines, Thailand, and Uruguay.

further in 1990 [GATT, 1989].) Therefore, the United States may not be specifically obliged to write a bill that reduces the trade-distorting effects of its domestic agricultural policies, but writing legislation that moved those policies in the opposite direction would probably be viewed internationally as a sign of bad faith. The U.S. voice has, after all, been prominent among the numerous voices calling for reform of national agricultural policies. To ignore the possibility of such an agreement in the GATT when writing the 1990 agricultural bill would send a strong negative signal to other GATT members about U.S. commitment to agricultural trade reform that would ultimately have adverse effects on the U.S. agricultural sector and very probably far beyond.

The interests of developing countries are a special case. Those countries that are importers of agricultural products gain from world market surpluses. But while ready access to low-priced commodities for import and subsidized or donated products provide short-term benefits for developing country consumers and governments, in the long run they may act as disincentives to the development of local production capacity. Ultimately, developing countries need access to markets in other countries for their products, both agricultural and industrial or manufactured, so that they can raise their own income levels and thus become even better markets for the products of the industrial nations.

The Economic Well-being of Rural Areas

Finally, it is likely that rural economic development will continue to draw attention in the context of the new agricultural legislation, despite considerable evidence that the tendency to equate agricultural with rural is erroneous. Farming and agricultural service industries together contributed only about 8 percent of the personal income in nonmetropolitan areas in 1986. In 1984 only about 9 percent of rural employment was in farming—compared to 40 percent in manufacturing, 16.5 percent in services and trades, and 13 percent in government—and farm families earned more than 60 percent of their income from off-farm jobs (Swanson, 1989). Therefore, raising the farm prices for a select set of agricultural commodities is an inefficient and ineffective way to promote rural economic activity. Conversely, however, economically diversified rural communities can offer greater employment opportunities to farmers who wish to continue farming but who also have difficulties meeting their financial commitments.

That the Food Security Act seems to have done little to stimulate rural development over the past three years appears to have prompted policymakers to look more closely at alternative approaches for promoting economic activity in rural America. The Senate has written a separate rural development bill (S. 1036), and at the time of this writing, the House Agriculture Committee's Subcommittee on Rural Development, Conservation, and Credit was working on a companion bill (Freshwater, 1989). But, as Freshwater points out, despite these moves to develop a separate rural development policy, most federal funds for rural development will continue to flow through farm programs. Thus, rural development proponents will likely have to face the reality that funds for rural development will continue to compete with funds for commodity programs and the political support for commodity programs tends to be more organized and stronger than that for rural development.

What is most needed for rural America is a new view of the rural economy, not as an economy and society separate and distinct from the rest of the country but as an integral part of it. Swanson (1989) makes this point, noting that, in fact, many of the problems experienced by rural communities are remarkably similar to those now plaguing inner cities—deteriorating education and health care systems and high crime rates. Agricultural enterprises will likely continue to be important economic activities in rural America for the foreseeable future, but perhaps the more appropriate way to address the issue of rural development in relationship to the next agriculture bill is to ask how revitalizing rural communities might help relieve farmers of their dependence on price and income supports, not how price and income supports will help revitalize rural communities.

SOME MODEST GUIDELINES

It has often been said that agricultural policy changes are evolutionary rather than revolutionary, and there is little reason to expect a revolution to occur before the next U.S. agriculture bill is passed. Any changes will likely be made at the margin, in response to new and pressing objectives of the groups already influencing agricultural policy or, possibly, of new ones yet to be recognized, with the budget continuing to provide constraints. Marginal changes can, however, yield significant results in the long run, so identifying and articulating possible new goals and interests can provide guides to possible deviations from the general direction of agricultural policy.

The FSA made relatively marginal changes in the farm programs, but it did set a new course; lower market prices made U.S. agricultural commodities more competitive, but at the same time income supports were kept high. The result has been a very expensive bill; in the future there will be considerable pressure to reduce the cost of all agricultural programs. Policymakers will have to face the difficult trade-off between keeping consumer and export prices low and maintaining farm income, with a smaller total budget available.

In the short term, farm price and income support programs will and should continue to focus on farms and farm incomes. To shift the focus or to eliminate the support programs completely now, after more than half a century of institutionalized support, would create severe hardship in many areas of the economy, not just in agriculture.

Changes in policies change the rules of the game for participants. In any change there will be winners and losers. Therefore, transition programs may be needed to help ease the losers into new roles. One way to facilitate transition is to allow flexibility in decision making. For farmers, flexibility in production decisions allows them to respond to market signals. Policy and program flexibility is also important to allow adjustment in programs within the lifetime of legislated bills (see American Farm Bureau Federation, 1988; Tutwiler and Carr, 1988).

One of the more difficult challenges for policymakers will be to identify what agricultural policies in general and commodity programs in particular can and cannot do. Congressman Dan Glickman (D-Kan.) recently addressed this challenge and suggested that what is needed is for the policy community to recognize what it does not want the commodity policies to address (Glickman, 1988). Currently, the commodity programs are seemingly expected to do it all—support or raise farm income, stabilize farm prices and incomes, expand exports, ensure a safe and plentiful food supply for Americans, protect soil and water resources, and promote rural development. Glickman suggests that the policies would be less blunt instruments if they were tailored to accomplish narrower goals. Both approaches have advantages and disadvantages. The more one set of policies tries to do, the less well it achieves any of the objectives, at least at an acceptable cost. On the other hand, a program that meets the narrowly defined goals of a single-issue bill may never materialize because the bill fails to draw sufficient support in the Congress.

The long-term goal should continue to be the promotion of a healthy, competitive, and diverse agricultural sector and viable, diverse rural communities. But any goals will have to be reconciled with the constraints set by the budget limitations and environmental, consumer, agribusiness, and other policy and societal objectives. Efficiency and equity are performance criteria that society deems desirable, but they often conflict. The balance between them is a political decision.

The U.S. agricultural community must begin to broaden its focus. It can no longer look primarily at the commodity programs as the main factor determining the sector's fortunes. The sector is an integral part of the U.S. and world economy. It stands to benefit greatly from public and private research and development, yet the agriculture community pays far less attention to state and federal agendas for scientific research and higher education and where agriculture fits into these than to land set-asides and target prices. A similar argument can be applied to the development and release of new technologies; unless the U.S. agricultural community plays an active role in the review, evaluation, and regulatory process for new technologies, U.S. agriculture could see promising opportunities lost to other countries whose agricultural sectors are on the ball. Environmental policies and regulations will increasingly affect U.S. agriculture. So far, much of the attention of environmental interest groups has been focused on relatively minor changes in commodity legislation. In the future, they may not feel so constrained and may seek to pass separate environmental legislation that could have a much greater impact on the entire food and fiber system. Many other U.S. policies—for example, tax, monetary, banking, health, trade, labor, immigration, and foreign—will continue to influence U.S. agriculture's well-being.

A sound economic base, a strong educational system, well-funded public research and development institutions, and a well-developed rural infrastructure were some of the factors that contributed to the United States' competitiveness in agriculture for much of the twentieth century. It would seem foolish to forget the importance of those factors in a rush to regain market shares and support returns to overpriced factors of production.

ACKNOWLEDGMENTS

The author gratefully acknowledges constructive criticism and input from the members of the Food and Agriculture Committee, David Trechter, and Roger Hitchner and intellectual stimulation from colleagues at the National Center for Food and Agricultural Policy. All errors or omissions are, of course, the sole responsibility of the author.

REFERENCES

American Farm Bureau Federation, 1990 Farm Program Study Committee. 1988. *1990 Farm Program Study,* August.

Browne, William P. 1988. "The Fragmented and Meandering Politics of Agriculture," in M. Ann Tutwiler, ed., *U.S. Agriculture in a Global Setting: An Agenda for the Future* (Washington, D.C., Resources for the Future).

Carr, A. Barry. 1988. "Agricultural Information and Technology: A Continuum of Change," in M. Ann Tutwiler, ed., *U.S. Agriculture in a Global Setting: An Agenda for the Future* (Washington, D.C., Resources for the Future).

Freshwater, David. 1989. "Rural Development and the 1990 Farm Bill," paper presented at the 1989 annual meeting of the American Agricultural Economics Association, Louisiana State University, Baton Rouge, July 30–August 2.

General Agreement on Tariffs and Trade (GATT). 1989. "Completion of Montreal Package," News of the Uruguay Round of Multilateral Trade Negotiations, Information and Media Relations Division of the General Agreement on Tariffs and Trade, mimeo, April 12, p. 6.

Glickman, Dan. 1988. "The Outlook for U.S. Commodity Policy in the Next Decade," *Looking Ahead* vol. 10, no. 4, pp. 1–4 (Washington, D.C., National Planning Association, August).

Kramer, Carol S. 1988. "Consumer Demands: A Balancing Act," in M. Ann Tutwiler, ed., *U.S. Agriculture in a Global Setting: An Agenda for the Future* (Washington, D.C., Resources for the Future).

Schneider, Keith. 1989. "Stores Bar Milk Produced by Drug," *The New York Times,* August 24, p. A1.

Sugarman, Carole. 1988. "Testing . . . Testing . . . ," *The Washington Post,* September 28, p. E1.

Swanson, Louis E. 1989. "The Rural Development Dilemma," *Resources* no. 96 (summer), pp. 14–16.

Thompson, Paul B. 1988. "The Philosophical Rationale," in M. Ann Tutwiler, ed., *U.S. Agriculture in a Global Setting: An Agenda for the Future* (Washington, D.C., Resources for the Future).

Tutwiler, M. Ann, and A. Barry Carr. 1988. "An Agenda for the Future," in M. Ann Tutwiler, ed., *U.S. Agriculture in a Global Setting: An Agenda for the Future* (Washington, D.C., Resources for the Future).

U.S. Department of Agriculture (USDA). 1984. *History of Agricultural Price-Support Programs, 1933–84* AIB 485 (Washington, D.C., Economic Research Service, December).

U.S. Department of Commerce, Bureau of the Census, and U.S. Department of Agriculture (USDA). 1988. *Current Population Reports: Farm Population,* Series P–27, no. 16 (June).

U.S. House of Representatives. 1985. "Conference Report on H.R. 2100, Food Security Act of 1985," *Congressional Record* vol. 131, no. 175, part II (December 17), p. H 12249.

Winters, L. Alan. 1988. "The So-Called 'Non-Economic' Objectives of Agricultural Policy," OECD Working Paper (Paris, Organization for Economic Cooperation and Development, April).

PART 1

CURRENT U.S. AGRICULTURAL POLICIES

AN INTRODUCTION TO U.S. AGRICULTURAL PROGRAMS

JOSEPH V. KENNEDY and JON VISSER

Federal farm legislation consists at present of a wide variety of commodity and conservation programs that pursue different and often contradictory goals. This chapter outlines the underlying structure of these programs and attempts to provide the reader with a basic understanding of each of the major components of the federal government's agricultural policy. Its primary purpose is to provide a background for some of the policy concepts discussed in the rest of this volume.

The implementation and operation of government programs is primarily the responsibility of the U.S. Department of Agriculture (USDA), headed by the Secretary of Agriculture. The primary financial arm of the USDA in supervising farm programs is the Commodity Credit Corporation (CCC). The CCC is a multibillion-dollar corporation owned and controlled by the federal government. To a large extent it acts as a financial clearinghouse for the government's commodity operations. Actual staffing and decision making are handled by a number of other agencies within the USDA, each with its own responsibilities. The most important to the commodity programs is the Agricultural Stabilization and Conservation Service (ASCS). This agency employs most of the professionals who actually implement and oversee the commodity programs. The key agency for implementing many of the government's agricultural conservation programs has been the Soil Conservation Service (SCS).

Although national agricultural policy is determined primarily in Washington, D.C., the vast majority of the decisions that most immediately affect individual farmers are made in thousands of federal offices across the country. Almost every county in the nation has a federal office, usually housing representatives of both the ASCS and SCS. These offices make basic determinations such as whether or not a farmer and his or her land qualify for various government programs, the farming history of the land including crop acreage and yields, and compliance with program requirements. Farmers participating in government programs quickly become familiar with their county offices. These offices, in turn, acquire an intimate knowledge of both the program requirements and the local farming community.

The current agricultural legislation, the Food Security Act of 1985, contains numerous options and alternatives, many of which are left to the discretion of the Secretary of Agriculture. As a result, much of the structure of today's programs, while based in the authorizing legislation of the act, is the result of administrative decisions made within the USDA and published as regulations or administrative guidelines.

PRICE AND INCOME SUPPORT PROGRAMS FOR FOOD GRAINS, FEED GRAINS, AND COTTON

The most important commodity programs in terms of government expenditures, acreage planted, and farmer income are those for the five major crops: wheat, feed grains (primarily corn—the largest program crop—but also oats, barley, sorghum, and rye), soybeans, cotton, and rice. The programs for each of these commodities are similar, although there are significant differences. Primary among these differences are the absence of target prices for soybeans and rye and the existence of mandatory marketing loans for cotton and rice farmers.

These government programs are voluntary in the sense that each farmer is allowed to assess the benefits and costs of participating in the government program and choose whether or not to enroll. A farmer opting not to enroll may generally plant any crop he or she wishes and sell it to the highest market bidder, but in doing so, the farmer forgoes current and future benefits. If the farmer plants nonprogram crops on base acreage he or she "loses" crop base (see below) and hence future payments. If the farmer plants additional program crops he or she may also become ineligible for government price and income supports for program crops.

Eligible Production

One of the most important concepts of the present farm programs is "eligible production." All the major commodity programs base payments to farmers on units of production: pounds for cotton, hundredweight for rice, and bushels for wheat and feed grains. The level of a farmer's support is based on his or her planting history. Farmers only receive payments for eligible production.

Eligible production is defined as a farm's program acreage multiplied by its program yield. The crop acreage base is equal to the average acreage planted and considered planted to the program crop during the last five years. There are many cases where land left idle or planted to other crops is considered planted to the program crop for the purpose of maintaining the crop acreage base in future years. Considered planted acreage includes land idled under Acreage Reduction Programs, land left unplanted under the 0/92 or 50/92 provisions, and, in certain cases, land devoted to other crops. Acreage Reduction Programs usually require farmers to leave a certain percentage of this land idle. Farmers who exceed their program acreage (less any required idled acreage) are ineligible for government support payments on any of their production of that crop.

Program yield equals the average crop program yield on the farm between 1981 and 1985 inclusive, excluding the highest and lowest annual program yields. Generally, the program yields for 1981 through 1985 equaled the average actual yield for the previous five years, excluding the highest and lowest actual yields. Thus, the program yield for 1986 was a weighted-average of actual yields dating back to 1976. Later legislation provided that the 1986 program yield could not fall below 90 percent of the 1985 program yield. Farmers are not prohibited from exceeding their program yields, but any excess production is not eligible for government payments. Although the Secretary has the authority to update farm yields on the basis of information from the most recent years, this authority has not been exercised to date. As a result, program yields have not increased with actual yields since 1986. This freeze in program yields has forced most farmers to look at market prices rather than the level of government support when deciding whether to add inputs in order to increase yields on their program acreage.

As with most payment programs, government payments are made only for actual production. Thus, a farmer who is prevented from planting a crop because of poor weather conditions is not eligible for deficiency payments on the unplanted crop. A special feature of the government programs creates an exception to this rule, however. Under the 0/92 and 50/92 provisions of the current farm legislation, grain farmers may choose to leave unplanted all or any portion of their crop and still receive 92 percent of the deficiency payment they would have received had they grown a full crop. Growers of cotton and rice must plant at least 50 percent of their crop to receive 92 percent of the benefits. Farmers are also eligible for deficiency payments if their crops fail after planting.

Acreage Reduction Programs

To control production, raise market prices, and lower government outlays, the government usually requires that farmers participating in support programs idle a certain percentage of their acreage base. The Food Security Act sets maximum and minimum limits on the extent of this requirement. These vary among crops but generally range from 0 to 35 percent. The percentage of land to be withdrawn from production is determined annually by the Secretary of Agriculture within the ranges set by the act.

The Secretary may also order a voluntary paid land-diversion program under which farmers can choose whether to idle additional acres in exchange for a direct payment from the government.

Farmers are generally required to conserve idled acreage by planting a cover crop on the land or otherwise protecting the land from erosion. The Secretary may allow farmers to plant nonprogram crops on the land, but this authority has generally not been exercised.

Acreage Reduction Programs (ARPs) are less than 100 percent effective in controlling crop production for two reasons. First, farmers generally devote their least fertile land to the ARP requirement. Second, by reducing production, ARPs raise market prices. This, in turn, may cause farmers to use the production costs saved on the retired acreage to purchase additional inputs for the remaining land. The net result then is a "slippage" in the effectiveness of the ARP requirement as the retired land produces a less-than-proportional decrease in overall production and farmers substitute other inputs for the land retired under the program. The fact that any additional production owing to yield increases does not qualify for deficiency payments reduces but does not eliminate this problem.

Target Prices

The total level of government support for each of the crops covered by these programs, except soybeans and rye, is determined primarily by the target price, which sets a minimum income support on all eligible production. Farmers participating in the programs receive a deficiency payment from the government for each unit of eligible production. This payment is generally equal to the difference between the target price and the higher of the loan rate or the average market price. As a result, participating farmers are assured of receiving the target price as their gross amount of return on each unit of production up to the program yield. Target prices thus act as income supports by supplementing the cash receipts a farmer can obtain from the commodity markets. Except to the extent that they affect production decisions, target prices do not affect the market price of a commodity.

Loan Rates

Current farm legislation specifies the formula for determining loan rates for each crop. Eventually, the loan rates will be set equal to 75–85 percent of the average market price for the preceding five years, not including the highest and lowest annual prices. However, the Food Security Act of 1985 placed limits on the maximum amount by which the loan rates for grains could drop from year to year. Minimum loan rates were specified for soybeans, cotton, and rice. These limits have resulted in loan rates that are higher than the formula price.

The Secretary of Agriculture has the authority to lower the loan levels as much as 20 percent if it is determined that the statutory levels are making U.S. production uncompetitive in world markets. This authority has frequently been exercised since enactment of the Food Security Act of 1985. It automatically increases the potential level of deficiency payments since it increases the difference between the target price and the loan rate. Unlike normal deficiency payments, payments made as the result of this authority (called Findley payments) are not included in the $50,000 payment limit discussed in the subsection entitled "Payment Limits."

Once a crop is harvested, participating farmers may use it as collateral to obtain a nonrecourse loan from the government. The loan amount is equal to the loan rate multiplied by the number of units of production placed under loan. Unlike deficiency payments based on target price, a farmer's entire production is eligible for protection under the loan programs.

The loan rate works as a price support program, essentially providing a floor for market prices. The loans for most crops mature in nine months. At or before this time, the farmer may repay the loan with interest, receive the pledged crop, and sell it at the market price. Alternatively, the farmer may default on the loan and keep the loan proceeds. Because the program involves nonrecourse loans, the government's right to recovery is limited to the crop used as collateral for the loan. It is this feature that causes the loan program to act as a price support for the commodity market. At prices close to or below the loan rate, farmers will find it more profitable to default on their loans and transfer title of the crop to the government than to sell the crop at current market prices.

Loan rates are especially important to the soybean program, since soybeans are the only major crop without a target price. For soybeans it is the loan rate, rather than the target price, that represents the guaranteed minimum payment a farmer can expect to receive on his or her production.

The loan rate also provides an important signal for production in excess of a farm's program yield. Because excess units of production are eligible for the loan program but not for deficiency payments, it is the loan rate, together with the expected market price, that provides the relevant information to farmers when making production decisions.

Marketing Loans

During the early to mid-1980s loan rates acted as floors on domestic commodity prices that kept them well above world prices. This, combined with a strong U.S. dollar, resulted in the United States losing market share to foreign competitors. The Food Security Act initiated a mandatory program for cotton and rice called the marketing loan. Under the marketing loan, the USDA publishes an estimate of the world price for cotton and rice each week. Farmers then have the option of repaying their commodity loans at this lower world price rather than at the original loan rate. The loan is considered paid in full even though only a portion of the loan amount is actually repaid. Supporters of this program hoped it would make cotton and rice more competitive in world markets by allowing prices to fall below the domestic loan rate and, by decreasing loan forfeitures, reduce government stocks.

The 1985 Food Security Act gives the Secretary of Agriculture the authority to initiate a marketing loan program for wheat, feed grains, and soybeans, but to date this authority has not been exercised.

Cross Compliance and Offsetting Compliance

The 1985 Food Security Act does not allow the imposition of strict cross-compliance requirements as a condition for participation in the commodity programs.

Under strict cross compliance, a farmer is prohibited from participating in the government program for a commodity unless he or she also participates in the commodity program for every other crop grown on the farm. However, the Food Security Act does permit the Secretary of Agriculture to prohibit farmers who participate in the program for wheat and feed grains in years in which an Acreage Reduction Program is in effect from exceeding their crop acreage base for any other crop grown on the farm. Farmers are not required, however, to comply with other provisions of the programs such as ARPs. Thus, although farmers do not have to participate in other crop programs, the amount of land they can plant to other crops is limited to the acreage base of those crops. Any production of other program crops is ineligible for government support unless the farmer formally participates in the programs for those crops. The Food Security Act allows the Secretary to extend this requirement to the programs for cotton and rice.

Under offsetting compliance, a farmer may not participate in a government program for a given crop unless all of the farms operated by that individual are enrolled in the program for that crop. This prevents a farmer from offsetting mandated reductions in production on one farm with expanded production on other farms outside the program. The 1985 Food Security Act prohibits the imposition of offsetting compliance as a condition for participating in the programs for rice and cotton. Imposition for wheat and feed grains is left to the Secretary's discretion.

Payment Limits

The Food Security Act of 1985 imposes a limit of $50,000 per year on direct payments to each person. However, the limit does not apply to many types of government outlays, particularly the value of government loans and crop purchases, gains realized from repayment of a marketing loan at a price below the original loan level, disaster payments, and any Findley payment made as a result of the Secretary's decision to lower the loan levels.

Congress and the USDA have had difficulty developing an effective definition of what constitutes a "person" for the purposes of applying the limitation. Anecdotal evidence suggests that as some farms approach the payment limitation, additional people have been added to the management of the farm, thus increasing the number of persons eligible to receive payment. This practice was severely restricted by the Budget Reconciliation Act of 1987, which tightened the definition of what constitutes a "person" for payment purposes. Persons are now eligible for government payments only if they are "actively engaged" in farming through participation in the management of the farm or contributions of land or labor toward its operation. In addition, no individual may receive payments from more than three farms. Finally, a separate $250,000 payment limit has been established, which includes deficiency payments and other payments such as storage subsidies, disaster payments, rental payments from the Conservation Reserve Program, and the difference between loan receipts and repayments under a marketing loan program.

MARKETING ORDERS

Federal marketing orders are used to regulate the marketing of a wide variety of agricultural products, including fluid milk (see below), fruits, and vegetables. They are administered by the Agricultural Marketing Service of the USDA. Each order is designed to cover a specific commodity grown within a designated region. Marketing orders must be approved by the growers within the geographic area covered. Usually two-thirds of those voting must approve; orders are then legally binding on all producers and/or handlers of that commodity.

The Capper-Volstead Cooperative Marketing Act of 1922 allows farmers to use marketing orders to influence the price and quality of the commodities they produce. Marketing orders can contain a number of different provisions, including (1) limits on the total amount of a commodity allowed to enter the market, (2) site and grade standards to improve or standardize quality, (3) mandatory or discretionary check-off provisions to finance administrative costs and advertising campaigns, and (4) marketing schedules, which spread out the delivery of commodities to the market.

THE DAIRY PROGRAMS

Milk Marketing Orders

Marketing order provisions were first instituted under the Agricultural Marketing Agreement Act of 1937. Unlike most commodity programs, which are administered by the ASCS, the milk marketing orders are administered by the Agricultural Marketing Service of the USDA. Milk marketing orders are defined for each of several major markets of fluid milk in the country. Each order has two basic sets of provisions. The first sets the minimum prices for fluid milk that handlers must pay to producers. The second specifies the distribution of returns among producers for selling milk. The provisions of each order apply only to Grade A milk that will be used in a fluid state.

Milk marketing orders are defined geographically and are voted on by milk producers operating within the covered region. Once two-thirds of the producers within a region have approved a milk marketing order, it becomes legally binding on all processors operating in the regulated area.

Each marketing order designates a minimum price that milk handlers must pay to producers or their cooperatives. Most orders base this price on the average price of manufacturing grade milk in Minnesota and Wisconsin (M-W Price), adding a premium for Class I milk (milk used for fluid purposes) and a regional differential. The addition of the regional differential generally causes minimum Class I prices to increase with distance from Wisconsin. Because the M-W Price is supported by federal purchases of manufactured products, the prices set by these marketing orders are influenced by the federal price support purchases.

Price Support Purchases

The government also maintains a price support program for milk producers. The Food Security Act sets a minimum level at which the government supports the price that milk producers receive. Because milk is not a storable commodity, the government implements this program by purchasing processed dairy products such as butter, cheese, and nonfat dry milk. The price at which each product is purchased is calculated to ensure that under normal market conditions the processors from whom the government buys the products are able to offer the support price to the producers from whom they purchase milk, after subtracting reasonable production costs. Unlike commodity loan programs, the milk price support is a direct purchase program: the government is committed to buying the total amount of commodities offered to it at the purchase price.

Changes in the price support level are currently tied to the level of government purchases. In any year in which the projected level of government purchases is expected to exceed 5 billion pounds in milk equivalent, the support price is lowered 50 cents per hundredweight. If the projected level of purchases is below 2.5 billion pounds, the support price is raised 50 cents per hundredweight. On occasion, however, Congress has passed legislation to avoid a scheduled reduction in the support price.

Dairy product imports into the United States are limited by quotas.

The Dairy Termination Program

High support levels and rapidly increasing yields per animal have combined to create chronic surpluses of dairy products. Unlike the programs for grains and cotton, the dairy program does not impose production controls on producers. The Food Security Act of 1985 attempted to deal with the surplus by initiating the Dairy Termination Program.

In the early 1980s the government implemented a diversion program under which producers were paid to dump surplus milk rather than sell it. The diversion program temporarily reduced surpluses, but rapid expansions in production including increases in output per animal eventually restored the problem of surpluses.

The Dairy Termination Program attempted to achieve a more permanent solution to the problem of surpluses. Under the Food Security Act and regulations issued by the USDA, dairy producers wishing to participate in the program submitted bids to the department. Producer bids were based on dollars per hundredweight of the producer's historical fluid milk production. Successful bidders agreed to slaughter or sell for export their entire population of cows, heifers, and calves and to refrain from producing milk during the next five years. The Food Security Act linked the termination program to the price support level by providing that the USDA could not lower the support price of milk in subsequent years unless it accepted enough bids into the termination program to reduce milk production by at least 12 billion pounds during the eighteen months of the program (April 1, 1986 to September 30, 1987). The Food Security Act gave the Secretary of Agriculture authority to conduct future termination programs if conditions warrant.

THE SUGAR PROGRAM

Sugar production in the United States is subsidized by a system of price supports and import quotas. The structure of the sugar program places most of the cost for this subsidy—roughly $3 billion each year—on the consumers and users of sugar rather than the government. As a result, the budget pressure that has recently led to cuts in the target prices and support levels for other programs has had little impact on the sugar program.

The Food Security Act requires the Secretary of Agriculture to support the price of raw cane sugar at a minimum of 18 cents per pound. The price of beet sugar is set administratively at a level that is fair and reasonable in relation to the support price of cane sugar. This support is offered as nonrecourse loans to refiners who process raw sugar into its refined form.

The act also requires the Secretary to operate the program in such a way that there is no net cost to the U.S. government. The Secretary must maintain the domestic price of sugar at a level high enough to prevent refiners from forfeiting their loans to the government. Domestic sugar producers do not face production restrictions, and the United States is currently a net importer of sugar. There are, however, barriers to sugar imports into the United States. U.S. trade law currently imposes a tariff of 2.5 cents per pound of sugar imported. But this is not high enough to overcome what has historically been a large difference between the U.S. and world prices of sugar. As a result, import quotas have been necessary. The authority for these quotas is contained in a headnote to the U.S. trade laws. The quota is determined by the Secretary each year and is set equal to the maximum amount of imports that the domestic market can absorb while still keeping the domestic price of sugar at a level high enough to prevent forfeitures of sugar to the government.

The total import quota is divided among countries that have historically exported sugar to the United States. Countries with quota levels include both developed and developing nations, although there have been periodic attempts in Congress to transfer quotas from developed countries such as Australia to less developed countries such as those in the Caribbean Basin and the Philippines.

The sugar program has resulted in a significant reduction in the domestic demand for sugar. High domestic prices have caused many manufacturers—the most important among them being soft drink producers—to switch to alternative sweeteners. Primary among these alternatives are high-fructose sweeteners made from corn starch. These have increased the domestic demand for corn, making corn producers beneficiaries of the sugar program.

In response to a complaint from the Australian government, a recent panel of the General Agreement on Tariffs and Trade (GATT) determined that the U.S. sugar program violated GATT agreements. This determination has been accepted by the U.S. government. As a result, the United States will either have to reform its sugar program to bring it into compliance with the GATT or offer compensation to Australia in the form of reduced tariffs on other goods. It is unclear which alternative the government will propose or, if the program is reformed, when and how this will be implemented.

THE TOBACCO PROGRAM

The tobacco legislation is contained in permanent legislation, not the Food Security Act. Because authority for the tobacco program does not expire in 1990, new legislation is not needed. It is therefore possible that the program will not be a subject of the 1990 farm debate.

Tobacco is one of the few commodities in which production is based on a national quota system. Unlike most other programs, individual farmers do not have the right to decide whether or not to participate in the program. The programs for major types of tobacco are based on farm level poundage marketing quotas that represent historical production. Quotas are attached to individual farms and convey the right to plant and market a certain amount of tobacco each year. The amount of quota is adjusted according to anticipated demand; however, the land to which quotas are attached is fixed. Thus, new farmers may not market tobacco unless they purchase or rent land with a quota allocation from an existing holder in their own county.

The government has a separate program for each type of tobacco grown in the United States. Every three years, quota owners or growers vote to determine whether to have a program for the coming three years. If at least two-thirds of the farmers agree to a program, the government will run one. The program is then mandatory for all growers, and no farmer is allowed to market that type of tobacco without a quota allocation. Growers of some types of tobacco have voted against having programs.

The permanent legislation requires the USDA to set minimum price levels for each of the two major types of tobacco, burley and flue-cured. Each year the Secretary surveys domestic and foreign demand to determine the level of domestic production that can be sold at the minimum price level. The right to grow this amount is then allocated proportionally to quota holders. Tobacco growers typically market their crop through auctions. At these auctions the government enforces the price support by lending the growers' association the funds needed to purchase any tobacco not sold at the support price. A quota assessment is placed on all tobacco marketed. The proceeds from this assessment are used to reimburse the Commodity Credit Corporation for any losses associated with these loans.

THE PEANUT PROGRAM

The price support program for peanuts is run in a manner similar to the tobacco program. A farmer referendum was held in January 1986 to determine whether a program would be implemented. Because two-thirds of the farmers voting favored such a program, the Secretary will implement one for each crop year from 1986 to 1990. The authority from this referendum expires in 1991.

Under the government program, the Secretary determines an annual national poundage quota for peanuts for domestic edible, seed, and related uses. The Secretary may not set a quota below 1.1 million tons. This national quota is apportioned among farms based on historical production. Each farm is allocated a poundage quota. Farmers may sell or lease this quota to other farmers in their county.

The production and marketing of peanuts are divided into quota and "additional" peanut categories. Both are entitled to price protection under the program, but the support price for quota peanuts is much higher than that for additional peanuts. Quota peanuts are defined as those eligible for domestic edible use. Domestic edible peanuts include those used to produce domestic food peanuts, seed, and peanuts used on the farm. Total sales of quota peanuts from a farm may not exceed the poundage quota that the farmer has been given by the government or has leased from another farmer. The total amount of quota peanuts is determined by the Secretary on the basis of the forecasted demand for these peanuts at the legislated loan rate. Additional peanuts are defined as all peanuts sold from a farm in excess of the farm's poundage. They may be purchased only for exporting or crushing.

The Food Security Act imposes a marketing penalty on all peanuts marketed for domestic use in excess of a farm's poundage quota.

The price support for peanuts for domestic use is set legislatively and is indexed to changes in the cost of production, excluding land. The support level for additional peanuts is set at the level the Secretary determines appropriate based on the demand for peanut oil and meal, the prices of other vegetable oils and protein meals, and the demand for peanuts in foreign markets. The market price for additional peanuts is usually well above this loan level.

The peanut program allows buyers to purchase additional peanuts at the price support level for domestic-use peanuts. The proceeds from these sales are placed in a pool for distribution to growers within the region. As a result, the domestic-use loan level acts as both a floor and a ceiling, since farmers may sell to the government at the support level but buyers may purchase additional peanuts whenever the market price rises above the support level.

THE WOOL AND MOHAIR PROGRAM

Authority for government support of wool and mohair production is contained in the National Wool Act of 1954. The 1985 Food Security Act extended this authority through 1990. Under the program, the Secretary of Agriculture must support the price of wool and mohair through loans, purchases, payments, or other operations. The support rate for shorn wool is based on an index of prices paid by farmers for commodities and services including interest, taxes, and farm wages. The support price for both pulled wool and mohair must maintain normal marketing relationships between pulled and shorn wool and maintain approximately the same percentage of parity for mohair as for shorn wool.

THE HONEY PROGRAM

The government offers nonrecourse loans to domestic honey producers. Under the Food Security Act, the loan level must at least equal 75 percent of the average annual price level in the previous five years, excluding the highest and lowest annual prices. The Secretary has implemented a marketing loan for honey producers, allowing them to repay their loans at lower prices. The total value of these loans is limited to $250,000 per individual.

CROP INSURANCE AND DISASTER PAYMENTS

Farmers may purchase crop insurance against losses on their farm. Insurance is available against different risks for a wide range of commodities. Farmers participating in the program may choose to insure 50, 65, or 75 percent of their normal established yields from the farm. They also have a choice as to the market price at which they wish to purchase insurance. The insurance premiums are subsidized by the federal government.

Crop insurance is not designed to cover all of a farmer's losses, and most farmers have historically chosen not to purchase insurance. In 1988 and 1989, Congress passed legislation mandating disaster payments for farmers who suffered losses because of a variety of weather conditions, primarily the droughts of those years. These disaster payments are usually based on a percentage of what the farmer normally would have produced. Total disaster payments are limited to $100,000 annually per person.

TRADE-EXPANDING PROVISIONS

The debate on the 1985 Food Security Act was conducted amid rapid declines in both the volume and value of U.S. agricultural exports and sharp increases in the level of U.S. government stocks. A number of the trade provisions in the act were therefore aimed at moving U.S. commodities onto world markets. These include export subsidies, export promotion, credit guaranteed for export purchases, and commodity donations. Many of the payments under these programs are made in the form of government-owned commodities.

Although most of the programs are funded from the CCC account, the majority are administered by the Foreign Agricultural Service rather than ASCS.

The Export Enhancement Program

The Export Enhancement Program (EEP) was first initiated in May 1985. The Food Security Act broadened the program's scope and mandated a minimum level of payments. Under the EEP, government subsidies are offered to domestic exporters selling specific commodities to specific countries. Originally these commodities and countries were to be carefully selected in order to counter instances of unfair subsidization by other exporters, primarily the European Community (EC). The fact that this program subsidized sales to nontraditional U.S. customers but not to traditional purchasers of these commodities, including the Soviet Union and China, created tension and pressure to expand the program. After much debate the program's authority was enlarged to include these countries and lost much of its original targeted nature.

Although the EEP has been used to support sales of a large variety of commodities, the vast majority of its initiatives have been to subsidize exports of wheat and wheat flour, the commodities with which the EC competes most aggressively with the United States.

EEP proposals are approved by an interagency review group. Proposals are specific to both country and commodity, entitling the recipient country to purchase a stated volume of a given commodity. The country then solicits offers from U.S. exporters for the commodity it is authorized to purchase. Once the exporter and the purchasing country have agreed on the price and terms of the sale, the exporter submits its contract to the USDA for approval together with a request for the specific amount of subsidy needed to make the sale. If the USDA believes that the contract price does not fall below what it considers to be the world price, and the amount of the requested bonus does not exceed the amount the USDA thinks is necessary to make the sale, it will approve the contract and pay the exporter the requested bonus. In recent sales to the Soviet Union, the department has provided exporters with the flat amount per ton that it was willing to offer as a subsidy rather than allowing exporters to request their own levels.

The EEP thus provides payments to companies that export commodities abroad. Bonus payments are made in the form of commodity certificates redeemable for CCC-owned commodities. These payments enable exporters to offer the purchasing countries a lower price than they otherwise would.

Much of the debate on the wisdom of this program involves the concepts of diversion and additionality. Diversion occurs when subsidized sales to a country fail to result in a reduction in the absolute volume of EC sales of the commodity but instead divert sales to other countries. The volume of U.S. exports to these other markets may then be lowered, countering at least part of the increased imports resulting from the EEP and diminishing the program's effectiveness. In addition, if the EEP is used to subsidize sales that would have otherwise taken place even if unsubsidized, the program does not result in additional exports of U.S. agricultural products (that is, there is no additionality) and instead causes a loss of revenues. Any diversion of EC exports to other countries increases this lack of additionality.

Targeted Export Assistance

The Food Security Act of 1985 created the Targeted Export Assistance Program as a new export promotion program. The act allows the Secretary of Agriculture to use CCC funds or commodities to counter the effects of unfair trade practices by foreign countries. The USDA has implemented this provision by creating a program that provides government support to commodity organizations engaging in export promotion activities abroad. Much of the funding from this program has gone to commodities or products not eligible for other government programs such as the EEP. This includes high-value horticulture crops such as fruit and nuts.

Organizations that wish to participate in the program apply to the USDA for funding. Successful applicants receive generic certificates to fund agreed-upon marketing activities aimed at promoting U.S. products in foreign markets. Participants in the program include farmer-owned agricultural trade organizations, state agencies, and private corporations. Under the program's regulations, these organizations must demonstrate that (1) their product has been the target of unfair trade practices, (2) an adequate supply of the product exists, and (3) at least 50 percent of the product originates in the United States. Preference is given to products that originate fully in the United States. Participants must also submit a marketing plan outlining how they will use the funds from the program.

Credit Guarantees

The United States offers both short-term (up to three years) and medium-term (from three to ten years) loan guarantees to foreign purchasers of U.S. agricultural products. These guarantees are intended to help buyers finance the purchase of U.S. commodities. The government annually allocates guarantee authority among countries. When these countries purchase U.S. products, they generally finance the purchase with the exporter. The exporter typically assigns the product to a U.S. financial institution, which collects payments on the debt. If the loan is guaranteed, the financial institution may file a claim with the CCC if payments on the loan are not made. Upon receipt of a claim, the CCC will assume the payments. The CCC charges borrowers a fee, which is less than 1 percent of the loan value, to cover administrative expenses and possible defaults.

The Food for Peace Program

The Food for Peace program, or Public Law 480, named after the 1954 legislation that first established the program, provides surplus agricultural commodities to low- and middle-income countries. These shipments may take the form of donations or subsidized sales.

Under Title I of the program, foreign governments may purchase U.S. agricultural commodities with long-term concessional credit. The purchasing country usually generates domestic currency by selling these commodities in its domestic market. It may then use the funds for certain developmental or other purposes specified in P.L. 480. The revenue generated by this program may not be used to fund projects aimed at increasing the production of agricultural commodities that would compete in world markets with U.S. agricultural exports.

Title II of P.L. 480 authorizes the U.S. government to donate surplus commodities to other governments, private voluntary organizations (PVOs), or the World Food Program (WFP). The minimum amount of annual donations is set at 1.9 million tons for each of fiscal years 1987 to 1990. Of this amount, at least 1.425 million tons must be channeled through PVOs or the WFP. A PVO may apply for authority to sell up to 5 percent of any commodities it receives, provided it details the intended use of these funds.

Section 416 and Food for Progress

Section 416 of the Agricultural Act of 1949 authorizes the Secretary of Agriculture to make domestic or overseas donations of CCC stocks. The 1985 Food Security Act significantly expanded the list of commodities eligible for donation under this program and for the first time set a minimum level of annual donations equal to the lesser of 500,000 tons or 10 percent of the CCC's uncommitted stocks of grains and oilseeds.

The Food Security Act authorized a new commodity program to assist countries committed to the implementation of market-oriented agricultural policy reforms. Under the program, called Food for Progress, the USDA is to distribute at least 75,000 tons annually by authority of section 416. No sales that would displace U.S. commercial sales are allowed.

TRADE RESTRICTIONS: SECTION 22

Apart from the commodity-specific trade restrictions discussed in the previous sections on the commodity programs, probably the most important trade restriction is that commonly referred to as section 22. Section 22 of the Agricultural Adjustment Act of 1933 allows the president to impose quotas and tariffs on the imports of an agricultural commodity or product whenever the imports are materially interfering with the operation of the government commodity programs. No finding of unfair competition is required for action to be taken. Section 22 would normally violate the GATT, but since 1955 the United States has obtained a waiver for this section.

COMMODITY STOCKS

Existing farm legislation contains a variety of programs aimed at taking stocks off the market. Most of these programs originated to make it easier for farmers to withhold their crops from the market until prices rose. By allowing farmers to spread their sales more evenly across the marketing year, these programs are designed to even out seasonal and yearly fluctuations in prices and production.

Farmer-Owned Reserve

Under the Farmer-Owned Reserve (FOR), the Secretary of Agriculture makes multiyear loans to producers of wheat and feed grains to maintain reserve stocks of these commodities. The Food Security Act set maximum and minimum levels on the amount of grain that could be brought into this program. Under current legislation, whenever the amount of wheat in the FOR is less than 300 million bushels and the market price does not exceed 140 percent of the nonrecourse loan rate, the Secretary must encourage participation in the FOR by offering increased storage payments and loan levels, interest waivers, or other incentives. The corresponding minimum level for feed grains is 450 million bushels.

The upper levels for participation in the program are set at 30 percent of the estimated use of wheat and 15 percent of the estimated use of feed grains. The Secretary has the authority to increase these maximum levels to 33 and 16.5 percent, respectively, if necessary for the effective operation of the FOR. Whenever enrollment threatens to exceed these levels, the Secretary must close the reserve to further crops.

A farmer who enrolls crops in the FOR receives a three-year loan from the government. The farmer also receives an annual storage payment for maintaining the stored crop. In return, the farmer agrees not to sell the crop on the market unless the market price exceeds the maximum of the target price or 140 percent of the loan rate for that year's crop. Farmers who violate this agreement are subject to penalties, including the return of storage payments. When the market price exceeds the maximum of the target price or 140 percent of the loan rate in effect for the year in which the crops were placed under loan, the farmer may market the grain held in storage without penalty. Farmers do not have to sell the crop, but they no longer receive storage payments.

CCC Stocks

As a result of both forfeitures under the commodity loan programs and direct market purchases, the CCC has in the past acquired large stocks of surplus commodities. Legislation usually prevents the CCC from selling these stocks on the open market except at high prices. This prevents the CCC from directly competing with farmers for market sales.

A number of programs permit the Secretary to make donations of CCC stocks to various domestic and foreign programs. In addition, government payments under some programs are made in the form of generic or commodity certificates. Commodity certificates can be redeemed for CCC-owned stocks. The restrictions on government commodity sales do not apply to releases under a certificate program. Thus, certificates provide the CCC with a way to release its stocks onto the market at current prices.

Generic Commodity Certificates

The Food Security Act authorized the use of generic commodity certificates. The certificates are denominated in dollars and can be redeemed for cash or used to purchase a certain amount of government-owned commodities at the government-determined market price. The CCC periodically publishes an inventory of the commodities it owns. Certificate holders may select commodities from this catalog using certificates to pay for them at various locations around the country. Farmers may also use the certificates to redeem their crops out of the CCC loan program. The certificates are negotiable, so farmers can sell them to other individuals if they wish.

The price at which certificates may be redeemed for crops is known as the posted county price (PCP). It is set by the CCC on the basis of the estimated market price in each area of the country. Although the government attempts to follow market prices, the PCP can also affect market prices by controlling the flow of government stocks onto the market.

Certificates often trade above their face value. The use of certificates allows traders to purchase certificates in one area of the country and redeem them for commodities in a different area without paying transportation costs for shipping the commodities. In addition, large traders are able to consolidate purchases and fit the timing and location of redemption into their other business operations. Finally, when the market price is below the loan rate, certificates allow a farmer to gain the full benefit of the loan program without having to pay for storage during the loan period. This is accomplished by placing the crop under loan and then immediately buying it back from the government at the PCP with certificates.

CONSERVATION PROVISIONS

Current farm legislation contains a number of provisions aimed at encouraging the conservation of farmland, especially in fragile environments. These programs are generally of two types. Under the first, the government makes payments available to producers who agree to follow certain practices ranging from the adoption of better conservation measures to the retirement of land from farming. The second denies farmers the right to participate in commodity programs if they engage in certain environmentally destructive behavior.

The Conservation Reserve Program

The Conservation Reserve Program (CRP) was created by the 1985 Food Security Act. Under the CRP, the Secretary is required to retire between 40 million and 45 million acres of highly erodible farmland by 1990. Although the program is aimed mainly at controlling erosion, the Secretary has the authority to enroll land with soils that have excess salinity or pose an off-farm environmental threat. The legislation established a timetable under which roughly 10 million acres are to be enrolled in the CRP each year. To date, 33 million acres are participating in the program.

The CRP operates under a bid system. The USDA periodically solicits bids to bring additional land into the program. Landowners may then submit bids to the Secretary listing the annual payment they will accept in exchange for retiring a specific number of acres. Successful bidders are required to enter into 10- to 15-year contracts with the government. During this time the land may not be farmed for any purpose without the Secretary's approval. In addition, the farm's crop base or allotment is reduced in proportion to the amount of the farm placed in the CRP. In return the farmer receives the bid amount as an annual payment from the government. The government also pays for 50 percent of the cost of planting an eligible cover crop on the land. Land entered into the CRP must generally have a certain erosion rating to be eligible for participation in the reserve. Annual payments under the program are subject to a separate $50,000 limit per person.

Sodbuster, Swampbuster, and Conservation Compliance

Since the enactment of the 1985 Food Security Act, any person who produces a commodity on highly erodible land (that is, engages in "sodbusting") is ineligible to participate in most government agricultural assistance programs, including the price and income supports, crop insurance, disaster payments, and Farmers Home Administration loans. If the land was used to produce a commodity or fulfill an acreage reduction requirement during the 1981–1985 crop years, ineligibility is delayed until the later of January 1, 1990, or the time that the land is classified under the land capability classification system by the SCS. Once each farm has been categorized according to its erosion potential, the land must be farmed according to a conservation compliance plan approved by the local SCS committee or the Secretary. The farmer must be in full compliance with this program by January 1, 1995, to remain eligible for government support.

Also since the enactment of the 1985 Food Security Act, any person who produces an agricultural commodity on wetlands that are converted to farm use (that is, engages in "swampbusting") becomes ineligible for the same government programs listed in the sodbuster provisions. For both the sodbuster and swampbuster provisions, this ineligibility extends to all farms operated by that person.

CREDIT PROGRAMS

The federal government operates a number of programs providing either direct credit or credit guarantees to farmers. These programs attempt to reduce the cost of medium- and long-term credit to farmers.

The key institutions involved in credit programs are the Farm Credit System and the Farmers Home Administration. Also important is the Rural Electrification Administration, which issues direct and guaranteed loans to finance utility projects in rural communities.

Farm Credit System

The Farm Credit System (FCS) is a family of member-owned cooperatives that provides credit and related services to farmers, ranchers, and related businesses. Although it is not formally a government agency, it is sponsored by the federal government and is thus able to obtain financing at favorable rates from the marketplace.

In the late 1970s and early 1980s the FCS issued loans to farmers on the basis of individual net worth statements that reflected rapidly rising land prices. When these prices dropped dramatically, the FCS and many private bankers found that the value of their collateral no longer covered the amount of loans outstanding. When rapidly dropping prices caused many farmers to fall behind on their loan repayments, agricultural lenders found themselves facing large losses.

In 1987 Congress passed the Agricultural Credit Act, which restructured the FCS system. The legislation established a new entity, the Federal Agricultural Mortgage Corporation ("Farmer Mac"). Farmer Mac will establish underwriting standards for agricultural mortgages. These mortgages will then be pooled and sold in a secondary market. A subordinated participation interest equal to 10 percent of the principal value of the pool will be required to cover defaults on the underlying mortgages. Farmer Mac will guarantee timely payments of principal and interest on the pools and will cover defaults after the 10 percent reserve is exhausted.

Farmers Home Administration

The Farmers Home Administration (FmHA) makes direct loans to eligible farmers. It also guarantees loans made by other agricultural lenders. FmHA serves as a "lender of last resort," making loans to family farmers who, for one reason or another, are unable to obtain credit at reasonable rates and terms.

FmHA's direct loans often supplement borrowing from other sources. They are usually intended to serve as a temporary bridge for individuals who are just beginning to engage in farming or are attempting to recover from setbacks. In addition to lower qualifying standards, FmHA loans usually carry lower interest rates and easier servicing requirements. In the past the agency has been reluctant to foreclose on farmers who experience difficulty repaying their loans. Instead, FmHA has usually attempted to restructure the loan in a way which allows the individual to continue farming the land.

During the 1980s FmHA experienced large losses on its loan portfolio as a result of the deterioration in farm income. Congress has generally pressed the agency to be lenient with borrowers and to continue extending credit in order to help producers stay on the farm.

Recently the agency has attempted to concentrate on guarantees rather than direct loans. Under this program FmHA will agree to guarantee the repayment of principal and interest on loans made by private institutions to qualifying borrowers.

ACKNOWLEDGMENT

The authors gratefully acknowledge constructive comments and input from David Trechter, Roger Hitchner, Daniel Sumner, and Larry Salathe.

SELECTED READINGS

Food Security Act of 1985, P.L. 99–108, 99th Congress, 1st session (Washington, D.C., Government Printing Office).

Glaser, Lawrence K. 1986. *Provisions of the Food Security Act of 1985* AIB–498 (Washington, D.C., U.S. Department of Agriculture, Economic Research Service, April).

Heimlich, Ralph E., and L. L. Langer. 1986. *Swampbusting: Wetland Conversion and Farm Programs* AER–551 (Washington, D.C., U.S. Department of Agriculture, Economic Research Service, August).

Hiemstra, Stephen W., S. R. Koenig, and D. Freshwater. 1988. *Prospects for a Secondary Market for Farm Mortgages* AER–603 (Washington, D.C., U.S. Department of Agriculture, Economic Research Service, December).

McDowell, Howard, A. M. Fleming, and R. F. Fallert. 1988. *Federal Milk Marketing Orders: An Analysis of Alternative Policies* AER–598 (Washington, D.C., U.S. Department of Agriculture, Economic Research Service, September).

U.S. Department of Agriculture. 1985. *Sodbusting: Land Use Change and Farm Programs* AIB–536 (Washington, D.C., Economic Research Service, June).

U.S. Department of Agriculture. 1985. *Agricultural-Food Policy Review: Commodity Program Perspectives* (Washington, D.C., Economic Research Service, July).

U.S. General Accounting Office. 1988. *Farm Programs: An Overview of Price and Income Support, and Storage Programs* RCED–88–84BR (Washington, D.C., February).

A MARKET CONTEXT FOR THE 1990 FARM BILL DEBATE

PATRICK O'BRIEN, WILLIAM H. MEYERS, and ANDREW S. MORTON

Farm bill debates typically start off with prognostications about the market environment likely over the life of the new legislation. This starting point reflects the importance of the market setting in shaping both the legislative agenda and the options open to policymakers. It also reflects the difficulties involved in forecasting market movements several years in advance with any degree of reliability.

The problems facing the agricultural sector and the options available to policymakers are seen very differently if the market setting over the life of new legislation is expected to be bullish or bearish. The 1981 farm bill debate, for example, took place amidst bullish market expectations; it focused heavily on expanding capacity fast enough to meet demand and raising income supports fast enough to keep pace with increasing production costs and market prices. Conversely, the 1985 debate took place amidst bearish expectations; it focused heavily on restricting capacity and reducing supports to discourage excess production and expand use.

Unfortunately, neither prognosis proved particularly insightful. The bearish market environment of the first half of the 1980s differed substantially from the setting projected in 1981. And the setting over the last half of the 1980s proved less bearish than projected in 1985. Simply put, the major commodity markets are volatile enough that short-term changes in weather, policy, or the macroeconomy can overturn the best longer-run projections.

A look at farm income and farm program costs illustrates this point. A 10 percent swing in commodity prices—a range well within past year-to-year movements—can generate a 30 percent swing in farm income. Similarly, a 10 percent swing in program commodity prices when prices are near support levels can generate a 35 percent swing in farm program costs (U.S. Department of Agriculture, 1989).

Keeping in mind the difficulties inherent in making projections, we speculate in this chapter on the market setting likely in the first half of the 1990s using three different perspectives. The first draws on the more qualitative approach to projections used before advances in computer science and econometrics made large-scale modeling possible. It focuses on evaluating agriculture's performance over the last several decades to identify upturns and downturns in the market, analyze the factors behind these market movements, and speculate on how these factors might influence future performance.

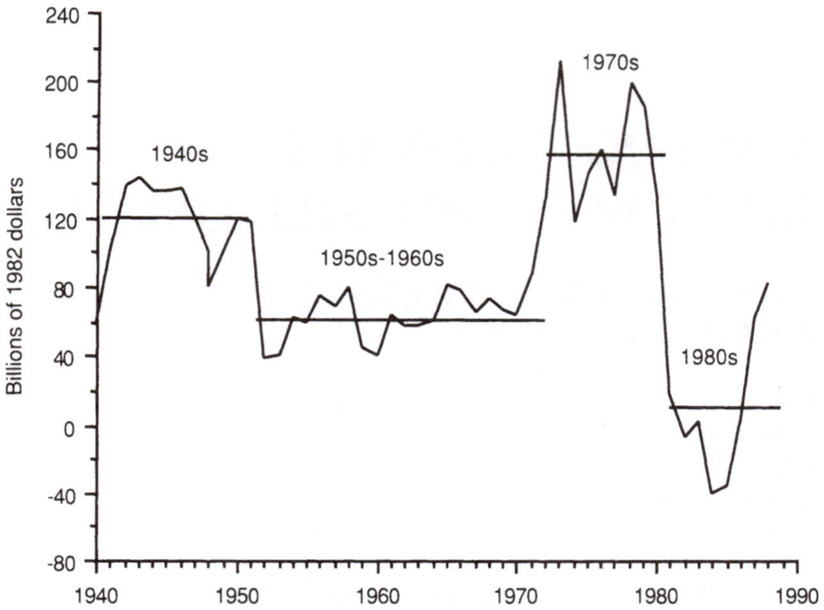

FIGURE 1. Real composite agricultural sector returns, 1940–1988. *Notes:* Returns are calculated as net cash income and value of change in equity. Horizontal lines are returns averages for the 1940s (1940–1951; mean = $120 billion), the 1950s and 1960s (1952–1972; mean = $63 billion), the 1970s (1973–1980; mean = $158 billion), and the 1980s (1981–1988; mean = $11 billion). *Source:* Data from U.S. Department of Agriculture, various 1988 and 1989 issues, *Economic Indicators of the Farm Sector* (Washington, D.C., Economic Research Service).

The second perspective draws heavily on large-scale simulation modeling to provide a more quantitative picture of the early 1990s. Past performance notwithstanding, many of the issues being debated involve complex trade-offs that cannot be evaluated without longer-run commodity supply, demand, and price indicators and sector performance measures. Sufficient numerical detail is included for a base scenario from two different sources to support benefit–cost analysis of policy options.

The last perspective focuses on the total budget and its influence on agricultural programs, recognizing that real concern about the federal deficit could be of overriding importance in the upcoming debate.

Combined, the three perspectives provide a view of what the early 1990s could hold for the agricultural sector.

A QUALITATIVE ASSESSMENT OF THE OUTLOOK FOR THE EARLY 1990s

How has the agricultural sector fared over the last several decades? What does its historical performance record show regarding market upturns and downturns? What does this imply for policymakers focusing on the 1990s?

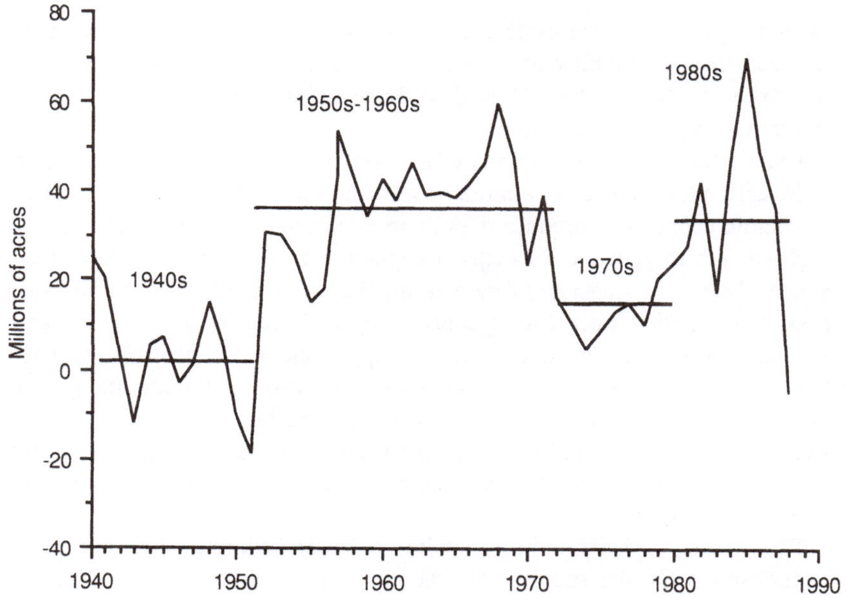

FIGURE 2. Excess capacity in U.S. agriculture, 1940–1988. *Notes:* In acreage equivalents. Horizontal lines are excess capacity averages for the 1940s (1940–1951; mean = 3 million acres), the 1950s–1960s (1952–1972; mean = 37 million acres), the 1970s (1973–1980; mean = 13 million acres), and the 1980s (1981–1988; mean = 35 million acres). *Source:* Dvoskin (1988).

In a sector as diverse as agriculture, it is difficult to develop any single measure of performance without overgeneralizing. Therefore, figure 1 uses a deflated composite income measure to track sector performance. The composite combines net cash income (cash income from all sources minus cash expenses for all items) with changes in the sector's equity position (changes in asset values minus changes in debt). As the plot indicates, the sector has gone through four more-or-less distinct market upturns and downturns since 1940. The 1940s and the 1970s were generally bullish periods, while the 1950s–1960s and the 1980s were more bearish. It is also noteworthy that within each of these periods there were shorter-lived upturns and downturns in composite returns.

While this financial information helps to identify broad movements in the market, it does little to identify the cause-and-effect relationships needed to analyze past performance or to speculate about the 1990s. Figure 2 uses an alternative data set from a recent study by the Economic Research Service on excess capacity that reflects the same market movements but provides a stronger framework for analysis (Dvoskin, 1988). The excess capacity data shown in the figure measure the difference between the sector's capacity to produce with its established natural resource and technology bases and effective market demand for its products here and abroad at support prices. Potential supply includes diverted production—the product that could have been produced with set-aside acreage in use—and the excess stocks accumulated from commodity price and income support operations. Commercial use excludes disappearance linked to domestic and foreign food assistance programs and subsidized exports.

Underlying this second measure of performance is a sense that postwar U.S. agriculture has faced a more-or-less chronic problem of excess capacity—more capacity to produce than there is demand for its product at prices high enough to generate "socially acceptable" farm incomes. Alternatively stated, these excess capacity data measure the extent of government intervention in the market since excess capacity by definition would not exist in a free market setting. In either case, as supply shocks or demand windfalls here or abroad work to reduce excess capacity, markets tighten and returns turn up. The converse is also true. Note that the bullish 1940s and 1970s were periods of sharp and sustained downturns in excess capacity, while the bearish 1950s–1960s and 1980s coincided with sharp and sustained upturns in excess capacity.

This excess capacity indicator points to several simple cause-and-effect relationships. On the supply side, the sector's capacity to produce can be broken down in the simplest terms into resource and productivity components. Effective demand can be broken down into factors affecting the large stable domestic market and those affecting the smaller but more volatile export market. The severity of the sector's excess capacity problem, and in turn the market environment, can be seen as the changing balance between these supply and demand factors.

Agriculture's Capacity to Produce

Figure 3 traces the resource and productivity factors shaping the supply side of the excess capacity measure over the postwar period. The figure uses data from the U.S. Department of Agriculture (USDA) on cropland use as a proxy for the larger natural resource bundle used in agriculture. USDA's all-factor productivity index is used to measure the pace at which farmers have been able to increase output per unit of natural and man-made input over time.

The sector has enjoyed steady growth in productive capacity of roughly 2 percent per year. As the cropland use plot suggests, the sector's resource base has actually declined fractionally since 1940, with year-to-year fluctuations around an essentially constant 375 million to 400 million acres related largely to changes in various acreage control programs. Hence, growth in productivity has accounted for all of the sector's postwar 150 percent growth in capacity.

Equally important, these productivity and natural resource patterns have not been sensitive to changes in producer returns. As figure 1 indicates, returns have fluctuated widely, but these swings have not been reflected in the movement of natural resources—as opposed to labor and capital—into or out of agriculture or changes in productivity growth. While this resource rigidity relates in part to the acreage-based farm support programs in effect

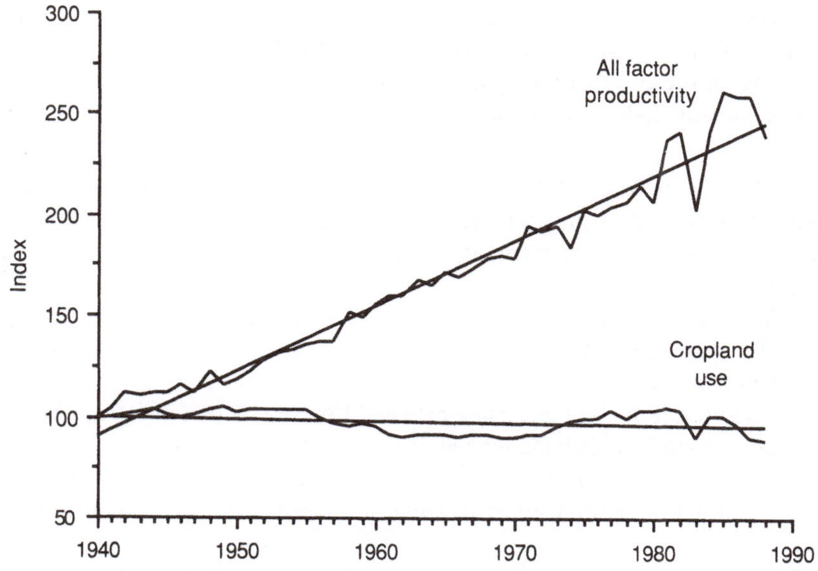

FIGURE 3. Productive capacity components of U.S. agriculture, 1940–1988, 1940 = 100. *Source:* Data from U.S. Department of Agriculture, various 1988 and 1989 issues, *Economic Indicators of the Farm Sector* (Washington, D.C., Economic Research Service).

for much of this period, it makes economic sense if the cost of resource entrance or exit was high. It also makes sense if productivity growth depended on longer-term technological advances generally independent of market movements and adoption decisions based on producers' drive to improve returns at the margin rather than on overall sector returns.

Prospects for Productivity Growth in the 1990s. Will these productivity and resource use patterns continue in the 1990s and support trend increases of 2 percent per year in productive capacity?

A number of recent studies—most notably the Office of Technology Assessment's lengthy report on agricultural technologies—point toward continued and possibly accelerated growth in productivity (U.S. Congress, Office of Technology Assessment, 1986). These studies tend to focus both on the backlog of technologies awaiting adoption and the new technologies coming on line. Many also point out that past productivity assessments typically have underestimated, rather than overestimated, the potential for growth. In the 1990s, however, environmental concerns about the use of a broadening range of manufactured inputs could affect productivity adversely to a far greater extent than ever before.

For example, federal and state activities have been or are being launched to counter surface and groundwater contamination by strengthening restrictions on the use of agricultural chemicals. If strictly enforced, these programs could affect the use of productivity-enhancing fertilizers and pesticides, slowing future productivity growth and possibly eroding past gains. Programs already under way in the water-quality area alone include—

- The Clean Water Act of 1972, as amended in 1987, mandating state water-quality control programs and requiring that nonpoint source pollution of the type generated by many agricultural production activities be included in the programs;
- The Safe Drinking Water Act of 1974, as amended in 1986, requiring that the states develop plans to protect wellhead areas from contaminants including agricultural chemicals;
- The Federal Insecticide, Fungicide and Rodenticide Act of 1972, providing more extensive pesticide regulation and restrictions on use;
- The Resource Conservation and Recovery Act of 1976, as amended in 1984, dealing with solid waste management and hazardous water disposal; and
- The Toxic Substance Control Act of 1976, providing for testing and additional controls on substances including agricultural chemicals.

The use of agricultural chemicals also comes under the purview of the Super Fund Program established as part of the Comprehensive Environmental Response, Compensation, and Liability Act of 1986 and the Environmental Protection Agency's groundwater protection activities.

Proposed legislation on food safety could have an equal or even larger impact than these water-quality acts. All six of the major food safety bills introduced in Congress since January 1989 call for expanded government intervention to license and otherwise regulate the use of fertilizers and pesticides at the farm level and chemical additives and entire production and marketing processes downstream from the farm gate to the consumer.[1]

Questions about the approval and adoption of new biotechnologies often touted as a major source of potential productivity growth in the 1990s are also being raised. The debate over livestock hormones, such as bovine somatotropin, is a particularly graphic example. The rising concern about food safety points to public interest in understanding the full effects of technological change and a willingness to slow or prevent change where it is proved to be or is widely perceived to be hazardous.

How can this potential for slowed growth in productivity be quantified? Clearly, any precise measure would be little more than speculation. But for want of anything better, a standard deviation on the historical productivity growth data suggests a range of 1–3 percent per year around the postwar 2

[1]The Kennedy-Waxman Bill (S. 722 and H.R. 1725); the Dorgan Bill (H.R. 1387); the Sikorski Bill (H.R. 1508); the Smith Bill (H.R. 604); the Kolter Bill (H.R. 240); and the Moakley Bill (H.R. 2051).

percent average. Assuming that rigorously enforced environmental regulation pulled productivity growth toward the bottom end of this range, annual increases could be 1 percent or less.

Prospects for Resource Use in the 1990s. Recent legislation and proposals before Congress affecting land use also suggest that resource use patterns could change significantly in the 1990s. The Food Security Act of 1985, for example, included provision for—

- The Conservation Reserve Program, with up to 40 million acres of erodible land retired through ten-year leases.
- Conservation cross-compliance requiring the Soil Conservation Service to work with farmers to develop conservation plans to reduce soil losses to acceptable levels on erodible soils by 1990. Dicks (1989) estimates that as many as 20 million to 25 million acres could be affected by this provision.
- Swampbusting and sodbusting restrictions that slow future expansion in acreage that would otherwise have offset regular acreage losses to nonagricultural uses. Dicks (1989) suggests that possibly 1 million acres each year could be affected.

Several other less visible conservation-oriented programs could also have an increasingly significant impact on resource use. For example, the Resource Conservation and Recovery Act could have a considerable effect on land use through its mandate to identify national conservation priorities. And while the Forestry Incentive Program affects only 1 million to 2 million acres per year in its current form, its expansion is under discussion both in Congress and USDA.

With rigorous enforcement these various programs could shrink agriculture's land base by 60 million acres or more by 1995, with at least two-thirds of this coming from cropland. This amounts to long-term retirement of roughly 15 percent of the sector's 375-million to 400-million-acre cropland base before any annual Acreage Reduction Program is implemented.

Implications for Capacity Growth. While historical trends point to continued strong growth in capacity, it is clear from the preceding discussion of environmental quality regulations that significant change in agricultural productivity growth and resource use are possible and even probable. But what are the orders of magnitude?

Contrasting environmentalists' concerns with the Office of Technology Assessment's technology prognosis and factoring in potential changes in land use suggests that capacity to produce could increase at the trend pace of 2 percent or more per year or essentially stagnate. This range is wide enough, if extended through the mid-1990s, to generate a full 15 percent difference in capacity by 1996 between these high and low prognoses before other market forces—such as demand pressures and changes in commodity prices—are taken into account. While it is difficult to draw specific policy implica-

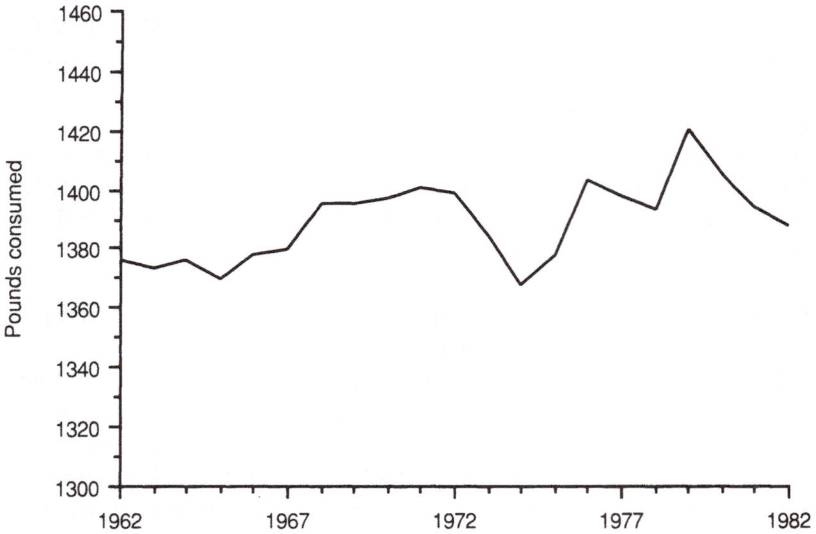

FIGURE 4. U.S. per capita food consumption, 1962–1982. *Source:* Data from U.S. Department of Agriculture, various 1986–1988 issues, *Food Consumption, Prices, and Expenditures* (Washington, D.C., Economic Research Service).

tions from this broad a range or to calculate probabilities associated with any one prognosis, the range is clearly broader than in past debates with greater risk of slower, rather than faster, growth in capacity.

Demand Prospects in the 1990s

On the demand side, the excess capacity issue becomes one of how slow growth in the large stable domestic market combines with potentially faster but more volatile growth in the smaller export market.

Domestic Demand Prospects. As figure 4 indicates, domestic use of farm products has essentially stagnated over the last two decades at 1,375–1,400 pounds per capita. Changes in income growth and employment can add price strength in a few selective markets. But even wide swings in these conventional demand drivers make little difference in volume terms. Moreover, the shift within the 1,375–1,400 pound total toward more livestock products requiring several pounds of feed per pound of meat, milk, and eggs produced has also slowed. Meat supplies of 210–220 pounds per capita appear to have saturated the market. With population growth slowing to 0.8 percent per year and little growth in alternative uses for agricultural commodities on the horizon, prospects are for less than 1 percent growth per year in the domestic market in the 1990s.

Export Demand Prospects. In the export area, historical data suggest that two issues are in question—the pace of growth in the world market and changes in U.S. market share. As figures 5, 6, and 7 indicate, growth in the world market has been variable but is stable when compared to fluctuations in the United States' export market share.

Although highly simplistic, developments in the world market can be linked to two sets of factors: shock factors and longer-term growth factors. Shock factors—production shortfalls and windfalls and demand shortfalls and windfalls—tend to dominate year-to-year movements in world trade. But important as these shocks may be, they are essentially impossible to project. Hence, qualitative projections are generally limited to recognizing that "normalized" projections will almost certainly be too high or too low for any one year but, it is hoped, are valid over a series of years.

Among the longer-term factors that can be projected, economic growth is critical. The affluence-related drive in a broadening circle of countries to upgrade diets, both by increasing caloric intake and substituting livestock products for crop products, underlay much of the postwar growth in imports. With resource or productivity constraints, or both, on domestic production, many countries had to turn increasingly to the world market, first for foodstuffs and eventually for feedstuffs and livestock products.

More recently, economic growth in the middle-income countries has become critical, particularly when growth in domestic income and employment has been linked to growth in export earnings or foreign borrowing with which to finance foreign purchases. As the situation in sub-Saharan Africa demonstrates, population growth in and of itself counts for little in the world market.

Adjustments in farm and trade policies in the importing countries have also been major world market drivers. The shift toward liberalized trade policies and less protectionist agricultural policies in many countries in the 1970s encouraged import expansion. Conversely, policy retrenchment in the 1980s generated market contraction.

It is important to recognize how strongly these two sets of demand factors interacted to generate upturns or downturns in the world market. Growth spurts in import demand coincided with accelerating economic activity, expanding export earnings, and policy liberalization in the 1970s; the most pronounced slowdowns coincided with economic contraction, losses in export earnings, the debt crisis, and policy retrenchment in the first half of the 1980s.

Changes in exportable supply have also driven growth in world trade. Despite significant expansion in domestic demand, a core of exporters have been able to supply a growing world market—at real prices down by half since 1960 and a quarter since 1980.

What are the prospects for these longer-term import demand and export supply factors in the early 1990s? Is a return to the postwar trend of 4–5 percent growth in world trade likely after the ups of the 1970s and the downs of the early 1980s?

Most macroeconomic forecasters expect the early 1990s to be a period of expanding economic activity more bullish than the early 1980s but not as

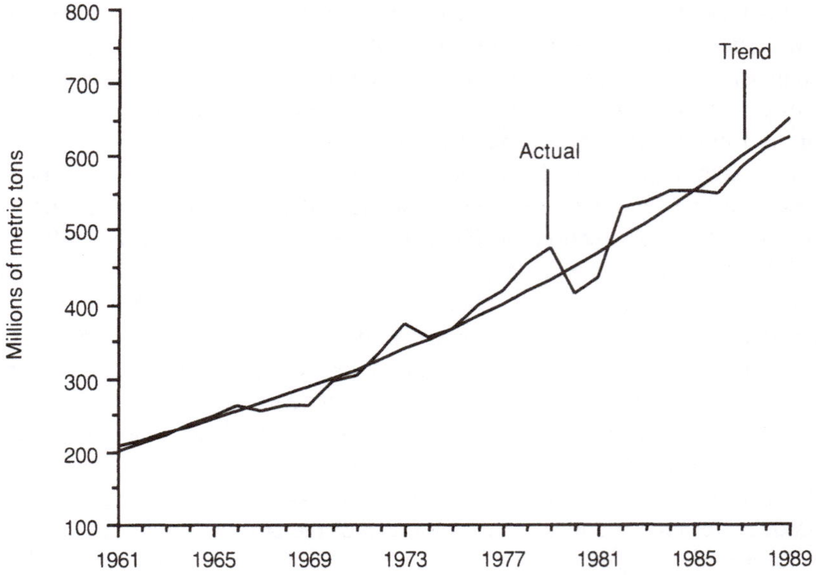

FIGURE 5. World agricultural trade, 1961–1989. *Source:* Data from United Nations Food and Agriculture Organization, various 1980–1988 issues, *Trade Yearbook,* Rome.

strong as in the last half of the 1980s (Monaco, 1989). This should provide a firmer base for growth in demand for farm products in the food deficit countries to be met with either expanded local production or imports. Many analysts question whether growth in production in key importing countries will continue in the 1990s at the accelerated pace of the 1980s after a decade of low commodity prices discouraged investment in agricultural production capacity. There may thus be some pickup in import demand compared with the early and mid-1980s but slower growth than in the expansionist 1970s. Continued policy reform in the agricultural and trade areas in many importing countries should also add strength to import demand—both by accelerating economic growth and by opening the economies of these countries to products from abroad.

What about export availabilities? Given the proven capacity of the major exporters and any continuation of their policies encouraging overproduction, exportable supplies should be plentiful enough to meet substantially larger world import demand at least through the mid-1990s with little nominal increase—and possibly continued real decreases—in commodity prices.

Given these import demand and export supply prognoses, it is difficult to argue for any strong deviation from the longer-term 4–5 percent annual growth trend.

How would trade liberalization affect this outlook? Liberalization along the lines under discussion in the General Agreement on Tariffs and Trade (GATT) would change world trade volumes and patterns significantly (Ron-

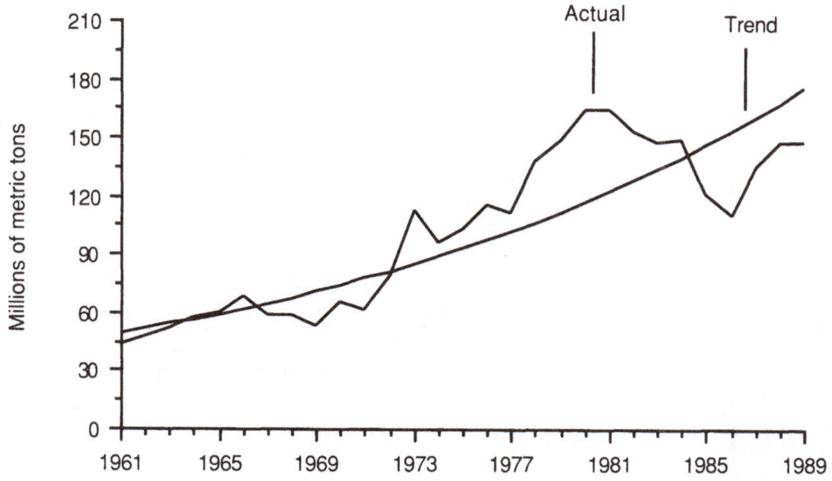

FIGURE 6. U.S. agricultural exports, 1961–1989. *Source:* Data from United Nations Food and Agriculture Organization, various 1980–1988 issues, *Trade Yearbook,* Rome; U.S. Department of Agriculture, various 1980–1989 issues, *Foreign Agricultural Trade of the United States* (Washington, D.C., Economic Research Service).

ingen and Dixit, 1989). There is general agreement that trade prices would be somewhat higher; increased import demand in the importing countries and the loss of export subsidies would work to raise prices and offset the impact of eliminating producer supports. There is also agreement that world trade volume would be somewhat higher. At issue, however, is the small probability of something approaching full-scale liberalization and the timing of reform. Given the tenor of talks to date, full-scale liberalization appears less and less likely. Moreover, the impact of even far-reaching reform agreed to in the next twelve to eighteen months would be quite small through the life of this next round of farm legislation.

U.S. Market Share. What about the U.S. share of the world market, the more volatile of the two trade factors? In the most simplistic terms, U.S. volume share of world agricultural trade has swung from a high of more than 40 percent to a low of less than 20 percent since 1961, largely in response to changes in the ability of the United States to offer quality products and associated services at competitive prices (figure 7). Changes in political relationships with the centrally planned economies and short-lived constraints on exportable supply have occasionally been significant factors. But by and large, traditional price issues have dominated. This is not surprising, given the nature of the world market and the United States' role as a large supplier but with less than a monopoly hold on the market. The world market has generally been a buyers' market, with a broad range of products—many at

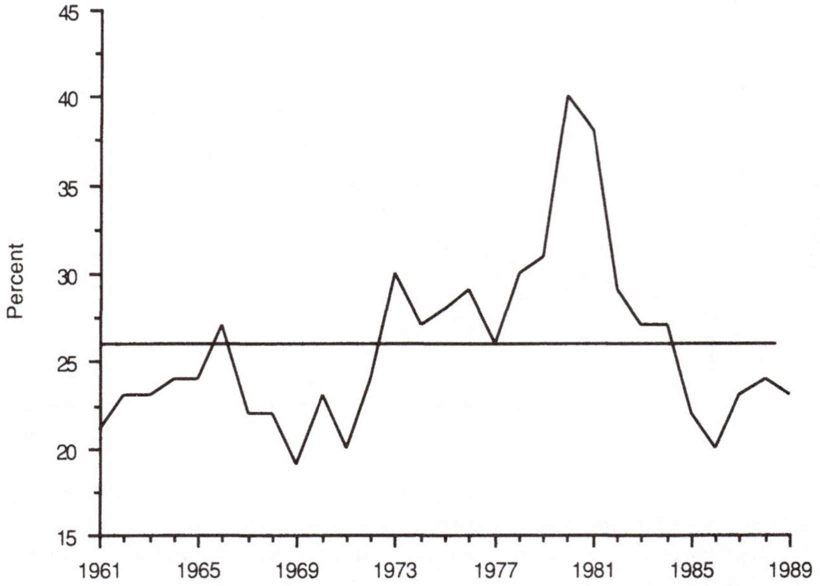

FIGURE 7. U.S. volume share of world agricultural trade, 1961–1989. *Source:* Data from United Nations Food and Agriculture Organization, various 1980–1988 issues, *Trade Yearbook,* Rome.

least partially substitutable—available from many sellers in an information-rich marketplace. These characteristics make it extremely price competitive.

For example, the world wheat market tends to be insensitive to general price levels: a given change in wheat export prices across suppliers seldom generates much of a change in import demand. However, a relatively small change in an individual supplier's prices relative to other suppliers can lead to large shifts in importer purchases to alternative sources. Similarly, questions of grain quality, reliability as a supplier, and an exporter's willingness to match competitor subsidies and delivery terms can also shift purchases. However, with contract specifications generally well defined, these considerations tend to be absorbed into the overriding question of landed local currency cost.

Figure 8 uses wheat export price data to make this point. After a decade of competitive pricing reinforced by the declining value of the dollar in the 1970s, U.S. wheat export prices moved above competitor prices in the early and mid-1980s. Domestic support prices were raised in the 1981 farm bill and worked in turn to raise U.S. export prices in a weakening world market. Dollar appreciation eroded U.S. competitiveness further and boosted the appeal of competitor supplies. The impact was immediate. As U.S. prices moved $30–$40 per ton above the export prices offered by the European Community (as well as by Canada, Australia, and Argentina), U.S. market share fell from 46

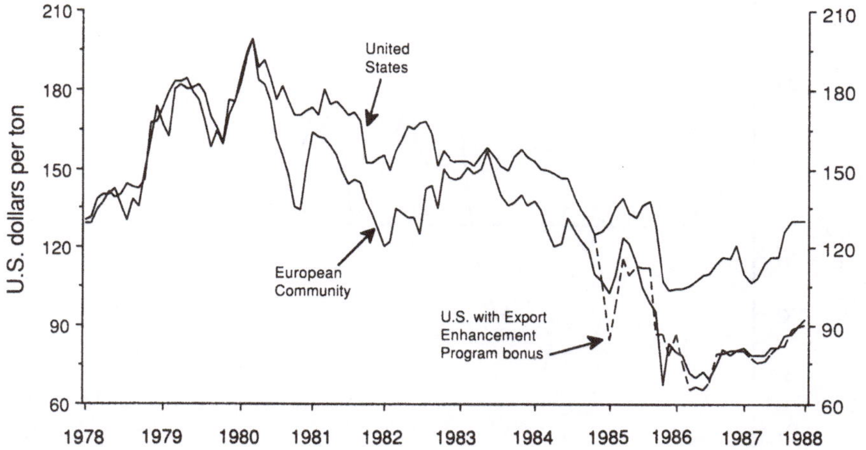

FIGURE 8. U.S. and European Community wheat export prices, July 1978–July 1988. All July years. *Source:* Unpublished USDA reports.

percent in 1979 to 26 percent in 1985. Similar problems with the other major commodities helped drop the U.S. share of the overall world market for farm products from a peak of 40 percent in 1980 to 20 percent in 1986.

The reduced support levels and expanded export marketing programs provided for in the Food Security Act of 1985 have helped to ease this price competitiveness problem, as has depreciation of the dollar. But as figures 7 and 9 suggest, U.S. share is still below the average for the last two decades and well below the peak of the early 1980s.

What does this perspective on market share suggest for U.S. exports in the 1990s? Increased U.S. concern with pricing, including the 1985 act's more flexible support programs and expanded export programs, will help keep U.S. prices competitive—particularly with a stable or weakening dollar. However, this is not likely to be enough to push U.S. share back up to the upper end of the historical range (around 35–40 percent). Peak U.S. export share has depended on the combination of rapid growth in the world market and a strong U.S. competitive position. Hence, moderate growth in world imports combined with continued pressure on the United States to be competitive point to a share in the 25 percent range similar to the long-term average. A 25 percent share of a steadily expanding world market would push U.S. exports up 4–6 percent per year.

Demand Implications. What do these combined domestic and export perspectives imply for overall growth in demand for U.S. farm products? Given a roughly one-third/two-thirds volume split between export and domestic de-

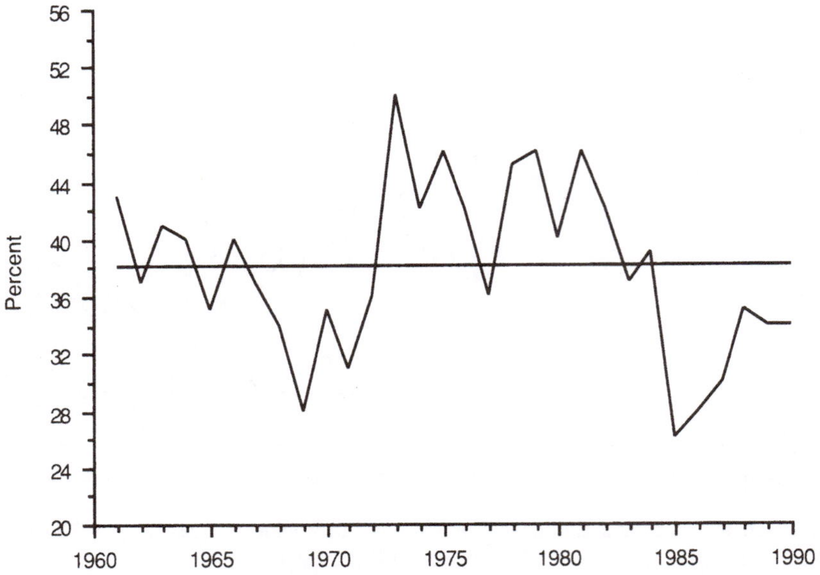

FIGURE 9. U.S. share of the world wheat market, 1961–1990. *Source:* Data from United Nations Food and Agriculture Organization, various 1980–1988 issues, *Trade Yearbook,* Rome. 1989–1990 estimates from USDA.

mand, a 0.8 percent annual increase in domestic demand combined with 4–6 percent growth in the export market suggests a 2–3 percent annual pace overall. This is comparable to the pace for the postwar period as a whole—well below the fast-growth 1970s but well above the 1-2 percent pace of the 1980s.

Implications for the Market Environment in the 1990s

What do these productivity, resource, domestic demand, and export demand factors suggest for changes in excess capacity and, in turn, the market environment in the early 1990s? For one thing, they suggest that growth in production capacity could prove slower and less certain, unless offsets to environmental pressures on resource and input use are found. One such offset could take the form of accelerated substitution of more environmentally positive technologies. Biotechnology breakthroughs, for example, could ultimately supply agriculture with environmentally positive technologies and better resource management techniques. But such a transition is likely to stretch well beyond the life of the next farm bill. This offset could also take the form of a political decision to weaken or delay implementation of environmental programs if it becomes clear that they are a serious drag on supplies and prices.

These factors also suggest that despite the far larger size of the domestic market, export performance will continue to be the major demand concern of the 1990s. With more normal growth in world trade and a competitive U.S.

export stance, steady gains in exports are possible. This suggests that the bearish experience of the United States in much of the 1980s could prove to be as serious an aberration as the export boom of the 1970s. Combined domestic and export demand growth of 2–3 percent per year could more than offset growth in capacity and gradually work to reduce the excess capacity—both surplus stocks and excess acreage—the sector will carry with it into the 1990s.

Hence, while excess capacity is unlikely to disappear completely in the next five years, gradual reductions could shift the sector closer to balance. It is important to note how fragile this conclusion is. It abstracts from short-lived supply and demand shocks. It assumes away trade wars with the Europeans, a retreat to high price supports and a closed agricultural economy, or a repeat of the world recession of the 1980s. Any one of these events could weaken U.S. export performance and expand, rather than shrink, excess capacity. However, the probability of this bearish outcome appears small. Given the unique mix of weather, macroeconomic, and policy factors that combined to push excess capacity toward zero in the 1970s, a more bullish short supply alternative is also possible but not highly probable.

The next section looks at quantitative estimates, from two sources, of a number of variables that are important in assessing the performance of the agricultural sector. Both sets of estimates are more precise than the foregoing assessment, and are, therefore, more sensitive to the underlying assumptions.

A QUANTITATIVE ASSESSMENT OF THE OUTLOOK FOR THE EARLY 1990s

Projections of the outlook for agricultural markets will vary according to the models being used, the underlying macroeconomic and policy assumptions, and the judgments of the analysts involved in a particular study. The more comprehensive the coverage, the more numerical detail and questions about approach and technique can distract the user's attention from the analysts' most important conclusions. Comparing general results from different studies for key variables and the cause-and-effect relationships underlying them can help users focus on critical issues rather than on specific estimates, which are of more limited value.

Two sets of projections are compared here. One was prepared at the Food and Agricultural Policy Research Institute (FAPRI) and the other at the Congressional Budget Office (CBO). Treatment is limited to reporting on the major commodity markets, land use, and farm program costs. But more detailed results are available in publications available from both organizations (Food and Agricultural Policy Research Institute, 1989; U.S. Congress, Congressional Budget Office, 1989).

The materials that follow start with a comparison of basic macroeconomic and policy assumptions. Supply, demand, and price projections are included for the wheat, corn, soybean, rice, and cotton markets, along with a comparison of land use indicators. Because CBO does not report net returns and

TABLE 1. Comparison of Macroeconomic Assumptions Used by the Food and Agricultural Policy Research Institute and the Congressional Budget Office

Variable	1988	1989	1990	1991	1992	1993	1994
U.S. GDP/GNP (percent change from previous year)							
FAPRI (GDP)	4.0	2.9	2.2	3.0	2.7	2.7	2.7
CBO (GNP)	3.9	2.8	1.7	2.3	2.5	2.5	2.5
Implicit GNP deflator							
FAPRI	3.4	4.8	4.8	4.3	4.4	4.2	4.4
CBO	3.5	4.5	4.4	4.3	4.3	4.3	4.3
U.S. exchange rate (percent change from previous year)							
FAPRI	−5.8	5.4	−5.0	−3.0	−2.7	−2.1	−0.1
CBO	−4.2	6.8	−6.4	−6.0	−4.1	−3.7	−2.9
Three-month T-bill rate (percent change from previous year)							
FAPRI	6.7	8.6	8.2	7.4	7.1	6.8	6.7
CBO	6.7	8.2	7.2	6.8	6.5	6.3	6.1
Regional GDP (percent change from previous year)[a]							
World	4.0	3.5	3.1	3.3	3.3	3.3	3.7
Africa	2.3	2.3	2.0	2.5	3.5	3.8	3.7
Latin America	0.3	−0.5	3.5	4.2	2.4	3.3	4.2
Pacific Basin	8.0	6.3	5.5	6.2	6.1	6.1	6.0
Western Europe	3.5	3.3	2.4	2.6	2.5	2.8	3.2
Centrally planned economies	5.0	3.9	4.1	3.6	4.1	4.1	4.9

Notes: Data for 1988 are actual values. GDP, gross domestic product; GNP, gross national product.

Sources: For FAPRI assumptions: Wharton Econometric Forecasting Associates, 1989, "World Economic Outlook" (Bala Cynwyd, Pa., July); CBO assumptions: U.S. Congress, Congressional Budget Office, 1989, "The Economic and Budget Outlook: An Update" (Washington, D.C., August).

[a]CBO does not have an explicit foreign sector, so no comparative data are available.

farm income, reporting for these indicators is drawn from the FAPRI materials.

Assumptions

Macroeconomic Assumptions. The FAPRI projections use macroeconomic assumptions published by Wharton Econometric Forecasting Associates (the WEFA Group) in July 1989. CBO's assumptions were provided by the CBO Fiscal Analysis Division. Both sets of projections have U.S. gross domestic product (GDP) or gross national product (GNP) growth rates declining in 1989 and 1990 (table 1). The WEFA forecasts have the GDP growth rate declining in 1990 and recovering quickly in 1991, whereas CBO expects a greater slowdown in GNP in 1990 and a slower recovery thereafter. Both have the value of the dollar continuing to decline throughout the projection period after the upward bump in 1989, although the exchange rate measure used is not exactly the same for both organizations. The intermediate-term interest rate projections as reflected in the three-month Treasury-bill rate show similar trends under the two alternatives, but the CBO projected rates are consistently lower.

The CBO does not have an explicit foreign sector in its modeling system, so no comparable data are available for economic growth and exchange rates for major foreign countries or regions. The WEFA Group forecasts have most of the developed countries' and world GDP growth slowing down in 1989 and 1990, then recovering to more moderate growth rates in the later years.

Pacific Basin growth rates are slightly lower relative to the late 1980s but continue to be higher than growth elsewhere in the world. Projected growth rates for Africa and Latin America in general are better for the 1990s than in the latter part of the 1980s. The improved rates of economic growth in Africa and Latin America are still very low or even negative on a per capita basis. The regional averages shown also mask very large differences between countries with very low growth rates and those with better-than-average economic performance.

U.S. Agricultural Policy Assumptions. CBO and FAPRI both assumed that the major provisions of the Food Security Act of 1985 would continue. Both projections hold target prices constant at 1990–1991 levels for the projection period 1991–1995. Although the target price assumptions are the same, and both CBO and FAPRI continue to have the crop loan rates follow the rules established in the act, some other program provisions are slightly different (table 2). FAPRI assumes higher Acreage Reduction Program (ARP) rates for corn and wheat from 1992 to 1993 and for rice and cotton from 1992–1993 onward. CBO assumes a 35-million-acre cap on Conservation Reserve Program acres, while FAPRI assumes 40 million acres. In both studies, the dairy price reduction formula was continued, allowing the dairy support price to fall $0.50 in each year that net removals exceed 5 billion pounds. In both analyses this led to a decline in dairy support prices until they reach $9.10 in 1992. CBO allows them to drop one more step to $8.60 in 1993.

Commodity Market Projections

Both sets of projections use commodity supply, demand, and price data from the summer of 1989 as a starting point. CBO estimates were made in August, while the FAPRI projections were done in September. Hence, the FAPRI results include somewhat more recent information from the USDA market reports. Projection results are summarized below for wheat, corn, soybeans, rice, and cotton.

Wheat. With the reduction of ARP rates from 27.5 percent to 10 percent in 1989–1990 and to 5 percent in the succeeding years, wheat acreage increases 10 million acres to more than 76 million acres in 1989–1990 and ranges from 74 million to 79 million acres for the following years (table 3). FAPRI estimates 82 million acres for 1990–1991 because of the new program an-

TABLE 2. Comparison of Policy Assumptions Used by the Food and Agricultural Policy Research Institute and the Congressional Budget Office

Variable	1988–1989	1989–1990	1990–1991	1991–1992	1992–1993	1993–1994	1994–1995
Crop ARP rate (percent)							
Corn FAPRI	20.0	10.0	10.0	10.0	12.5	12.5	10.0
Corn CBO	20.0	10.0	10.0	10.0	10.0	10.0	10.0
Wheat FAPRI	27.5	10.0	5.0	5.0	10.0	10.0	5.0
Wheat CBO	27.5	10.0	5.0	5.0	5.0	5.0	5.0
Rice FAPRI	25.0	25.0	15.0	20.0	20.0	20.0	20.0
Rice CBO	25.0	25.0	15.0	15.0	12.5	12.5	10.0
Cotton FAPRI	12.5	25.0	10.0	15.0	20.0	20.0	20.0
Cotton CBO	12.5	25.0	20.0	15.0	15.0	15.0	15.0
Dairy policy							
Net removals (billions of pounds)							
FAPRI (calendar year)	8.9	11.5	11.2	8.6	4.9	2.6	1.4
CBO (fiscal year)	9.1	9.8	9.5	8.3	6.2	5.4	n.a.
Support price (dollars per hundredweight)							
FAPRI (calendar year)	10.7	10.1	9.6	9.1	9.1	9.1	9.6
CBO (calendar year)	10.6	10.1	9.6	9.1	8.6	8.6	n.a.

Notes: ARP, Acreage Reduction Program; n.a. means not available.

Sources: Food and Agricultural Policy Research Institute, 1989, "Summary of FAPRI Baseline," mimeograph (Columbia, University of Missouri, and Ames, Iowa State University, September 22), used with permission; U.S. Congressional Budget Office, 1989, "The Economic and Budget Outlook: An Update" (Washington, D.C., August).

nounced in August 1989 (after the CBO analysis). Production rises accordingly, and prices gradually fall from the highs of 1989–1990.

The differences in the two projections are relatively small and do not lead to a different interpretation of the market outlook. In both cases, exports fall from the high levels of 1987–1988 and do not return to those levels until after 1992–1993. This represents a return to normal rather than a change in the export path. The high export levels in 1987–1988 were driven by abnormally high imports by the USSR and China and abnormally low exports by Australia.

Corn. Because of the reduction in the ARP rate from 20 percent to 10 percent in 1989–1990, projected corn acreage increases about 5 million acres in 1989–1990 and ranges between 72 million and 75 million acres for the next five years (table 4). With the higher acreage and the assumption of normal weather, corn production increases dramatically and prices are projected to fall 40–50 cents per bushel in 1989–1990 and remain between $2.00 and $2.20 per bushel for the next five years. The CBO price projections are generally 5–10 cents per bushel lower throughout the projection period than those of FAPRI. This can be explained largely by the lower FAPRI production estimates for 1989–1990 and the higher FAPRI assumptions of ARP rates (and lower production) in some later years.

TABLE 3. Comparison of Projections of the Food and Agricultural Policy Research Institute and the Congressional Budget Office, for Wheat

Variable	1988–1989	1989–1990	1990–1991	1991–1992	1992–1993	1993–1994	1994–1995
Price (dollars per bushel)							
FAPRI	3.72	3.90	3.24	3.01	3.31	3.46	3.50
CBO	3.72	3.92	3.50	3.26	3.19	3.24	3.28
Participation rate (percent)							
FAPRI	85.3	77.0	93.6	77.8	79.4	70.8	70.6
CBO	85.8	77.0	74.3	75.8	77.0	76.4	76.4
Acreage planted (millions of acres)							
FAPRI	65.5	76.8	82.0	77.9	73.9	75.7	79.0
CBO	65.5	76.8	77.1	77.5	77.3	77.3	77.5
Production (millions of bushels)							
FAPRI	1,811	2,064	2,648	2,538	2,438	2,527	2,663
CBO	1,811	2,120	2,511	2,559	2,593	2,633	2,678
Domestic use (millions of bushels)							
FAPRI	978	975	1,004	1,014	1,024	1,048	1,062
CBO	963	1,022	1,045	1,087	1,110	1,123	1,147
Exports (millions of bushels)							
FAPRI	1,425	1,335	1,343	1,426	1,480	1,562	1,636
CBO	1,440	1,254	1,361	1,419	1,496	1,547	1,554
End stock (millions of bushels)							
FAPRI	694	469	789	908	862	799	784
CBO	694	559	687	762	772	757	757
Target price (dollars per bushel)							
FAPRI/CBO	4.23	4.10	4.00	4.00	4.00	4.00	4.00
Loan rate (dollars per bushel)							
FAPRI	2.21	2.06	1.95	1.91	1.99	2.05	2.00
CBO	2.21	2.06	2.02	2.09	2.24	2.24	2.13

Sources: Food and Agricultural Policy Research Institute, 1989, "Summary of FAPRI Baseline," mimeograph (Columbia, University of Missouri, and Ames, Iowa State University, September 22), used with permission; U.S. Congressional Budget Office, 1989, "The Economic and Budget Outlook: An Update" (Washington, D.C., August).

For the corn market, the differences between the two projections are again relatively small and explained largely by differences in the policy assumptions and the size of the 1989–1990 crop. The greater amount of acreage planted, larger production, and lower prices in the CBO projection lead to higher levels of stocks being estimated throughout the projection period.

Soybeans. Soybean acreage is projected to increase by 2 million to 3 million acres in 1989–1990 in response to the high drought-induced prices in 1988–1989 (table 5). The lower prices in 1989–1990 resulting from higher production levels lead to lower acreage planted in subsequent years. Prices are projected to remain in the $5–6 range through 1994–1995. The FAPRI price projections for soybeans are about 50 cents per bushel higher than CBO's in the later years.

TABLE 4. Comparison of Projections of the Food and Agricultural Policy Research Institute and the Congressional Budget Office, for Corn

Variable	1988–1989	1989–1990	1990–1991	1991–1992	1992–1993	1993–1994	1994–1995
Price (dollars per bushel)							
FAPRI	2.55	2.18	2.17	1.99	2.13	2.20	2.17
CBO	2.54	1.99	1.97	1.99	2.09	2.17	2.23
Participation rate (percent)							
FAPRI	87.1	79.8	82.8	81.1	79.7	74.3	71.4
CBO	87.0	80.0	80.0	78.0	73.0	71.0	69.0
Acreage planted (millions of acres)							
FAPRI	67.6	72.3	74.5	73.9	71.4	72.4	73.9
CBO	67.6	72.8	75.6	75.2	75.5	75.3	75.0
Production (millions of bushels)							
FAPRI	4,921	7,321	7,973	8,028	7,903	8,114	8,369
CBO	4,921	7,507	7,957	8,081	8,273	8,388	8,512
Domestic use (millions of bushels)							
FAPRI	5,255	5,754	5,848	5,934	5,923	5,932	5,943
CBO	5,257	5,565	5,958	6,130	6,220	6,259	6,276
Exports (millions of bushels)							
FAPRI	2,075	1,980	1,972	2,031	2,112	2,243	2,378
CBO	2,102	1,827	1,956	2,063	2,138	2,202	2,272
End stock (millions of bushels)							
FAPRI	1,855	1,445	1,601	1,666	1,536	1,478	1,529
CBO	1,827	1,945	1,991	1,883	1,801	1,733	1,700
Target price (dollars per bushel)							
FAPRI/CBO	2.93	2.84	2.75	2.75	2.75	2.75	2.75
Loan rate (dollars per bushel)							
FAPRI	1.77	1.65	1.57	1.49	1.41	1.34	1.30
CBO	1.77	1.65	1.56	1.49	1.50	1.51	1.54

Sources: Food and Agricultural Policy Research Institute, 1989, "Summary of FAPRI Baseline," mimeograph (Columbia, University of Missouri, and Ames, Iowa State University, September 22), used with permission; U.S. Congressional Budget Office, 1989, "The Economic and Budget Outlook: An Update" (Washington, D.C., August).

Both projections show some growth in exports from the drought-induced low level of 1988–1989. However, even the CBO export estimates, which are higher, do not project a return to the levels of exports achieved in 1986 and 1987. The poor export prospects are related to slow growth in total world demand for soybeans and a slight decline in the U.S. share of the export market.

Rice. FAPRI participation rates and ARP rates for rice remain high in the projection period, so planted acreage changes relatively little (table 6). CBO eases ARP rates beginning in 1991–1992, which results in higher production, lower prices, and higher use and stock levels. Both projections have prices declining from near $8.00 per hundredweight in 1989–1990 to about $6.50 per hundredweight in 1991–1992. CBO price projections remain near that

TABLE 5. Comparison of Projections of the Food and Agricultural Policy Research Institute and the Congressional Budget Office, for Soybeans

Variable	1988–1989	1989–1990	1990–1991	1991–1992	1992–1993	1993–1994	1994–1995
Price (dollars per bushel)							
FAPRI	7.35	5.78	5.26	5.76	5.97	5.87	6.02
CBO	7.33	5.63	5.30	5.24	5.31	5.28	5.30
Acreage planted (millions of acres)							
FAPRI	58.9	60.5	57.6	56.5	58.4	59.6	60.0
CBO	58.9	61.7	59.0	59.1	59.2	59.7	60.3
Production (millions of bushels)							
FAPRI	1,539	1,889	1,913	1,899	1,978	2,038	2,072
CBO	1,539	1,968	1,918	1,941	1,961	1,999	2,040
Domestic use (millions of bushels)							
FAPRI	1,156	1,199	1,264	1,295	1,330	1,368	1,396
CBO	1,137	1,192	1,229	1,255	1,275	1,300	1,323
Exports (millions of bushels)							
FAPRI	530	584	598	621	634	650	669
CBO	549	639	668	684	694	711	724
End stock (millions of bushels)							
FAPRI	155	262	313	296	310	330	337
CBO	136	273	293	294	286	274	266
Loan rate (dollars per bushel)							
FAPRI/CBO	4.77	4.53	4.50	4.50	4.50	4.50	4.50

Sources: Food and Agricultural Policy Research Institute, 1989, "Summary of FAPRI Baseline," mimeograph (Columbia, University of Missouri, and Ames, Iowa State University, September 22), used with permission; U.S. Congressional Budget Office, 1989, "The Economic and Budget Outlook: An Update" (Washington, D.C., August).

level, but FAPRI prices return to about $7.50 per hundredweight because of the lower production in later years.

Cotton. The outlook for cotton prices is relatively static (table 7). In both projections, cotton prices remain in the range of 58–64 cents per pound during the next six years. CBO also projects a relatively static production and use pattern. FAPRI projects greater growth in exports and domestic use, but this is accompanied by higher production stimulated by slightly higher prices. The wider range of estimates in the cotton projections indicates a greater disparity between FAPRI and CBO in the behavior of the cotton sector models.

Sector Aggregates

Planted Area. Planted area is estimated by FAPRI to increase 15 million acres in 1989–1990 as a result of reduced acreage reduction requirements for wheat and feed grains (table 8). The CBO estimate of the increase in planted acreage is 16 million acres, owing to CBO's higher estimate for soybean plantings.

TABLE 6. Comparison of Projections of the Food and Agricultural Policy Research Institute and the Congressional Budget Office, for Rice

Variable	1988–1989	1989–1990	1990–1991	1991–1992	1992–1993	1993–1994	1994–1995
Price (dollars per hundredweight)							
FAPRI	6.71	8.11	6.74	6.57	7.03	7.58	7.46
CBO	6.75	7.75	7.00	6.40	6.30	6.40	6.50
Participation rate (percent)							
FAPRI	94.1	95.1	85.3	91.0	92.0	90.0	87.7
CBO	94.0	94.0	93.0	93.0	95.0	93.0	95.0
Acreage planted (millions of acres)							
FAPRI	2.93	2.77	3.17	2.98	2.88	2.87	2.91
CBO	2.93	2.77	3.05	3.05	3.21	3.15	3.30
Production (millions of hundredweight)							
FAPRI	159.5	152.3	172.4	165.6	162.3	163.9	167.5
CBO	159.5	153.1	169.8	170.8	180.8	179.3	188.5
Domestic use (millions of hundredweight)							
FAPRI	81.0	84.1	85.5	86.4	87.8	88.9	90.1
CBO	86.2	89.7	92.9	96.1	99.6	102.7	105.5
Exports (millions of hundredweight)							
FAPRI	87.0	78.0	84.3	85.9	80.9	80.0	81.0
CBO	76.0	73.0	75.2	78.0	81.8	83.8	87.2
End stock (millions of hundredweight)							
FAPRI	26.9	22.2	29.8	28.1	26.7	26.7	28.1
CBO	32.4	26.8	32.7	33.6	37.3	34.6	34.9
Target price (dollars per hundredweight)							
FAPRI/CBO	11.15	10.80	10.71	10.71	10.71	10.71	10.71
Loan rate (dollars per hundredweight)							
FAPRI/CBO	6.63	6.50	6.50	6.50	6.50	6.50	6.50

Sources: Food and Agricultural Policy Research Institute, 1989, "Summary of FAPRI Baseline," mimeograph (Columbia, University of Missouri, and Ames, Iowa State University, September 22), used with permission; U.S. Congressional Budget Office, 1989, "The Economic and Budget Outlook: An Update" (Washington, D.C., August).

Planted acres are projected to remain steady near these levels for the succeeding years. The estimates of the combined planted and idled acres are very similar for the two sources, since the higher CBO plantings are approximately offset by lower idled acres. Most of the difference in idled acreage is explained by a 5-million-acre difference in the Conservation Reserve Program acreage.

Commodity Credit Corporation and Conservation Reserve Program Outlays. Estimated net outlays by the Commodity Credit Corporation (CCC) for the coming years are all lower than the $12 billion levels for fiscal years 1988 and 1989 (table 8).[2] The cost estimated by FAPRI for fiscal year 1990

[2]In September 1989 the U.S. Treasury estimated the fiscal year 1990 CCC outlays adjusted for CBO accounting to be $10.4 billion.

TABLE 7. Comparison of Projections of the Food and Agricultural Policy Research Institute and the Congressional Budget Office, for Cotton

Variable	1988–1989	1989–1990	1990–1991	1991–1992	1992–1993	1993–1994	1994–1995
Price (dollars per pound)							
FAPRI	0.56	0.63	0.59	0.58	0.60	0.61	0.64
CBO	0.57	0.61	0.59	0.58	0.58	0.59	0.60
Participation rate (percent)							
FAPRI	89.0	84.9	75.7	80.2	78.7	75.8	73.0
CBO	88.0	87.0	80.0	76.0	78.0	78.0	78.0
Acreage planted (millions of acres)							
FAPRI (all)	12.5	10.5	12.6	11.9	11.6	11.6	11.7
CBO (upland)	12.0	10.1	10.7	10.2	10.3	10.2	10.2
Production (millions of bales)							
FAPRI (all)	15.4	12.3	16.2	15.5	15.4	15.6	16.0
CBO (upland)	15.1	11.7	13.4	13.1	13.4	13.6	13.9
Domestic use (millions of bales)							
FAPRI (all)	7.6	7.7	7.8	8.0	7.9	8.0	8.1
CBO (upland)	7.5	7.2	7.0	6.9	7.1	7.2	7.3
Exports (millions of bales)							
FAPRI (all)	6.2	7.8	7.4	7.5	7.7	7.8	8.1
CBO (upland)	6.3	7.7	6.6	6.4	6.4	6.5	6.7
End stock (millions of bales)							
FAPRI (all)	7.1	3.9	4.9	4.9	4.9	4.8	4.8
CBO (upland)	7.2	4.2	4.2	4.0	4.1	4.1	4.1
Target price (dollars per pound)							
FAPRI/CBO	0.80	0.73	0.73	0.73	0.73	0.73	0.73
Loan rate (dollars per pound)							
FAPRI	0.52	0.50	0.50	0.50	0.50	0.50	0.50
CBO	0.52	0.50	0.50	0.50	0.51	0.54	0.54

Sources: Food and Agricultural Policy Research Institute, 1989, "Summary of FAPRI Baseline," mimeograph (Columbia, University of Missouri, and Ames, Iowa State University, September 22), used with permission; U.S. Congressional Budget Office, 1989, "The Economic and Budget Outlook: An Update" (Washington, D.C., August).

dropped more than that estimated by CBO primarily because of FAPRI's higher market prices. For the period beyond the 1985 act, estimated costs begin at just above $11 billion and decline from there. In the FAPRI projections, costs decline by about $4.4 billion in two years, from fiscal years 1992 to 1994, compared with a decline of $1.8 billion during the same period in the CBO projection. Much of this difference is explained by higher prices and lower deficiency payments in the FAPRI estimates associated with higher ARP rates assumed in the analysis.

CRP costs were moved out of the CCC budget in fiscal year 1988. Nevertheless, these costs are important to agriculture and are expected to reach $2 billion by fiscal year 1992 and remain stable at that level thereafter.

TABLE 8. Comparison of Projections of the Food and Agricultural Policy Research Institute and the Congressional Budget Office, for Sector Aggregates

Variable	1988–1989	1989–1990	1990–1991	1991–1992	1992–1993	1993–1994	1994–1995
Land use (eight crops) (millions of acres)							
Planted							
FAPRI	241	256	262	257	251	256	261
CBO	241	257	265	264	265	265	265
Planted and idled							
FAPRI	319	315	314	316	319	319	319
CBO	319	315	319	317	317	317	317
Net CCC outlays (billions of dollars) (net of the Conservation Reserve Program spending)							
FAPRI	12.3	6.3	11.4	11.1	8.2	6.7	6.8
CBO	12.6	9.8	11.8	11.2	10.3	9.4	n.a.
CRP costs (billions of dollars)							
FAPRI	1.4	1.6	1.9	2.1	2.0	2.0	2.0
CBO	1.6	1.8	2.1	2.0	1.9	1.9	n.a.
Crop net returns (dollars per acre)							
Corn							
Participants	86.88	135.00	137.37	136.37	133.05	129.75	127.92
Nonparticipants	67.16	83.58	94.53	74.70	90.43	93.95	84.03
Wheat							
Participants	55.61	61.36	77.42	76.92	72.55	71.33	74.15
Nonparticipants	66.95	63.60	56.71	48.36	59.37	63.12	62.11
Soybeans	128.27	109.96	102.52	120.11	126.06	120.51	123.60
Rice							
Participants	206.33	181.70	189.16	171.54	168.03	160.85	145.86
Nonparticipants	66.03	118.44	33.06	18.16	36.41	53.81	29.52
Cotton							
Participants	153.00	110.61	144.63	133.10	121.54	115.64	108.21
Nonparticipants	60.31	91.68	80.39	72.66	78.42	80.47	85.33
Total (eight crops) (aggregate, participants, and nonparticipants) (billions of dollars)	23.52	24.87	26.00	25.49	25.98	25.21	24.92

Note: CCC, Commodity Credit Corporation; CRP, Conservation Reserve Program; n.a. means not available.

Sources: Food and Agricultural Policy Research Institute, 1989, "Summary of FAPRI Baseline," mimeograph (Columbia, University of Missouri, and Ames, Iowa State University, September 22), used with permission; U.S. Congressional Budget Office, 1989, "The Economic and Budget Outlook: An Update" (Washington, D.C., August).

Returns. Net crop returns are reported as a means of looking at the well-being of crop producers. CBO does not provide estimates comparable to those of FAPRI. Ignoring the abnormal drought year of 1988–1989, corn net returns decline slightly over the projection period and the gap between participant and nonparticipant returns narrows. Wheat participant returns are projected to remain fairly stable, while nonparticipant returns are expected to increase and narrow the gap between the two. Soybean net returns vary substantially from year to year, since prices are more variable than for other commodities and there are no deficiency payments to stabilize receipts. For rice, the gap between participant and nonparticipant net returns continues to remain very large throughout the period. Cotton participant net returns also decline, and returns for nonparticipants remain more stable, leading to a smaller gap between the two.

Net returns for the eight program crops are calculated by adding the net returns of participants and nonparticipants to develop an aggregate measure. Declining net returns per acre for participants are somewhat offset by higher plantings and higher returns for nonparticipants. Thus, total net returns to crops during the next six years remain between $25 billion and $26 billion. This is $3 billion to $4 billion below the average for the 1986–1987 to 1987–1988 period (that is, the period prior to the drought of 1988–1989).

Implications of Extending Current Programs

The agricultural outlook under the assumption of continuation of the 1985 Food Security Act's major program provisions and favorable macroeconomic conditions is moderately optimistic. Crop shortfalls in North America and the Soviet Union during the last two years reduced excess U.S. crop inventories sharply and have made it possible to continue 1985 programs at lower costs than in the first few years of the act. Average annual costs in 1991–1994 are projected to be $5 billion to $6 billion lower than during the 1987–1989 period. Moreover, a larger proportion of future costs is expected to be in direct payments rather than in commodity loans.

Idled acreage is expected to remain between 50 million and 60 million acres under a five-year extension compared with an average of 62 million acres during 1986–1989. Equally important, two-thirds of this smaller total is expected to be in the longer-term Conservation Reserve Program compared with only one-third or less under the early years of the 1985 act. This would reduce the cropland readily available to offset a drought such as occurred in 1988 or an export demand windfall. In general, the reserve supplies represented by crop inventories and land idled as part of the annual Acreage Reduction Program are expected to be substantially lower than they were in the second half of the 1980s, implying greater potential for price variability.

Producers of program crops are expected to receive more from the market and less from government payments in the future. Payments are projected to average 8 percent of total crop cash receipts compared with 16 percent on average during the 1985 act. In terms of farm income, lower payments to crop producers are somewhat offset by higher market prices and higher plantings. Program participation is expected to decline only slightly, since program benefits would still be attractive, particularly with somewhat lower acreage reduction requirements and increases in downside risk as loan rates continue to decline.

Uncertainty still remains a central element of the outlook. While the scenarios reported here suggest a "soft landing" as the sector moves toward more market orientation, a succession of bumper crops or another export slump could quickly increase program costs and reduce farm income. Another major drought in the United States or abroad could have the opposite effect. A major macroeconomic downturn could also reduce demand for farm products and raise government costs. These kinds of uncertainties will be of equal or greater importance than in the past. Another source of uncertainty is that of the effect of the overall budget deficit, and attempts to reduce it, on agricultural programs.

FEDERAL BUDGET ISSUES AND THE 1990 FARM BILL

Many observers of U.S. agricultural policy do not see much pressure for program change in 1990 coming from within agriculture. Because crop exports and farm income are relatively strong and surplus stocks have largely disappeared, the agricultural community appears to support continuation of the programs provided for in the Food Security Act of 1985. Most observers argue that pressure for changes will come largely from forces outside the agricultural community—from groups concerned about the federal budget deficit, agricultural chemicals, and food safety.

This section focuses on one of these concerns—the federal budget—and assesses prospects for change in farm programs arising from pressure to trim the deficit. Three questions are central to this issue:

1. How large are the projected deficits relative to the targets specified in the Balanced Budget and Emergency Deficit Control Reaffirmation Act of 1987 (commonly known as Gramm-Rudman-Hollings or GRH) and how much of any deficit reduction will have to be absorbed in farm programs?
2. Assuming commodity program spending will be reduced, what will it be reduced from? What policy assumptions underlie CBO and Office of Management and Budget (OMB) baseline budget projections?
3. If proposed reductions are large, can the agriculture committees use budget gimmicks to avoid painful program cuts?

Deficit Projections

The spending and revenue projections that will form the basis for the fiscal year 1991 budget resolution will not be released until January 1990. However, barring significant changes in the macroeconomic outlook over the next three months, recent projections by CBO and OMB are reasonable proxies for the deficit situation the federal government will face during the 1990 farm bill debate.

The Outlook Under Current Policies. Under a continuation of the policies in place as of August 1989, large budget deficits will persist for the next five

TABLE 9. Deficit Projections and Targets Under the Balanced Budget and Emergency Deficit Control Reaffirmation Act of 1987 (Gramm-Rudman-Hollings)

(outlays by fiscal year, in billions of dollars)

	1990	1991	1992	1993	1994
August 1989 Congressional Budget Office baseline deficit[a]	141	144	141	143	128
August 1989 Congressional Budget Office baseline deficit[a] assuming full implementation of the 1990 summit agreement	118	127	124	128	112
Deficit targets under the Balanced Budget and Emergency Deficit Control Reaffirmation Act of 1987	100	64	28	0	0

Source: U.S. Congress, Congressional Budget Office, 1989, "The Economic and Budget Outlook: An Update" (Washington, D.C., August).

[a]The 1990 deficit includes savings from shifting the military pay date ($2.1 billion), advancing farm deficiency payments ($850 million), and writing off uncashed food stamps ($477 million).

years. CBO's updated projections of future revenues and outlays under this scenario show a federal deficit that is nearly flat in nominal dollars for several years—$141 billion in fiscal year 1990 and $143 billion in fiscal year 1993. In fiscal year 1994, the projected gap between spending and revenues drops to $128 billion (see table 9).

The Outlook Under Implementation of the Summit Agreement. The baseline projections do not take into account implementation of the fiscal year 1990 budget summit agreement announced by President Bush and Congress in April 1989. At present, reconciliation and appropriation bills have not been enacted that would implement the summit agreement. While spending reductions under the GRH began October 16, 1989 (5.3 percent for nondefense programs, including agriculture), many observers believe that deficit reduction legislation will be enacted and that across-the-board reductions will be temporary or that deficit legislation will at least cause OMB to reduce the size of across-the-board reductions.

If the summit agreement is fully implemented, CBO estimates that the deficit will be $118 billion in fiscal year 1990 and $127 billion in fiscal year 1991, before falling to $112 billion in fiscal year 1994. Even when full implementation of the summit agreement is assumed, the projected deficits far exceed the GRH targets.

The Projected 1991 Deficit Relative to the GRH Target. In 1991, the initial fiscal year that would be affected by the 1990 farm bill, the projected deficit, assuming full implementation of the summit agreement, exceeds the GRH target by $63 billion. Even this scenario may be overly optimistic. Although Congress is currently working on deficit reduction legislation, in accordance with the summit agreement, out-year savings targets are not binding, so the actual savings in 1991 and later years may be smaller than assumed. If deficit reduction legislation is enacted that saves nothing in 1991 and beyond, the 1991 estimated deficit would exceed the GRH target by $81 billion. Thus, the projected 1991 deficit will probably be somewhere between $63 billion and $81 billion above the target. CBO's views, however, do not determine the necessity for across-the-board cutbacks. The Supreme Court found the original version of the GRH unconstitutional, because it gave a key role to a legislative branch officer (the Comptroller General of the General Accounting Office) in carrying out an executive function. As amended, the GRH assigns the task of determining compliance with the targets and ordering any across-the-board cutback to the Director of the Office of Management and Budget.

OMB's current view of the deficit is more optimistic than CBO's because OMB assumes somewhat greater economic growth and lower interest rates than CBO and estimates lower costs for the recent savings and loan legislation. Using OMB's July 1989 projections (adjusted for savings and loan and farm disaster assistance legislation), and with the same caution as above regarding the out-year effects of deficit reduction legislation, the projected deficit is $31 billion to $49 billion over the 1991 GRH deficit target.

Potential Cuts Facing the Commodity Programs

Most budget analysts agree that farm programs will be a target for cuts, despite CCC spending that is projected to remain well below the peak of the mid-1980s. However, the amount of fiscal year 1991 deficit reduction that Congress decides must be shouldered by commodity programs is very difficult to project.

A clue to the size of future cuts may be found in the recent summit agreement. The agreement called for legislative changes that would reduce the fiscal year 1990 deficit by $24.3 billion—$5.7 billion from higher revenues and $18.6 from lower spending. The agreement also assumed a few elements that did not require legislation but were instead carried out administratively and resulted in savings of $1.3 billion. These administrative measures have been incorporated into the CBO August 1989 baseline. Agricultural programs were to be reduced by $1.9 billion in fiscal year 1990. Of this total, $0.9 billion was obtained administratively when the Secretary of Agriculture moved payments equal to this amount into 1989 by increasing the size of 1989 crop deficiency payments. The remaining $1.0 billion in program cuts were to be obtained legislatively with $0.6 billion of that amount coming from commodity programs and the remainder from moving federal assistance to the Farm Credit System off-budget.

Unless the GRH deficit target is modified, the 1991 budget resolution will likely call for a deficit reduction of 30–100 percent more than called for in the 1990 summit agreement. Even if the increase in deficit reduction between 1990 and 1991 is accomplished by raising revenues, commodity programs

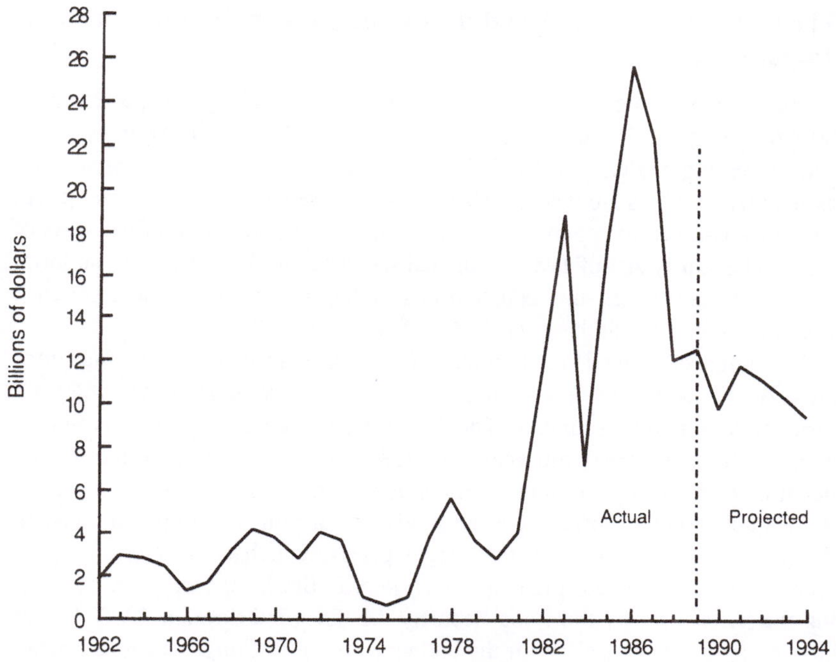

FIGURE 10. Commodity Credit Corporation outlays, 1962–1994. *Source:* Unpublished "actual" data from U.S. Department of Agriculture; "projected" data from Congressional Budget Office (1989).

could be asked to bear from $0.6 billion to $1.0 billion in spending reductions in 1991. If revenue increases account for less than all the increased deficit reduction, the agriculture committees may be asked to reduce CCC spending even more.

On the other hand, there are several arguments for limiting cuts imposed on farm programs for the purposes of deficit reduction. First, CCC spending is down from the very high levels of 1986 and 1987 (see figure 10). Second, commodity program interests in Congress view the current spending levels as appropriate. They are not embarrassed about the current and expected level of $10 billion to $12 billion per year. However, if spending was expected to stay around the $15 billion to $20 billion level, that might not be the case. Finally, there does not seem to be a groundswell of public or congressional opinion that farmers currently receive too much federal support.

Other candidates for cuts within the jurisdiction of the agriculture committees include the Conversation Reserve Program, the Federal Crop Insurance Program, Farmers Home Administration loan programs, and loans made to rural electrification and telephone systems.

The Outlook for CCC Spending and the Assumption of Constant Target Prices

Using the CBO August 1989 assumptions, CCC spending is expected to be $9.8 billion in fiscal year 1990 before rising to $11.8 billion in fiscal year 1991 and decreasing gradually to $9.4 billion by fiscal year 1994. These projections assume target prices are held at 1990 levels in subsequent years. If target prices had been assumed to decline at the same rate as legislated in the final years of the Food Security Act of 1985, estimated spending would be $0.5 billion lower in 1991. In later years, the reduction in spending would have been larger each year and savings would have reached $3.5 billion in 1994.

As Congress worked on its budget resolution earlier this year, questions arose over how the baseline should treat expiring provisions of the 1985 act's farm price support programs. The law currently governing such programs expires after the 1990 crop year. The GRH states that the baseline for the upcoming year should assume continuation of the programs at current rates; it does not, however, offer clear-cut guidance for longer-term projections. In the past, CBO has assumed that target prices, which help determine payments to farmers, would gradually decline after the 1990 crop year, continuing the pattern set by the Food Security Act for 1985 through 1990. But in developing its own baseline for the budget resolution, Congress chose instead to assume constant target prices after 1990. Because Congress prefers to use this assumption and because the GRH can be interpreted to support this approach, CBO has adopted it for its own new projections.

The baseline target price assumption is of great importance to the 1990 farm bill debate. While the current OMB baseline for CCC outlays assumes declining target prices, the administration has not yet addressed this issue. If both CBO and OMB ultimately adopt the constant target price assumption, the agriculture committees will be in a very strong position to keep target prices at or near 1990 crop levels.

Budget Gimmicks

Even if the budget resolution for fiscal year 1991 calls for sizable reductions in farm program spending, the agriculture committees may be able to avoid painful deficit reduction through the use of various budget gimmicks.

As mentioned earlier, the summit agreement called for legislation to reduce fiscal year 1990 commodity program spending by $0.6 billion. While CBO estimated that both House and Senate agriculture reconciliation bills would save the $0.6 billion minimum required under February 1989 baseline assumptions, it also estimates that the Senate bill will save only about $0.3 billion and the House bill only $0.2 billion against the more recent August 1989 baseline.

One type of gimmick is to take credit for spending reductions that probably would occur with or without legislation. For example, both bills reduce fiscal year 1990 funding for the Export Enhancement Program (EEP) below the level assumed in the February baseline. However, now that drought conditions have sharply reduced U.S. wheat supplies, few analysts expect EEP funding to be constrained by the committee bill levels.

Another type of gimmick is to take advantage of changing market conditions to offer legislation that would have saved money under assumptions used in the budget resolution, but would not do anything (and thus cause no pain to farmers) in the current market environment. For example, the Senate bill increases the maximum allowable Acreage Reduction Program percentage for the 1990 upland cotton crop by 5 percentage points from the current maximum 25 percent. The higher rate holds only if ending stocks for the 1990 crop year are projected to be above 7 million bales, which was the case in the February baseline assumptions package. However, because weather problems reduced the size of 1989 crop production and exports have been stronger than anticipated, current ending stock projections would not trigger the higher ARP.

While some gimmicks will likely be used to meet required 1991 CCC spending reductions, some real cuts are also likely. Real cuts will have to come from the direct payments to farmers that make up most of CCC spending. For example, this year's Senate agriculture committee bill would reduce deficiency payment rates for 1990 crops by 2.3 cents per bushel for wheat and corn, 0.515 cent per pound for cotton, and 5.15 cents per hundredweight for rice.

A Summary of the Effects of Budget Pressures

Farm programs seem destined for sizable spending reductions as the 1990 farm bill debate unfolds. The planned reduction could be 5–8 percent of the $11.8 billion in commodity programs spending currently expected for fiscal year 1991. On the other hand, the budget pressure could ease substantially if the current GRH deficit target of $64 billion is increased or if revenues are allowed to account for a much larger portion of deficit reduction than is currently expected.

Even if farm programs are targeted for relatively large budget reductions, the agriculture committees will probably minimize painful spending reductions through the use of budget gimmicks. However, the real cuts that are made will probably mean reduced direct payments to farmers. The degree to which budget gimmicks are used to satisfy budget constraints imposed on the 1990 farm bill will depend on several factors. The larger the cuts farm programs are faced with, the more likely it is that the agriculture committees will respond with gimmicks or spread spending reductions across other programs in their jurisdiction, but outside commodity programs. If CCC spending projections are reduced so that they contribute to an overall lessening of budget pressure, Congress will tend to be more tolerant of agriculture committee budget gimmicks. On the other hand, if CCC projections rise and add to overall budget pressure, Congress will be less tolerant of these gimmicks.

CONCLUSIONS

This chapter has presented qualitative and quantitative assessments of the economic and market environment in which the 1990 farm bill debate is likely to occur. Budget projections and prospective constraints have also been discussed. The policymaking environment in 1990 is expected to be somewhat benign compared with the dire circumstances under which the Food Security Act of 1985 was crafted. The scenarios presented here suggest that continuation of current programs with relatively minor adjustments is feasible unless budget constraints become more severe.

Uncertainties in the year-to-year variability around the most likely scenarios point to a need for continued or increased flexibility in program provisions. Short-term fluctuations in domestic production or export levels will be just as likely as in the recent past. Thus policies for the 1990s should—

- Look toward a gradually strengthening market but be flexible enough to be effective during short-term shocks in the market;
- Recognize the trade-offs between environmental concerns, including growing concerns about food safety throughout all segments of the food production and marketing system, and the sector's capacity to produce abundant supplies at falling real prices;
- Focus concern on ensuring U.S. competitiveness—providing a quality product that meets buyer specifications at the lowest possible price—in an expanding world market; and
- Focus on fine-tuning of current provisions rather than major revisions.

REFERENCES

Dicks, M. 1989. "Agricultural Production and Environmental Legislation: Implications for the Future." Working paper (Stillwater, Oklahoma State University, October).

Dvoskin, D. 1988. *Excess Capacity in U.S. Agriculture,* Agricultural Economic Report 580 (Washington, D.C., U.S. Department of Agriculture, Economic Research Service, February).

Food and Agricultural Policy Research Institute. 1989. "Summary of FAPRI Baseline," mimeograph (Columbia, University of Missouri, and Ames, Iowa State University, September 22).

Monaco, Ralph. 1989. "Macroeconomic Projections to 2000." Unpublished Economic Research Service Baseline Working Materials, October 13.

Roningen, Vernon, and Praveen Dixit. 1989. *Economic Implications of Agricultural Policy Reform in Industrial Market Economies.* Staff Report No. AGES 89-36 (Washington, D.C., U.S. Department of Agriculture, Economic Research Service, August).

U.S. Congress, Congressional Budget Office. 1989. "The Economic and Budget Outlook: An Update" (Washington, D.C., August).

U.S. Congress, Office of Technology Assessment. 1986. *Technology, Public Policy, and the Changing Structure of American Agriculture* (Washington, D.C., Government Printing Office, March).

U.S. Department of Agriculture. 1989. "Farm Income Forecast Errors," *Agricultural Outlook* AO/157 (Washington, D.C., Economic Research Service, October), p. 22.

Wharton Econometric Forecasting Associates. 1989. "World Economic Outlook" (Bala Cynwyd, Pa., July).

THE FOOD SECURITY ACT OF 1985 AS THE BASIS FOR FUTURE FARM LEGISLATION

ROBERT L. THOMPSON

There is widespread satisfaction among farmers and farm organizations with the basic structure of the Food Security Act of 1985. Members of Congress like it, too. In the debate on the drought disaster assistance bill during the summer of 1988, the House Committee on Agriculture exhibited great pride of authorship in the 1985 farm bill and protected it from being reopened. This author expects to see the members of the House and Senate agriculture committees, as well as the major farm organizations, push very hard to extend the 1985 farm bill with minor modifications. This chapter examines the Food Security Act of 1985 as a basis for the next major farm bill and looks at pressures that are likely to affect that debate. It also reviews needed modifications in the 1985 act if that act is to be the basis for writing the next farm bill.

FACTORS SHAPING THE 1990 LEGISLATION

Although the 1985 act will very probably provide the basis for the next farm bill, it is inevitable that it will be subject to some changes. At least two forces will necessitate change. The first of these, and probably the most important, is the pressure to reduce the federal budget deficit. Agriculture is viewed widely on Capitol Hill as having benefited greatly from government programs during the 1980s. Tens of billions of dollars of income have been transferred to producers through the commodity programs, a $4 billion Disaster Assistance Act of 1988, and a $4 billion Farm Credit System bailout. Members of Congress feel that they have been more than generous to agriculture and that it is now time to focus more resources on pressing needs of concern to liberal urban congressmen, who feel that their pet programs have been put on ice during the 1980s while agriculture has enjoyed great federal largess. The combined pressures to do something about other social problems while reducing the federal budget deficit will force agriculture to participate in the deficit reduction process by reducing the total cost of farm programs.

There are two large components in the cost of farm programs. One is the cost of acquiring and carrying stocks; the other is deficiency payments. The drought of 1988 and the growth of farm exports since 1986 have reduced the volume of stocks. The cost of carrying stocks, therefore, is not likely to be high. So, to reduce the cost of the programs much further, cuts will have to be made in deficiency payments, as discussed in more detail later in the chapter.

The second force shaping the next farm bill will be the international trade negotiations now under way under the auspices of the General Agreement on Tariffs and Trade (GATT). While it is impossible to predict the outcome of those negotiations at this time, it is noteworthy that this is the first GATT Round in which domestic agricultural policies that distort international trade are on the table along with border measures. The dynamics of the negotiations will inevitably interact with the dynamics of writing the next farm bill. It may have been advantageous to pass a one-year extension of the 1985 act that met the budget mark for fiscal year (FY) 1991 to permit completion of the GATT Round and to move writing the next farm bill into a nonelection year. The 1991 farm bill would then include the legislation implementing any policy changes necessitated by agreements in the GATT negotiations. Nevertheless, Congress seems determined to proceed with writing the next farm bill in 1990.

SPECIFIC AREAS FOR CHANGE

Let us turn now to the specific components of the Food Security Act of 1985 and how they might evolve in the next farm bill. The section starts with a discussion of the export crops—those crops covered by target prices, loan rates, deficiency payments, and marketing loans.

Export Crops

Loan Rates. The Bush administration will press to make sure that loan rates are maintained low enough that in normal years the market will clear at prices above the loan rate. Keeping loan rates low would ensure continued international competitiveness of American commodities while still providing a safety net. The formula included in the 1985 act seems to be working satisfactorily; loan rates are to be set at 75 percent of the five-year moving average market price, dropping the high and low observations. The law limited how quickly loan rates could get down to the level determined by the formula. The loan rate dropped 25 percent in the first year of the bill, and the law permits stepping it down 5 percent per year until it reaches 75 percent of the moving average market price. This approach has appeal for two reasons. First, it does provide a safety net but does not impede international competitiveness in a normal year. Second, using a formula removes politics from the issue of setting the loan rate. Any time that a country or group of countries has tried to run a buffer stock scheme, the acquisition price of the buffer stock has been politicized and set too high, with the result that stocks accumulate.

Establishing the loan rate on the basis of a percentage of the moving average market price was a major step toward removing politics from the establishment of the loan rate. With automatic adjustment of the loan rate, once the formula is reached everybody will be able to predict exactly what the loan rate will be as soon as the season average market price for the previous year is published. Farmers will not have to wait for the loan rates to be set and announced by the U.S. Department of Agriculture (USDA). The administration is likely to put a high priority on keeping that provision. There will be pressures from some corners to raise the loan rate as a means of lowering the budget cost of farm programs. I believe that a majority of the members of Congress were convinced by the experience under the 1981 act that it is necessary to keep loan rates low enough to ensure international competitiveness of U.S. exports. Otherwise, exports would drop and stocks accumulate, running up the budget cost once again.

Deficiency Payments. If the formula determination of loan rates is maintained, then there are only two other variables that can be manipulated to meet the budget constraint. One is the target price; the other is the volume of production eligible for deficiency payments. The cost of the program for any commodity equals the difference between target price and loan rate times the volume of eligible production. There will likely be some pressure to reduce target prices further over time, but actually doing so may be politically difficult. Therefore, the only remaining means of cutting the budget cost is by reducing the volume of production eligible for deficiency payments, either by reducing the area eligible for deficiency payments or by reducing the program payment yield as was done in 1985.

Under the 1985 act the cost of deficiency payments was reduced over time by two means. Target prices were cut by 10 percent over the last three years of the farm bill, from 1988 through 1990 (four years, from 1987 through 1990, in the cases of cotton and rice), and the program payment yield was reduced. Until 1985 program payment yields were based on a five-year moving average of proven yields on a farm. Under the 1985 act, in order to reduce the cost of deficiency payments, program payment yields were frozen at the average of the previous five years' program payment yields, each of which was itself the average of the previous five years' proven yields. This change, in effect, made the new program payment yield a weighted average of ten years' yields with the heaviest weight on 1981, with the weights declining back to 1976 and forward to 1985. When Congress realized what it had done, a "technical amendment" was passed to set a limit that no farmer could have his or her program payment yield cut by more than 10 percent from its 1985 level.

In addition, acreage bases were effectively frozen. Thus, the volume of production eligible for deficiency payments was reduced by roughly 10 percent relative to what it had been in 1985. The agriculture committees of Congress discussed reducing the fraction of the acreage base on which payments could be made. It was proposed.that each year USDA project total market needs (domestic plus export) for the coming year, divide that amount by the trend yield to determine how many acres of each crop would be needed to produce that quantity, and then calculate the fraction of the base that that would require. The rationale was that it makes no sense for USDA to pay deficiency payments on more acreage than it expects the market to need in the coming year. That proposal failed, but the same cost savings was accomplished by reducing the program payment yields.

In 1990 the only two ways to reduce the cost of farm programs further are to cut the target prices further or to reduce the volume of production on which deficiency payments are made. It is important not to raise the program payment yield. Freezing it at about 90 percent of 1985 levels removed the target price from its role as the supply-inducing price. When a producer decides whether to buy an additional unit of input, it now has to be profitable at the loan rate or world market price, whichever is expected to be higher. The producer no longer receives the target price for the last 10 percent or so of production. Therefore, if politics dictates against lowering target prices, the best way to meet the budget objective is to cut the fraction of each producer's base on which deficiency payments will be made. The so-called triple base has the great virtue of achieving both this objective and the need to provide greater flexibility (see Phipps, Rossmiller, and Meyers in this volume).

The 1985 act made a major commitment to reestablishing the export competitiveness of American agriculture. One means was reducing the loan rates, but there were two other measures built into that bill. One was the marketing loans that were mandated for cotton and rice and authorized for wheat, soybeans, and feed grains; the other was the Export Enhancement Program.

Marketing Loans. The marketing loan was created for cotton and rice because a 25 percent reduction in the loan rates for cotton and rice was nowhere near enough to get those two commodities back into an internationally competitive position. Cotton and rice have historically been successful at obtaining substantially higher levels of support than either wheat and feed grains. A drop of 25 percent in their loan rates would have left prices still well above the international market prices.

Cotton and rice producers have also frequently bumped up against the $50,000 payment limitation. The implied increase in deficiency payments from dropping the loan rate more than 25 percent would have been truncated by the payment limit. The marketing loan allows growers to take out their price support loan at the loan rate, but they have to repay it only at the world market price when the loan comes due. The difference between the loan rate and the world market price, or repayment rate, is a direct subsidy that is *not* counted against the $50,000 payment limitation. The marketing loan program thus did get U.S. cotton and rice back into an internationally competitive position again, while circumventing the $50,000 payment limitation to allow cotton and rice growers to receive even larger government payments.

The marketing loan appeared to work well for cotton in the first couple of years, as evidenced by the drop in Commodity Credit Corporation (CCC) cotton stocks from 9.4 million bales in 1985 to 5.0 million bales in 1986 (U.S. Department of Agriculture, 1989a). There exists a substantial range of substitutability of cotton for other fibers in the international textiles market. By dropping the price of cotton by a substantial margin, cotton's share of the world fibers markets increased from 7 percent in 1985 to almost 9 percent in 1987.[1] That is, by dropping the price as far as the United States did, cotton displaced man-made fibers and the backlog of cotton inventory was drawn down, moving onto the world market at a very competitive price. (U.S. cotton exports were also facilitated by a significant drop in world cotton production in 1986.) The market price later rose back close to the loan rate, with the result that the CCC again accumulated stocks, and the cost of the cotton program rose significantly. It thus appears that the marketing loan is not working nearly as well for cotton today as it appeared to be for a couple of years. Therefore the cotton program is likely to be revisited in the 1990 farm bill.

Rice marketing loans probably never worked as well as did those for cotton. One of the differences is that there is not as wide a range of potential substitutability of rice for other food grains as cotton for other fibers. No matter how much the price drops there is a limit to how much more rice people will eat in place of other food grains. As a result, the rice marketing loan was not as successful as the one for cotton in reducing U.S. inventories. It is, therefore, likely that marketing loans will be revisited in the 1990 farm bill debate. They are expensive relative to other farm programs. Cotton and rice are politically powerful commodities, but there is likely to be pressure to reduce loan rates on cotton and rice further, to bring them back closer to international competitiveness on the basis of market realities.

Export Subsidies. The Export Enhancement Program (EEP), a euphemism for export subsidies, was the other measure authorized in the 1985 farm bill to reestablish the export competitiveness of American commodities. The 25 percent drop in the corn loan rate, in combination with the decline in the value of the U.S. dollar, should have been sufficient to get U.S. corn back into the world market, but it was not quite enough to get wheat back in the market. As discussed earlier, it was nowhere near enough to bring U.S. cotton and rice prices down to the world market levels, but the marketing loan effectively served as an export subsidy for cotton and rice. While authorized for wheat, soybeans, and feed grains, the marketing loan was not implemented for them due to the huge potential budget exposure, especially for corn. Therefore, the EEP was implemented in order to make wheat internationally competitive again. Although wheat was the main beneficiary,

[1]Personal communication with Scott Sanford, Crops Division, Economic Research Service, U.S. Department of Agriculture, November 1989.

TABLE 1. U.S. Wheat Exports, 1985–1986 to 1989–1990

Year	Wheat exports (millions of bushels)
1985–1986	915
1986–1987	1,004
1987–1988	1,592
1988–1989[a]	1,423
1989–1990[a]	1,275

Source: U.S. Department of Agriculture, 1989, *Agricultural Outlook* AO-156 (Washington, D.C.), September.

[a]Estimates as of August 1989.

the EEP has also been used to assist exports of rice, poultry feed, poultry, dairy cattle, and eggs. The total value of EEP-assisted exports, as of May 1989, was $8.1 billion (U.S. Department of Agriculture, 1989b). Table 1 shows how wheat exports have risen since 1985.

As the United States moves loan rates down over time, toward the 75 percent of moving average market price formula, the EEP clearly becomes redundant. It is simply no longer needed to ensure that U.S. commodities are internationally competitive. Nevertheless, its continued use is justified by the need to fight fire with fire when the European Community (EC) uses export subsidies to take away former U.S. markets for wheat or other commodities. I expect that USDA will retain authority for the Export Enhancement Program, either as long as the European Community subsidizes agricultural exports, or until it is negotiated away in the GATT Round of trade negotiations, for example, in exchange for a limitation on EC export subsidies. In the meantime, as U.S. market clearing prices reside closer to world market levels, I would expect the EEP authority to be utilized less frequently.

Payment-in-Kind. Another important component in the 1985 act was the generic payment-in-kind (PIK) certificates. Contrary to all outward appearances, there really was some budget constraint when the 1985 farm bill was being written. But even the increase in the budget cost of the farm programs to $26 billion in FY 1986 was not nearly enough to satisfy some farm organizations or members of the agriculture committees. Therefore, in order to circumvent the budget constraint, payments-in-kind were authorized. These transfers were possible because giving away previously acquired assets of the federal government does not incur current outlays on the books of the U.S. government. As a result, significantly larger income transfers could be provided to farmers by paying part of them in kind. In fact, an estimated $17.5 billion net income transfer was made in kind during FY 1987, 1988, and 1989 above that made in cash.

Under the 1983 PIK program, payments-in-kind were made with physical commodities, but the cost and difficulty of making payments by means of physical products led to the generic payment-in-kind certificates as a paper substitute for moving commodities around. Using certificates is much cheaper and easier than making physical payments-in-kind to every individual producer. It is much more efficient to give individual producers certificates that can be cashed in for commodities but are fully negotiable. As a result, a large number of certificates end up in the hands of one farmer or grain trader who can then present a whole briefcase of them to the CCC warehouse and load out a whole unit train instead of having each individual farmer back a pick-up truck up to the warehouse. That is a slight exaggeration, but it illustrates the point.

There was also one other important motivating force in putting payment-in-kind certificates in place. The 1985 farm bill dropped loan rates by 25 percent in 1986, but the Senate and House agriculture committees refused to reduce the release price on CCC or Farmer-Owned Reserve (FOR) grain. The release price on grain in the FOR is 140 percent of the loan rate at which it entered the reserve, and the release price on CCC grain is 115 percent of that release price. So, when the U.S. loan rate and, in turn, the world market price structure dropped by 25 percent upon implementation of the 1985 act, the grain in the FOR and CCC inventories was totally insulated from the market. There was virtually no likelihood of its being released into the market because there was little likelihood that the market price would rise enough to trigger the release of that grain. In order to ensure that the market would remain liquid—even in the event of another year like 1983, when the United States had a significant acreage reduction plus a drought—USDA needed the flexibility to release CCC and FOR grain into the market at something less than the legislated release price. For this purpose the payment-in-kind certificate program was instituted. By these means, it would be possible to release government grain back into the market at current market prices rather than at the official release price. The release price problem with respect to CCC stocks is really over now because there is little grain in CCC hands that was acquired at the old, higher loan rates. The remaining problem, however, is with old FOR grain, and the FOR release prices still need to be reduced to the level of a reasonable relationship with current loan rates. A release price of 140 percent of *current* loan rates is not unreasonable for either FOR- or CCC-owned inventories.

PIK certificates are an issue that will probably be addressed in the next farm bill. Neither the Treasury Department nor the Board of Governors of the Federal Reserve System likes the fact that the USDA effectively obtained the right to issue currency in the 1985 farm bill. PIK certificates are dollar denominated, payable to bearer, fully negotiable, and—better than cash in your pocket—backed by real goods. Furthermore, they occasionally trade well above par. So, for a number of reasons, farmers love them, but the Federal Reserve and the Secretary of the Treasury are not enamored of USDA's issuing a form of currency, which really ought to be counted in Ml. With stocks low, generic certificates are unlikely to be used much in the near future.

Seventeen places in the Food Security Act of 1985 authorize payments-in-kind in lieu of cash. How many of these seventeen authorities are exercised by the Secretary of Agriculture, the fraction of payments that are made in kind versus in cash, and the timing of the release of those commodities were the most important determinants of short-term commodity market liquidity between 1985 and the drought of 1988. USDA decisions on payments-in-kind, therefore, were the most important determinants of short-run market price movements. Many people are nervous about government bureaucrats having so much control over short-run commodity market liquidity and price movements. The drought of 1988 has taken care of the situation for the present, but a number of observers are concerned that USDA not be placed in that role again. Therefore, on the basis of the currency issue with the rest of the administration and the concern over bureaucrats having too much control over market price movements, it would not be surprising if the authority for the USDA to issue generic certificates is eliminated in the next farm bill. In the meantime, payment-in-kind certificates are not likely to be used as frequently as during the 1986–1988 period.

Acreage Reduction. Another major feature of the 1985 act was its large commitment to acreage reduction in a number of forms. First, there was the authority for a 40-million to 45-million-acre conservation reserve. Second, large acreage reduction was required for program participation in the early years of the farm bill, and there was authority for paid land diversion, too. There were huge stocks overhanging the market. It was necessary to set aside significant amounts of land in the short run to reduce those stocks while the export market was rebuilt. Table 2 shows the quite dramatic increase in land set aside under the Food Security Act.

As market trends are evolving, achieving the authorized 40 million to 45 million acres for the conservation reserve should remove enough land from production that very large annual Acreage Reduction Programs (ARPs) will be unnecessary. The environmental movement may, however, press to expand the conservation reserve, perhaps to 65 million acres. In the face of growing world demand due to population and income growth, a conservation reserve of 65 million acres would very likely short the market if the land retired were productive cropland. The current goal of 45 million acres will be sufficient to get most of the highly erodible land and perhaps some wetlands out of production and to have sufficient land set aside so we would not need much in the way of annual Acreage Reduction Programs and certainly not paid diversion in the early 1990s for the purpose of production control.

One other feature of the Acreage Reduction Programs will likely be discussed at length in putting together the 1990 farm bill. The least satisfactory part of the 1985 farm bill today, from the standpoint of many producers, is the extreme inflexibility that the bill imposed on individual farmers' freedom of action in allocating their land among crops. For example, all the market signals in early 1988 were calling for more soybeans and oats. But corn growers could not afford not to plant their full historical corn acreage base to corn. That planting was necessary to protect the size of that extremely valuable asset—the corn acreage base—which entitled them to a future stream of government payments. The Disaster Assistance Act of 1988 provided

TABLE 2. Commodity Land Idled, 1982–1983 to 1989–1990
(millions of acres)

Year	Set-Aside[a]	Conservation Reserve Program[b]	Total
1982–1983	11.12	—	11.12
1983–1984	77.8[c]	—	77.84
1984–1985	26.69	—	26.69
1985–1986	30.74	—	30.74
1986–1987	44.28	8.2[d]	52.48
1987–1988	68.71	22.2	90.91
1988–1989[e]	68.03	30.6	98.63
1989–1990[e]	28.95	38.6–45.0[f]	67.55–73.95

Note: Dashes mean not applicable.

Source: Compiled from data from U.S. Department of Agriculture, various years, *Agricultural Outlook* (Washington, D.C., Economic Research Service).

[a]Includes diversion, PIK, and Acreage Reduction Programs.

[b]Given in cumulative figures.

[c]PIK year.

[d]First sign-up year.

[e]Estimates as of September 1989.

[f]38.6 million acres total as of October 26, 1989; 45.0 million is the goal amount by the end of 1990.

temporary relief by authorizing each corn grower to plant from 10 to 25 percent of his or her feed grains base to soybeans or sunflowers without the risk of losing his or her corn base. Some observers of the problem have suggested reverting to the normal crop acreage concept of the 1977 farm bill; others are touting the triple base concept. The latter has the additional virtue of producing budget savings.

The issue of inflexibility in the current bill could well lead to discussion of decoupling in the next farm bill debate. The issue concerns the appropriate basis for payments. One of the reasons for such inflexibility in the 1985 act was the need to reduce the volume of production eligible for deficiency payments. By freezing the acreage base and reducing the program payment yield, payments were decoupled from production at the margin. This point merits elaboration. There was a widely held view that target prices had induced farmers to expand production of crops that were in surplus. This view tied in neatly with the concerns about conservation, especially in wheat-growing areas. In the late 1970s and early 1980s a wheat grower could drop out of the program for a couple of years, break up sod in areas that never should have been plowed, and plant it to wheat to establish a larger wheat base. The annual average wheat acreage base increased from about 63 million acres in the 1970–1976 period to almost 78 million acres in the 1977–1983 period (U.S. Department of Agriculture, 1984). In fact, by the mid-1980s the national wheat base exceeded the largest acreage ever planted. When such farmers came back into the program two years later, many of them set aside that recently plowed-up land, with the result often being severe soil erosion on those acres.

There was widespread recognition that something had to be done about soil conservation. The conservation reserve addressed that problem, but there was also a feeling that the United States ought to reduce the incentive provided by the high target prices to expand acreage of crops that were already in surplus. First, the acreage base was effectively frozen. Then, the program payment yield was reduced by about 10 percent and frozen at that level. These changes meant that in future years about 90 percent of 1985's volume of production would be eligible for deficiency payments. Anything produced in excess of that amount had to be competitive at world market prices or the loan rates, whichever the farmer expected to be higher, because that is all one could get for the last units of production. So, a corn producer deciding whether or not to order one last tank of anhydrous ammonia must value that production in excess of 90 percent of 1985's program yield at the world market price or the loan rate, not at the target price.

In summary, the inflexibility arose from freezing acreage bases, imposing cross-compliance, and freezing program payment yields at less than their 1985 level. But a quandary remains. The cost of a small move in the direction of decoupling was this extreme rigidity in acreage allocation. Farmers clearly need more flexibility. The question is, Can we achieve greater flexibility by going back to something like the normal acreage concept, and yet meet the budget constraint and not provide an incentive through target prices to expand production of crops beyond the level that market realities would justify? It appears that the triple base concept has a better chance of meeting all these objectives than the normal crop acreage concept.

Over time it would be desirable if we could buy our way out of the bases and decouple payments completely from production in order to remove the present distortions from both production decisions and land prices. The capitalized value of the bases causes significantly more capital to be tied up in farmland than is necessary. Nevertheless, full decoupling is politically unpopular with farmers because it resembles welfare. While there will likely be flexibility in the next farm bill, the question is whether it can be accomplished without giving up some of the gains that were achieved in the 1985 act.

Conservation. The 1985 farm bill was one of the toughest pieces of conservation legislation in decades because of the marriage between the environmental movement and the farm organizations in support of the conservation reserve. Many, if not most, farm organizations supported the conservation reserve because they wanted more supply control; the environmental movement supported it because it wanted more soil conservation. The price of the environmental movement's support for more supply control via the conservation reserve was the farm groups' acceptance of "sodbuster," "swampbuster," and conservation compliance provisions—much tougher environmental constraints on production agriculture than in the past. Many farmers are hoping for some softening of the conservation compliance, swampbuster, and sodbuster provisions in the 1990 farm bill. This is unlikely. Agriculture may get even tougher environmental regulations. Water quality is one of the issues highest on the agenda. The main question is whether tougher environmental rules related to water quality are included in the next farm bill or in a separate omnibus water bill passed by Congress in the next four years.

Grain Reserves. Another area of debate in the next farm bill will be grain reserves. In the early to mid-1980s, there were huge stocks of grain in government hands (see table 3). There was no point in the private sector tying up capital to carry much inventory of grain because the government was willing to carry it at no direct cost to the grain companies. Large ARPs, export recovery, and the drought of 1988 have significantly reduced government stocks. The drought also rekindled the discussion about what role government reserves should play in the future. There will be strong support for carrying a humanitarian reserve in order to meet crisis situations such as famines or natural disasters anywhere in the world. A humanitarian reserve will certainly be maintained at some minimum volume, but how far beyond that will we go?

What role will buffer stocks play in market stabilization in future farm programs? In principle, a price support loan program is a buffer stock scheme. There is an acquisition price, which puts a floor under the market price, and a release price set at 140 percent of the acquisition price. When the market price reaches the release price, commodity stocks are automatically released back into the marketplace. This release of stocks protects the U.S. livestock industry, other domestic users, and foreign export purchasers against upside price risk.

In practice, however, acquisition and release prices often become politicized and are set too high, as occurred under the 1981 farm bill. When prices are set too high, stocks tend to accumulate without bound, requiring more and more supply controls until a crisis forces the necessary drop in the acquisition and release prices. The 1985 farm bill dropped the loan rate toward 75 percent of a moving average market price but left the release price on existing stocks unchanged. This stabilization function is unlikely to be given up in the next farm bill, even though the options market is fully prepared to provide insurance against price risk through puts and calls.

Some people will argue that the commodity price support loans should be converted into recourse loans that have to be fully repaid at maturity instead of using nonrecourse loans. Such a change in price support loans would eliminate the buffer stock role of U.S. commodity programs. This alternative is not likely to receive serious consideration. Rather the issues are likely to be (1) how large government inventories ought to be and (2) whether we ought to continue to have the Farmer-Owned Reserve.

It is not clear how the extendable three-year FOR contributes to the efficient functioning of markets. Nevertheless, the FOR is extremely popular

TABLE 3. Wheat, Feed Grain, and Rice Ending Stocks, 1980–1989

Year	Commodity Credit Corporation	Farmer-Owned Reserve	Free	Total	Stocks/use
Wheat ending stocks (millions of bushels)					
1980	255	360	374	989	0.43
1981	302	560	297	1,159	0.44
1982	257	1,061	197	1,515	0.63
1983	567	611	221	1,399	0.55
1984	553	657	215	1.452	0.55
1985	1,280	596	29	1,905	0.97
1986	1,066	632	123	1,821	0.83
1987	461	467	333	1,261	0.47
1988	216	287	197	694	0.29
1989[a]	160	150	282	592	0.26
Feed grains ending stocks (millions of metric tons)					
1980	16.9	0.3	27.0	44.2	0.20
1981	20.6	38.7	18.6	77.9	0.37
1982	37.4	58.2	13.0	108.6	0.49
1983	9.7	18.0	11.9	39.6	0.19
1984	27.4	15.3	14.7	57.5	0.26
1985	96.1	22.0	8.2	126.3	0.61
1986	111.2	43.1	−2.3	152.0	0.67
1987	60.9	32.8	39.8	133.6	0.57
1988	33.4	20.7	9.2	63.4	0.29
1989[a]	46.2	3.2	12.1	61.4	0.27
Rice ending stocks (millions of hundredweights)					
1979–1980	1.7	—	24.0	25.7	0.19
1980–1981	n.a.	—	16.5	16.5	0.11
1981–1982	17.5	—	31.5	49.0	0.33
1982–1983	22.3	—	49.2	71.5	0.54
1983–1984	25.0	—	21.9	46.9	0.38
1984–1985	44.3	—	17.0	64.7	0.53
1985–1986	43.6	—	33.7	77.3	0.62
1986–1987	8.7	—	42.7	51.4	0.32
1987–1988	0.1	—	31.3	31.4	0.21
1988–1989	0.0	—	32.4	32.4	0.20
1989–1990[a]	0.0	—	26.0	26.0	0.16

Note: Dashes mean not applicable; n.a. means not available.
Source: U.S. Department of Agriculture ASCS/CAD/July 1989, Washington, D.C.
[a]Estimated.

with farmers, who have earned a lot of money storing grain for the CCC, and it is unlikely to be eliminated in the next farm bill. One key concern, however, as noted earlier, should be to bring the release price for FOR grain into line with the *current* loan rate, rather than the loan rate at which the grain entered the reserve.

Other Commodities

Let us now turn to several other commodities, in particular, dairy and sugar, which are probably the most politically difficult with which to deal. The Washington lobbying capacities on behalf of dairy and sugar interests are among the most powerful of any, the defense industries included.

Dairy. In 1983 the Commodity Credit Corporation bought almost 17 billion pounds of milk equivalent (in the form of butter, cheese, and nonfat dry milk powder), 12 percent of all milk marketed in the United States that year (U.S. Department of Agriculture, 1984). Considering that about half of the milk produced in this country is consumed as fluid milk, that means the CCC bought 24 percent of all the butter, cheese, and powdered milk produced in the United States. In other words, about one pound out of every four of processing capacity in the United States was superfluous in 1983. The 1985 debate over how to reduce dairy surpluses and the cost of the dairy program was long and acrimonious. The dairy lobby wanted controls or a buyout, while the administration wanted to cut the support price in order to reduce the incentive to overproduce milk. In the end a deal was struck in which the dairy interests got the Whole-Herd Buyout Program, under which 8 percent of the nation's dairy herd was slaughtered or exported.

By this means, the farm bill set out to eliminate two-thirds of the excess capacity on dairy farms. The administration argued that there would still be an incentive for overproduction of milk if the support price was not also reduced. The dairy lobby argued that the Whole-Herd Buyout Program would suffice. Although they acceded to the whole-herd buyout, the administration insisted on fallback authority to cut the support price by fifty cents in each year that the projected government purchases exceeded 5 billion pounds of milk equivalent should the dairy lobby's claim prove to be wrong. The Whole-Herd Buyout Program did temporarily reduce milk production. In certain regions shortages arose. But at the same time that the program was implemented, feed costs dropped by 25 percent as a result of reducing feed grain loan rates. This drop in costs significantly increased the incentive to produce milk, and the drop in milk production from the buyout was very short lived.

By the beginning of 1987, projected CCC purchases were well above 5 billion pounds of milk equivalent and the first support price cut was triggered. Another two cuts in the support price were made, in October 1987 and in January 1988. The 1989 cut in support price was postponed by the Disaster Assistance Act of 1988, which actually allowed for an increase in the support price by fifty cents, although only for the second quarter of 1989. In lobbying for postponement of the 1989 price support cut, the dairy lobby reneged on the agreement it made with the administration in the 1985 farm bill deliberations and received more favored treatment than any other commodity group in the drought bill. The language in the Food Security Act requires the Secretary of Agriculture to reduce the support price again on January 1, 1990, if CCC net removals are projected to exceed 5 billion pounds of milk equivalent.

The "dairy problem" is, in fact, two problems. There is certainly an incentive for overproduction on many farms, but the larger problem is the excess capacity in the dairy processing sector, which is twice as large as that on dairy farms. (Recall that one of every four pounds of processing capacity in 1983 was superfluous.) So what was really needed in 1985 was a whole-milk-plant buyout, not a whole-herd buyout. I am convinced that if we are ever to solve the problem of very large dairy surpluses, we will have to attack it at the processing plant level, not at the farm level. The open-ended commitment by the CCC to buy at a guaranteed price any quantity of butter, cheese, and powdered milk offered sends precisely the wrong message to dairy processors. The infinitely elastic demand for these products at guaranteed prices is what keeps this system going. When the Whole-Herd Buyout Program caused regional shortages in the fall of 1986, some milk plants offered an incentive to farmers not in the buyout program. The milk plants paid a premium for every hundredweight of milk that farmers delivered that was in excess of what they had delivered the year before. That is, milk plants bid down their processing margins in order to maintain throughput in the plants—often to produce manufactured products to sell to the CCC. So, policies in 1986 were simultaneously signaling farmers to expand production while paying others to contract.

It appears that the dairy problem has been attacked at the wrong place by focusing only on the farm part of the program. Not surprisingly, the dairy cooperatives, which are the principal spokespersons in dairy policy debates, speak as processors, not as dairy farmers, when they lobby for the dairy program. There is no reason why there cannot be an efficient dairy industry in this country, but present policy has created a monster. It is not just the farm-level incentives but also, and maybe more important, the milk plant incentives that need to be addressed. Moreover, marketing orders carve up the United States into forty-three noncompeting regions in which "Class I differentials" raise the effective support price in high-cost regions to provide an incentive to increase milk production there rather than permitting fluid milk to be profitably shipped in from lower cost states. There is no economic justification for maintaining 1930s patterns of dairy production, given the huge changes that have occurred in our transportation and communications systems. There is no more justification for the Southeast to be self-sufficient in milk production than for Minnesota, Wisconsin, or New York to be self-sufficient in orange production.

Nevertheless, it will be very difficult to do anything at all about dairy policy. As noted, the dairy lobby showed the extent of its political power by being the only segment of American agriculture to tinker with its price supports and achieve a temporary price support increase in response to the drought of 1988 in addition to the feed assistance received by other segments of U.S. animal agriculture. In the 1990 farm bill, if not before, the dairy lobby will assuredly attempt to eliminate the automatic trigger for reducing the support price that it agreed to in 1985 in exchange for the whole-herd buyout. Some form of dairy supply control is likely to be proposed again, but I am not convinced that the "dairy problem" will be solved until the open-ended commitment by the CCC to buy butter, cheese, and powdered milk is limited.

Sugar. Sugar is the other very politically powerful commodity within agriculture and as a result also enjoys a very high rate of protection. The United States is a net importer of sugar. In effect, all the protection afforded the sugar growers is provided through import quotas. These quotas limit the imports of sugar to a small enough volume to ensure that the U.S. market price stays above the loan rate. This price enhancement ensures that sugar put under loan after harvest will never be forfeited to the CCC. By keeping the import quotas tight enough, the market price remains sufficiently above the loan rate to ensure that it is profitable to repay these loans with interest. Therefore, the CCC never incurs a long-term budget cost for the sugar program. This "no net cost" feature makes it popular with Congress, but the U.S. consumers bear the cost of the program by paying higher food prices.

This program also comes at tremendous political cost in terms of what it has done to the United States' reputation overseas. It also has a significant negative effect on a number of developing countries that might have grown into much better markets for products in which the United States has a comparative advantage. But in the 1980s we have supported the price of sugar at up to six to seven times the world market price.[2] The GATT panel finding of this summer—that the current U.S. sugar quota program is in violation of GATT rules—means that some changes will have to be made in the U.S. sugar program (Truell, 1989). It is most likely, however, that those changes will be made in the context of the GATT Round agreement, rather than in the 1990 farm bill.

Other Program Commodities. There are a number of commodities, including peanuts, wool, mohair, honey, and tobacco, that receive relatively small total amounts of commodity support, but they are important to some members of Congress who put a high priority on maintaining those programs. Those members are very astute at exchanging votes on issues important to the administration for support for those commodities. The Reagan administration talked a good line about reforming farm programs for these commod-

[2]It merits noting that the world market price of sugar was so low in large part due to the reduction in the United States' import quotas and expansion in the European Community's subsidized exports of sugar.

ities but was always willing to exchange support for crucial votes. This tradition is one of long standing, which probably will not change in the current administration. It appears that any significant reduction in supports for peanuts, tobacco, mohair, wool, and honey and any other commodities in this category would have to come about in the context of a GATT Round agreement that required *across-the-board* reductions in subsidies to all commodities simultaneously.

OTHER ISSUES

Payment Limits and Targeting

Another contentious point in the next farm bill debate will be targeting or limiting payments in the commodity programs. After much debate, the $50,000 payment limitation was left "unchanged" in the 1985 act. In reality, the payment limitation was effectively increased by a significant margin by writing a number of loopholes into the 1985 act. For example, an important objective of the marketing loan was to circumvent the $50,000 payment limitation on cotton and rice.

Nevertheless, there have been a number of media accounts of large payments to rich farmers, such as the $1 million deficiency payment to the Crown Prince of Liechtenstein and the more than $10 million payment to a large California cotton producer. When the public and members of Congress who view the programs as providing relief for financially stressed small farm operators hear reports of large payments going to rich farmers, they perceive that something is very wrong. The Disaster Assistance Act of 1988 perhaps provides a preview of how this sentiment might be expressed in the 1990 farm bill. (For a summary of the payment limitation-targeting provisions in the drought bill, see Sumner in this volume.) In addition, the budget constraint may provide further impetus for more stringent payment limits, although in practice they have not proven to be big budget savers.

Farm Credit

The Farmers Home Administration (FmHA) may again be an issue. This has become an issue because of the huge number of bad FmHA loans that will eventually have to be written off. In 1987 the FmHA had about 83,000 delinquent loans, representing a total loss of about $7 billion (U.S. General Accounting Office, 1988). I would expect the current administration to argue that the FmHA ought to be an agency that guarantees loans made by private banks or the Farm Credit System and that it should get out of direct lending completely. It is likely also to argue for focusing FmHA's attention on helping new farmers get into the business. But as long as some powerful congressmen continue to view FmHA as a granting agency, not a lender, major changes are unlikely to occur. There will be a lot of discussion, but

TABLE 4. Farm Income, 1980–1989

Year	Net cash farm income (billions of dollars)	Net farm income (billions of dollars)	Net farm income Deflated (billions of 1982 dollars)
1980	34.2	16.1	18.8
1981	32.8	26.9	28.6
1982	37.8	23.5	23.5
1983	36.9	12.7	12.2
1984	38.7	32.3	30.0
1985	46.6	32.2	28.9
1986	51.4	37.4	32.8
1987	57.7	47.1	40.0
1988	59.9	45.7	37.6
1989[a]	52–57	48–53	39–43

Source: Compiled from U.S. Department of Agriculture, 1989, *Agricultural Outlook* (Washington, D.C., Economic Research Service), August.
[a]Estimated.

only a modest movement in the direction of replacing direct lending with loan guarantees can be expected.

The Standard of Comparison for Farm Income

One factor that will have a significant influence on the next farm bill debate is the fact that net farm income reached record highs in 1987 and 1988 by a substantial margin (see table 4). Those numbers are likely to be held up for a long time to come as a new standard for comparison against which any future farm policy will be compared. Unfortunately, the 1987 net farm income was not in any sense a true measure of the long-term earning power of the U.S. farm sector. In fact, 1987 is an example of why the United States should have cut target prices for grains in the second year of the farm bill instead of waiting until the third. The delay created an artificially high expectation of the earning power of the U.S. farm sector by raising net farm income by such a large percentage over the previous record high. I anticipate the years 1987 and 1988 are likely to become the standard of comparison for future farm income, just as the 1910–1914 period has been in the past. Any farm program that reduces income below the 1987–1988 record will be viewed as a bad farm program by a lot of organizations, and that will have an effect on the debate.

CONCLUSION

This chapter has focused on the commodity program provisions of the 1990 farm bill. The debate over these provisions will probably be much less acrimonious than it was in 1985. In the background, the Uruguay Round of international trade negotiations will be going on at the GATT in Geneva throughout 1990. While the 1990 farm bill will be expected to reduce the budget cost of U.S. farm programs somewhat, it is important that the United States not unilaterally disarm in agricultural policy independent of the GATT. It is also important that Congress not tie the hands of the U.S. negotiators in Geneva, for there is hope that some multilateral reduction in trade-distorting agricultural subsidies can be achieved in this GATT Round. Much of U.S. agriculture would benefit from both a reduction in the incentive to overproduce in high-cost producing areas here and abroad and from greater access to foreign markets.

In conclusion, I predict that the commodity provisions of the 1990 farm bill will look much like those of the 1985 act, with a little more planting flexibility and a somewhat smaller budget cost. Much more time is likely to be spent on a "new agenda" of issues, including water quality, food safety, rural development, and animal welfare, than on the commodity program provisions.

ACKNOWLEDGMENTS

Journal Paper No. 12299, Indiana Agricultural Experiment Station. The author expresses his appreciation to Kristen Allen and Jon Visser for editorial assistance.

REFERENCES

Truell, Peter. 1989. "GATT Panel Finds U.S. Sugar Quotas Violate International Rules on Imports." *The Wall Street Journal,* June 1.

U.S. General Accounting Office. 1988. *Farmers Home Administration: An Overview of Farmer Program Debt, Delinquencies, and Loan Losses* (Washington, D.C., Government Printing Office), May.

U.S. Department of Agriculture. 1984. "Wheat: Background for 1985 Farm Legislation." Agriculture Information Bulletin No. 467 (Washington, D.C., Economic Research Service), September.

———. 1989a. *Agricultural Outlook* AO-156, table 17 (Washington, D.C., Economic Research Service), September.

———. 1989b. "The Export Enhancement Program: Review and Assessment." Unpublished draft (Washington, D.C., Economic Research Service).

U.S. Department of Agriculture, Agricultural Stabilization and Conservation Service. 1989. Supply and Use Tables. Unpublished mimeograph, July.

PART 2

ALTERNATIVE APPROACHES FOR U.S. AGRICULTURAL POLICY

DECOUPLING AND RELATED FARM POLICY OPTIONS

TIM T. PHIPPS, GEORGE E. ROSSMILLER, and WILLIAM H. MEYERS

The term *decoupling* has become a part of the language in farm policy debates and the General Agreement on Tariffs and Trade (GATT) negotiations. Unfortunately, not everyone agrees on the precise meaning of the term. In this chapter, we will define decoupling and decoupled programs with precision, mainly following the concept of these terms as used by Senators Rudy Boschwitz (R–Minn.) and David Boren (D–Okla.) in their 1985 farm bill proposal (the Family Farm Protection and Full Production Act of 1985) and by the U.S. administration in its July 1987 proposal for agricultural trade reform to the GATT. As will be seen, decoupling represents a free market ideal against which other farm policy mechanisms can be compared with regard to a variety of performance criteria. We will define decoupling; discuss the effects of decoupling on the economy, the environment, and various segments of the population; and conclude with a comparative analysis of alternative policies.

DECOUPLING DEFINED

Decoupling, in its purest sense, is defined as the replacement of the current U.S. system of enhancing the incomes of farmers, which is linked directly to a farmer's production level, with income assistance programs not based on production levels. As such, decoupling is a two-step process. The first step is the elimination of all farm programs that act by altering a farmer's production decisions relative to a free market. This step would entail ending the following: (1) any program that affects the prices received by farmers, such as price supports, target prices, quotas, tariffs, domestic demand enhancement, and export subsidies; (2) any program that affects the prices paid by farmers, such as subsidized credit, storage, or other input subsidies; and (3) any program that affects a farmer's choice of input mix, such as Acreage Reduction Programs. The second step is the creation of new income assistance programs not based on a farmer's production level. We henceforth refer to any such income assistance program as a "decoupled" program.

A slight digression is needed at this point. The definition of decoupling used above may be too restrictive to be relevant to policy decisions. Some existing programs can be considered to have such an indirect or small effect on production that they can be left aside or considered decoupled. Others may be deemed to have a wide enough social benefit that they should be retained even if they are not decoupled. Examples of the latter include government-supported agricultural research, international and domestic food aid, programs to improve environmental quality, extension and education, development of rural infrastructure, some grading and inspection programs, and other regulatory programs that protect health, safety, and welfare of the general population, including farm producers.

An additional digression is needed. Any transfer of income or other benefit—even a decoupled welfare payment, an unexpected tax refund, or a win in the lottery—will likely have some effect, however slight, on a farmer's production. At the very least, some resource allocation distortion will occur through the addition of resources to, or retention of resources in, the sector. Thus, there is no such thing as a perfectly decoupled income assistance program.

DECOUPLING AND ITS EFFECTS

Decoupling cannot be evaluated in isolation from current farm programs. In fact, the primary value of decoupling is that it would eliminate many of the economic and environmental problems that are created by current programs, while still providing income assistance.

Current farm programs seek to increase farm income by raising market prices (via price supports and supply control) and by subsidizing the prices received by farmers (via deficiency payments). Each of these policies causes certain economic and environmental problems.

Current farm policies cause three types of economic problems: (1) costs imposed on consumers in the form of higher prices for certain food products; (2) costs imposed on producers in the form of increased costs of production; and (3) costs imposed on the general economy in the form of increased taxes, reduced exports or increased imports, a reduced overall volume of world trade, and the maintenance of excess amounts of resources in agriculture relative to other sectors of the economy. These factors all combine to reduce the overall level of growth and employment in the economy.

The environmental problems of current farm policies are caused by the tendency of particular programs to encourage increased rates of application of fertilizers, insecticides, and herbicides; increased conversion of highly erodible soils or wetlands; and decreased provision of wildlife habitat; and to discourage the use of crop rotation techniques that enhance the soil and control pests. The environmental problems associated with current farm policies are discussed more fully by Reichelderfer (in this volume).

As noted by Reichelderfer, there are three ways to deal with the problems created by current farm programs: (1) create additional farm programs such as the conservation reserve or conservation compliance; (2) create additional environmental programs such as Environmental Protection Agency regulations; or fundamentally revise the way in which the U.S. government supports farm income. Decoupling is an example of the last option.

Because decoupled payments would not be based on a farmer's current production level, such a program would not create the same side effects as current programs. With decoupling, market prices would be allowed to seek their own level based on available supplies and effective demand. This shift to market orientation, of course, would require elimination of all access barriers against foreign suppliers to the domestic market, as well as elimination of all producer subsidies and consumer "taxes" in the form of high food prices.

The effects of decoupling can be summarized by looking at its effects on resource allocation and production, international trade, and consumers. As to resource allocation, with a decoupled income support program farmers would base their production decisions on the market price, not a government-supported price—that is, the market price would be the guide used by the farmer in deciding what to produce, how to produce it, and how much to produce. Resource allocation distortions caused by government interference with market prices and land use decisions would be eliminated.

Trade distortions would be eliminated automatically with the elimination of resource allocation distortions and the removal of trade barriers. The international market and the domestic markets would become parts of the same market with the same market price prevailing. With farmers producing for the market (rather than for a price administered through a government program), no incentive exists for farmers to produce surpluses in excess of the amount that can be sold at prevailing market prices.[1]

Consumers would also see the same market prices for agricultural products and would use those prices in determining what and how much to consume. Government-induced high food prices that result from market access barriers or producer subsidies would be eliminated. While U.S. consumers already face world market prices (the most notable exceptions being dairy products and sugar), consumers in other countries, such as the European Community and Japan, pay much of the cost of farm support through high food prices.

The effects of decoupling on the many segments of the food and agricultural system depend on whether it is accomplished unilaterally, or multilaterally through a process such as the GATT; whether short- or long-run effects are examined; and whether decoupling is applied to all agricultural commodities or to a subset, such as the food and feed grains sectors. Several quantitative studies of unilateral and multilateral trade liberalization may be used to infer some of the effects of decoupling, including both microeffects and macroeffects. These studies are described and their results summarized in the appendix to this chapter. It is important to remember, however, that trade liberalization represents only the first step of decoupling—the elimina-

[1]This does not mean that surpluses or shortages would not occur because of unforeseen circumstances, such as unusually good or bad weather.

tion of current farm subsidy programs—and does not include the replacement of current programs with decoupled income assistance.

AN ASSESSMENT OF THE FIRST STEP TOWARD DECOUPLING

Many effects of decoupling arise from shifts in the absolute and relative price levels of farm commodities. These shifts would bring about three types of changes in the agricultural sector. First, elimination of supports for program commodities such as wheat, feed grains, rice, cotton, and dairy products would lead to a reduction in the returns on those program commodities relative to nonprogram commodities. Farmers would therefore make the decision to shift resources into nonprogram commodities—such as livestock, fruits, and vegetables—on the basis of the relative returns from the various enterprises, not, as is now often the case, largely on the basis of the expected future stream of benefits from the price and income support programs.

Second, increased production of nonprogram commodities would also lower their returns. Falling returns throughout the sector should lead to a reduction in land rents and prices and in the use of marginally productive land for intensive agricultural enterprises.

Finally, the reduced cost of land relative to other inputs, combined with the elimination of the Acreage Reduction Programs for program crops, would lead farmers to substitute land for other inputs such as fertilizer and, to a lesser extent, pesticides.

The probable, short-term qualitative effects on U.S. agriculture of unilateral and multilateral elimination of agricultural commodity programs are summarized in table 1. Some of the predicted quantitative results of unilateral elimination of U.S. farm programs from three economic models are summarized in appendix table A-1 and for multilateral elimination in appendix table A-2. The tables show only estimates of the effects of the first phase of decoupling—the elimination of current farm programs. They do not include the fiscal and economic effects of decoupled income assistance, since at the time of writing no quantitative studies of these effects of decoupling were publicly available. In all of the scenarios, it is assumed that the Conservation Reserve Program is retained.

A Unilateral U.S. Move Toward Liberalization

Unilateral elimination of U.S. commodity price and income supports would lead to increased economic efficiency in the domestic economy, lower unit costs of production, and greater competitiveness of U.S. commodities on world markets. The three models of unilateral liberalization examined—FAPRI, POLYSIM, and NAC/BLS—all showed falling crop and livestock prices, an increased volume but lower value of trade, an increase in cultivated cropland, and a decrease in the acreage base (idled plus cultivated cropland). (See the appendix to this chapter for details.)

TABLE 1. Probable Short-Term Effects on U.S. Agriculture of Elimination of Agricultural Programs

	Unilateral elimination	Multilateral elimination
General economic efficiency	+	+
Unit costs of production	−	−
Land in cultivation	+	+
Acreage base (planted plus idled cropland)	−	?
Gross farm receipts	−	+
Sales	−	+
Government payments	−	−
Net farm income	−	−
Consumer prices	−	+
World prices	−	+
Government outlays	−	−
Land prices	−	?

Note: Pluses and minuses refer to the direction of change from current programs, not to whether the effect is good or bad. Question marks indicate that we are not able currently to predict the direction of the effect and that more research is necessary.

The volatility of producer and consumer prices would likely increase with unilateral elimination of programs. Although none of the studies of unilateral trade liberalization examined its effects on price volatility, the elimination of domestic target and support prices and government accumulation of surplus stocks would likely cause prices to become more sensitive to changes in world supply and demand. Increased volatility is not always bad. When supplies are low, higher prices help ration what is available. When supplies are high, low prices encourage people to consume more. In all likelihood, however, it would be necessary to continue some government holding of buffer stocks for national security and humanitarian purposes. And it is doubtful that price supports would be entirely eliminated. Low-slung price supports (price supports set below average market prices) could provide some downside price protection without significantly reducing the efficiency of production.

Effects on the environment would generally be positive. Land in cultivation would be farmed less intensively (lower per acre application rates of fertilizers and pesticides), and marginal land would be converted to forage or nonagricultural use. Elimination of support for program crops would lead to greater diversity of production and greater use of crop rotation techniques to control pests and augment the soil as the need to preserve a crop-specific base acreage is removed.

As with all policy changes there would be winners and losers. For example, the net position of the input supply industry would depend on whether lower per acre fertilizer and pesticide application rates were offset by the increased acreage in cultivation.

Removing supports for producing program crops would lead some farmers to shift more resources into nonprogram crops such as livestock, dry beans, sunflowers, fruits, and vegetables. An increased supply of nonprogram crops would lower their returns in most cases, although increased production may allow processors of certain commodities, such as sunflowers, to reap economies of size.

Although gross receipts to producers of crops and livestock would fall, the effect on the producers' net income is uncertain. Reductions in revenues would be offset partially by reduced costs of production.

In the longer run, the U.S. sugar, dairy, rice, and wool sectors would decline in size in the absence of government support. The wheat, feed grains, and soybean sectors would experience smaller income adjustment problems over time and would increase their volume of output because of their more favorable competitive positions in world markets. Processors and marketers of wheat, feed grains, and soybeans would gain from the increased volume of exports. Processors and marketers of dairy products, rice, and wool would lose as these subsectors contracted. The situation is more complex for sugar. Coastal sugar refiners may gain because of access to cheaper raw materials from abroad, while inland refiners would most likely decline.

Agriculturally dependent rural communities will be affected differentially, depending on the type of agriculture that dominates the region. Regions that produce sugar, dairy products, rice, and wool and regions with poor-quality land would see the relative importance of agriculture to their economic bases decline with a move toward program elimination. Regions that produce livestock, wheat, feed grains, and soybeans would see a smaller decline (or possibly none) in their agricultural economies, would experience fewer structural adjustments, and may end up being better off in the long run.

A Multilateral Move Toward Liberalization

A multilateral move toward agricultural trade liberalization would moderate many of the effects on U.S. agriculture of a unilateral U.S. move. The qualitative effects of multilateral liberalization are given in table 1. (See appendix table A-2 for quantitative effects.) With multilateral liberalization, reduced domestic production costs combined with elimination of foreign trade barriers should lead to greater U.S. exports of products such as wheat, feed grains, and livestock that are competitive internationally. World market commodity prices would be expected to rise as countries that now subsidize the sale of their surplus commodities on the world markets stopped doing so.

The amount of U.S. land in cultivation would be expected to rise, although the total acreage in cultivation would probably be less than the current acreage base (land in cultivation plus currently idled acreage). As in the case of unilateral liberalization, environmental quality could improve because of the elimination of the commodity program incentives to overapply pesticides and fertilizers. The improvement would not be as great as with unilateral decoupling because some of the gains could be lost because of higher prices and the greater acreage in production.

The net position of the input supply industry, again, would depend on whether lower fertilizer and pesticide application rates are offset by the increased acreage in cultivation. As with unilateral liberalization, processors and marketers of wheat, feed grains, soybeans, and livestock would gain from the increased volume of exports.

Increased export demand for competitive program crops would offset much of the reduction in current "coupled" government income support. Higher sales receipts are likely, but it is uncertain whether they would fully offset the decline in government payments. Compared with unilateral liberalization, projected multilateral liberalization results suggest that lower decoupled payments would be needed to offset fully reduced earnings from the market.

As with unilateral liberalization, rural communities in regions that produce sugar, dairy products, rice, and wool and regions with poor-quality land will be disadvantaged by the structural adjustment, although fewer areas will be disadvantaged than with unilateral liberalization. Regions that produce wheat, feed grains, and livestock may gain relative to their current position because of increased export demand.

Multilateral liberalization should lead to decreased volatility of world market prices as nations stop dumping their surplus production on the international markets and allow their domestic prices to become more responsive to world market signals. Currently, the European Community nations and Japan insulate their consumers and producers from changes in world prices by essentially fixing domestic prices. These higher domestic prices for food and agricultural products cause nations that have more open agricultural economies, such as Canada, New Zealand, Australia, and, in the case of food and feed grains, the United States, to experience much greater price volatility in response to market fluctuations than would occur without trade barriers.

Trade Liberalization with Decoupled Income Assistance

This section has so far addressed only the effects of the first step of decoupling—the unilateral or multilateral removal of price and income supports for farmers and all trade barriers. If decoupled income assistance were provided, the net effects of liberalization on net farm income and government payments would be indeterminate, a priori, since both would depend on the level of decoupled payments set by the administration and Congress and on whether or not producers of nonprogram crops were made eligible for decoupled payments. If decoupled payments were designed to offset exactly losses in net income to program commodity producers, government expenditures would be less with either unilateral or multilateral decoupling than with current programs. With income and price support programs, the government must spend more than a dollar for each extra dollar of income received by farmers (Just, Hueth, and Schmitz, 1982). With decoupled payments, as with any direct income transfer, each dollar of government expenditure would increase farm income by a full dollar.

POLICY ALTERNATIVES

As noted earlier, decoupling involves two steps: (1) the elimination of income support programs that are based on current production levels and (2) their replacement with income assistance payments not linked to current production. The preceding discussion shows that the first step of decoupling in its purest form would have beneficial effects for the economy as a whole as well as for the environment but would disadvantage producers of commodities that are not currently supported, producers and processors of commodities that are currently strictly protected from international competition, and rural areas that depend on commodities that would decline in importance after decoupling.

Although it is fairly clear that pure decoupling would be good for the economy as a whole and for the environment, the fact that it would disadvantage certain segments of the farm sector makes it unlikely that decoupling in its purest form is currently politically feasible. We might then ask, Is it possible to retain some of the benefits of decoupling and reduce the disadvantages if we move our farm programs toward decoupling without going all the way? Although the answer to that question is quite controversial, it is in the spirit of that question that we proceed. We will discuss several types of farm program mechanisms and then evaluate them on the basis of how well they achieve the objectives of decoupling. Alternatives are organized into three groups: (1) minor modifications of current policies that reduce the associated economic or environmental problems, (2) intermediate alternatives between minor modification and pure decoupling, and (3) different types of decoupled programs.

Minor Modifications of Current Policies

One option for reducing the economic and environmental problems of current farm programs while still keeping their basic structure intact is to modify some of the most troublesome features. Three approaches discussed here are modifications in the base acreage system, a phasedown of target prices, and placement of the Acreage Reduction Program (ARP) on a bidding basis. While none of these alternatives really decouples income support from production levels, they all provide farmers with more control over their planting decisions than do the current programs.

Modifications in the Base Acreage System. The current base acreage system limits the farmers' flexibility to respond to price changes or to practice crop rotation. Several alternatives exist that would give producers more flexibility while still retaining the base acreage system. Three options considered here are (1) allowing planting of rotation crops on a portion of the base acreage without losing program base eligibility for the original commodity; (2) granting temporary waivers to allow the planting of a particular alternative commodity, such as soybeans on corn base acreage, without decreasing the historical corn base; and (3) returning to a nonspecific commodity base.

Farmers could be permitted to plant up to a specified portion of their program base to certain nonprogram rotation crops without reducing their future program base. This option would increase farmers' flexibility to use certain crop rotations to improve the soil and control pests but would not increase their flexibility in responding to price changes in program crops. As such, this type of proposal would have positive environmental effects but would not significantly reduce the economic problems associated with the base system. This proposal would likely be opposed by current growers of nonprogram crops because of the increased competition it could potentially foster.

Another option used in the 1988 disaster assistance bill seeks to improve farmers' flexibility to respond to changes in the relative prices of program crops by granting temporary waivers to plant a particular crop that is in short supply on acreage designated for another program crop without losing the recorded base. This change would improve economic flexibility but not necessarily bring about environmental improvement.

A third option would be to return to a nonspecific commodity base instead of computing separate bases for each program commodity. This option would give farmers the opportunity to respond to price incentives. The problem is that as long as the current target price system is maintained (i.e., as long as there is the same relationship among target prices for various commodities), farmers would be responding to price incentives set by the government, not by markets. The greater flexibility in this option could also lead to greater price instability than is now experienced, with surplus in some years and shortage in others.

Phased Reduction in Target Prices and Acreage Controls. Another farm policy option, most closely associated with the Reagan administration position in the 1985 farm policy debate, is a phased reduction in target prices and acreage controls. Under this plan, target prices and acreage reductions would be gradually lowered over time until government price subsidies were eliminated. A complete phaseout is similar to the first step in decoupling: the elimination of the farm programs that alter farmers' production decisions. If loan rates continue to be set at levels below market prices, a phaseout of target prices and acreage controls would be practically the same as the first step of decoupling, including elimination of most of the economic and environmental problems caused by current programs. Under the Reagan administration option the second step of decoupling, the replacement of current income assistance programs with decoupled programs, would not necessarily have been taken.

Bid Acreage Diversion. Some of the problems with the current annual Acreage Reduction Programs could be reduced or eliminated by alternative program designs. The closer a land reduction program comes to mimicking the actions of a land market, the smaller will be the loss in efficiency due to the policy. An efficient land market operates at the margin of agricultural land use. As agricultural returns rise, marginal land is converted from idle or other uses to crop production. As returns fall, this land again leaves crop production. Current deficiency payment programs and Acreage Reduction Programs do not mimic a land market. Target or support prices set higher than market clearing prices encourage cultivation of poor-quality land that would not be cultivated otherwise.

One alternative that would be more efficient than the current acreage reduction system would be a movement to a strict paid diversion program based on a competitive bidding system. A bidding scheme, by operating similarly to a land market, would reduce the economic efficiency losses of land retirement. Economic efficiency losses would be reduced because the most productive land would remain in production and land of the worst quality would be idled. Budget costs, while higher than for the Acreage Reduction Program, would be lower than the current paid land diversion program because the government's acceptable bids could be determined by the productivity of each offered parcel, not by the value of the last (and best) accepted acre. There are, however, some important qualifications. A bid diversion program would require more land to be idled for a given reduction in production than the current system, because a bid system would draw the lowest-yielding land from a national pool of agricultural land, rather than requiring each farmer to idle a fixed percentage of his or her cropland. Another major disadvantage would be the high costs of administering such a program.

Intermediate Alternatives

Stabilization and Income Insurance. U.S. agricultural programs were originally represented as farm income stabilization programs and still include stabilization features. Deficiency payments help stabilize farm income since they move inversely with farm prices. Price support programs lead the government to acquire stocks when market prices fall below the loan rate and to sell stocks when the market price reaches a predetermined level. This process has a price stabilizing effect. The Farmer-Owned Reserve (FOR) also was designed to increase price stability by allowing wheat and corn farmers to extend their nonrecourse loans from one year to three to five years, with the government paying for on-farm storage. From its inception until 1980 the program was operated as a market stabilization device. Since 1980 the program has not been effective as a strict stabilization tool because in most periods both the entry and release prices have been set too high in relation to the market price. As such, the FOR has been more effective in reducing downward movements in prices than in reducing upward movements.

The FOR could be modified to make it once again effective in stabilizing rather than enhancing prices for farmers and consumers by reestablishing the original concept that the long-term expected size of the FOR is zero. This could be done by changing the trigger prices to allow farmers more flexibility in entering or leaving the reserve. As it exists today, the reserve is used more as a tool for supporting market prices than as a stabilization tool. Unfortunately, as with almost any stabilization program, political pressure will always exist to use it as a price support program.

Income insurance is an alternative for dealing with the problem of income instability. To begin with, if the goal of farm policy is to stabilize farm income, it is more economically efficient to approach that goal directly, rather than indirectly through price supports and deficiency payments. An income insurance program, if run in an actuarially sound manner, would avoid many of the economic and environmental problems of current policies because it would not work by altering the prices received by farmers.

Income insurance and price stabilization policies are technically not decoupling because they would still alter the allocation of resources between agriculture and the rest of the economy. Both programs would reduce risk in agriculture, which would encourage investment in agriculture and more production in areas that face weather or other types of risks. Both programs, however, would represent a sizable step toward decoupling.

A major problem with income insurance schemes is that they could easily become politicized, with the insurance payouts quickly being changed to income assistance. When and if that transition occurs, income insurance would create the same types of economic and environmental problems that the United States faces with its current programs.

Two-Price System. A two-price system, where the producer receives, for example, a high support price for the portion of his or her production that goes for domestic utilization and the world price for production above that level, comes fairly close to the concept of decoupling and in fact has been referred to as "decoupling at the margin." The incentive price viewed by the producer for all production greater than the domestic allocation would be the world price. Supply control measures, therefore, should not be necessary. Resource allocation and trade distortions would be minimal since the only incentive to overproduce comes from the extent to which the support may retain resources in agriculture that would move out in the absence of any support.

Two caveats are necessary about a two-price system of this type. First, the higher price for domestic consumption can be paid by either the taxpayer or the consumer. If the consumer pays, the quantity of agricultural products demanded will be less than if the taxpayer pays, and border restrictions would be necessary to prevent lower-priced foreign exports from entering the domestic market. Second, the amount of production eligible for support cannot be a simple percentage of a producer's total production; otherwise the producer will view his or her incentive price as a blend of the support and world prices and will attempt to increase his or her base production over time. The portion of production eligible for support on a given farm must be fixed and based on criteria unrelated to current or future production levels. A major drawback of this mechanism is that the existing producers receive the support entitlement, which becomes capitalized into land values, and thus entry into the industry becomes at least somewhat restricted.

Two proposals along these lines have been made recently. The first, by the International Agricultural Trade Research Consortium (IATRC) is known as the Producer Entitlement Guarantee (PEG). IATRC argues for the use of this approach multilaterally as part of the reform of domestic agricultural policies under a GATT agreement to liberalize trade. A PEG program would limit the quantity of an individual farmer's production eligible to receive support payments. The per unit payment could be based on historical levels of price supports. Governments would have the option of tying payments to production and therefore be free of the political objections often voiced against direct payments.

Each farmer's actual production would not be controlled. Farmers would be free to decide how much to produce above the quantity receiving support payments. In order to minimize trade distortions all existing border measures would be eliminated, as would all internal agricultural support measures except for payments on the specified PEG quantities. Consumers and users would pay the open market price for agricultural products, and farmers would receive this price for their marketings. The important point is that for a part of his or her production the farmer would realize returns of the market price plus the payment. But for the remaining production and any increases in production, the anticipated return would be determined by market prices.

The authors suggest that the volume of production eligible for support under a PEG program could be determined through GATT negotiations, but governments would have full discretion over to whom the initial PEGs are issued. To eliminate production inefficiency PEGs could be transferable among farmers or farms. If PEGs are not transferable, high-cost producers would be encouraged to continue farming in order to receive benefits under the program. With a transfer option they could sell their PEGs to low-cost producers or to the government.

Individual PEGs could be issued to existing farmers on the basis of production quotas currently held in Canada and the European Community or on land base and program yields in the United States. New entrants to farming would either purchase (or lease) PEGs from other farmers or obtain them directly from their governments. Otherwise, they would produce at world market prices and receive no government payments. PEGs could be tied to the land and transferred with the land. Alternatively, they could be given to individual farmers or farm workers (in which case the value of the PEGs would be reflected in the value of a certificate of eligibility if the government allowed PEGs to be transferable).

If governments wish to have a pool of PEGs for new entrants, then a fixed percentage of all private PEG transactions (farm-to-farm sales, parents to son or daughter, and so forth) could be automatically reclaimed to pass on to these entrants. Reclaimed PEGs could also be used to reduce the national quantity of production eligible for support to less than free-trade output levels. Government purchases of PEGs could also be used to transfer support eligibility to new entrants, eliminate (i.e., buy out) high-cost production causing trade distortions, or even phase out support payments completely.[2]

IATRC estimates that almost 98 percent of all present trade distortion would disappear if all countries would set production for price support eligibility at 80 percent of 1986 production levels. (For more discussion of the PEG in an international context, see Blandford et al., in this volume.)

A second proposal of this nature, known as the payment acres or triple base program, was recently analyzed by Abel, Daft, and Earley (1988). To provide producers the opportunity to respond to relative market prices without losing future support eligibility, price support eligibility would be established at less than 100 percent of a producer's planted base acreage. Acreage Reduction Programs (ARPs) would remain in force. The difference between total base acreage and the sum of payment acreage and ARP acreage could be planted to any crop, including the program crop, but would be without support eligibility. This proposal would provide program crop farmers some flexibility in cropping decisions without loss of program crop base but would also protect nonprogram crop farmers from major competition by program crop farmers.

Freezing Program Yields. The Food Security Act of 1985 froze program yields at their 1981–1985 level. Freezing program yields has been viewed by many as a significant step toward decoupling because it eliminates the incentive for farmers to apply more fertilizers and pesticides to increase their program yields to increase future deficiency payments. In a study of the U.S. wheat program, Tsigas, Hertel, and Preckel (1989) found that freezing program yields would significantly reduce the rate of application of nonland inputs such as fertilizers and chemicals.

Freezing both program yields and base acreage would be the same as decoupling since future deficiency payments would then be based purely on past production levels, and would not provide an incentive for increasing current production levels. Freezing both program yields and base acreage would also be the same as the PEG since the producer would receive the world market price for all production in excess of the program yield multiplied by the base acreage.

Since base acreage is not currently frozen, target prices still provide an incentive to increase the acreage under cultivation to increase future deficiency payments, so freezing program yields alone is not the same as complete decoupling. There are, however, significant economic disincentives in the Food Security Act not to increase the base acreage. Finally, until farmers believe that yields are permanently frozen, there may be incentives to apply

[2]The authors gratefully acknowledge the assistance of David Blandford, Harry de Gorter, Praveen Dixit, and Stephen Magiera in providing the foregoing description of the PEG.

more nonland inputs to keep recorded yields high in the hope that the program yields will be adjusted upward in the future.

Full Decoupling

Several options exist for fully decoupling income support payments from current production levels. The options differ mainly in the ways in which the decoupled payments are computed or disbursed.

Boschwitz and Boren Bill. The Boschwitz-Boren (the Family Farm Protection and Full Production Act of 1985) and Boschwitz, Boren, and Karnes (the Family Farm Protection Act of 1989) bills proposed to pay farmers annual transition or equity payments that declined over time. The payments would have been based on past acreage and yields so as not to influence current production decisions. In addition, the payments would have been subject to upper limits that increased less than proportionately to farm size, with an absolute upper cap of $200,000 per farm operator.

Other Options for Decoupling Income Payments from Current Agricultural Production. Another option for decoupling income assistance from current production levels is to link income assistance to the provision of public goods, such as open space, or to the adoption of environmentally beneficial practices, such as terracing or low input sustainable agriculture. The idea behind this approach, which is discussed further by Reichelderfer (in this volume), is that paying farmers to adopt socially beneficial practices could benefit both farmers and society in general. The problem with this type of approach is that it is very difficult to subsidize certain environmentally beneficial practices without indirectly influencing levels of production and market prices. This type of program would, therefore, be highly susceptible to political pressure to justify price enhancement policies on the basis of their contribution to environmental quality.

SUMMARY AND CONCLUSIONS

This chapter is not meant to be either an endorsement or a condemnation of any of the policy alternatives discussed. We have attempted to discuss problems with the current programs used to support farm income and to present alternatives that would correct some of those problems. The alternatives discussed range from minor modifications in current programs, such as the flexible base, to complete replacement of current programs with decoupled income support.

All of the alternatives discussed would improve the economic efficiency of U.S. agriculture. In general, the closer the alternative comes to a free market, the greater the potential gain in economic efficiency. A movement toward a free market would improve efficiency in two ways. First, removal of target prices and price supports at least at the margin would allow the prices that determine farmers' planting decisions to reflect world supply and demand conditions. A free market would not only eliminate surpluses but also allow farmers to respond more rapidly to shifts in relative prices. The greater efficiency would free resources to be used where they are most beneficial to the economy.

Second, removal of the Acreage Reduction Programs would allow farmers to choose the most economically efficient combination of inputs. This approach, in turn, would reduce average costs of production, which would benefit domestic consumers and lead to greater competitiveness of U.S. agricultural commodities on international markets. Eliminating Acreage Reduction Programs would also eliminate the incentives current programs provide to foreign competitors to expand their cultivated acreage.

Economic efficiency is extremely important, but it is only one dimension of policy. Policymakers are equally concerned with the distribution of the benefits and costs of alternative policies. As with all policy changes, some groups will gain and some lose. In general, a movement toward a freer market for agriculture would entail losses for those who are currently entitled to benefits under existing programs. It would also mean that some people in farming today would not be farmers in the future. As discussed earlier, losses could also accrue to producers who are not currently entitled to program benefits but would face increased competition from former program crop producers.

Society and the political system will decide how the U.S. agricultural sector will be supported. This decision should be based on an understanding of the trade-offs involved among economic efficiency, the distribution of gains and losses among different segments of the food and agricultural system and society at large, and environmental quality. In this chapter we do not pretend to have provided the basis for such a decision; rather, we have attempted to identify the types of trade-offs involved in formulating farm policies and summarize what is currently known from economic studies.

REFERENCES

Abel, Daft, and Earley. 1988. "An Analysis of Alternative Policies for Grains and Cotton" (Alexandria, Va.), December.

Just, R. E., D. L. Hueth, and A. Schmitz. 1982. *Applied Welfare Economics and Public Policy* (Englewood Cliffs, N.J., Prentice-Hall).

Tsigas, M. E., T. W. Hertel, and P. V. Preckel. 1989. "Program Yields and U.S. Crop Production: How Important Is the Freeze?" Draft report, Department of Agricultural Economics, Purdue University, West Lafayette, Ind.

APPENDIX

QUANTITATIVE ASSESSMENT OF POLICY REFORM

The quantitative analyses that come closest to evaluating the effects of unilateral and global decoupling are those that have evaluated unilateral and multilateral trade liberalization. As mentioned in this chapter, liberalization is only the first step of decoupling. These studies are relevant to the evaluation of decoupling, since the original U.S. proposal in the GATT called for elimination of current programs and their replacement with decoupled payments or programs. The various analyses estimate the effects of eliminating all current programs.

The general consensus of previous studies is that eliminating commodity programs multilaterally would increase world market prices and trade but would not raise prices to the level of current price and income supports in the developed countries. This section provides a summary of several recent studies that estimate the effects of eliminating current programs used to support agriculture in the United States and abroad and then compares these results with those from studies done in the early 1980s. Two types of models were used in these studies: partial equilibrium models and computable general equilibrium models. Partial equilibrium models focus on the effects of liberalization on the domestic and international agricultural sectors. The models compute the effects of liberalization on agricultural commodity prices, farm income, and agricultural trade flows. The computable general equilibrium models analyze the effects of liberalization on other sectors of the economy as well as agriculture and include estimates of the effects on domestic GNP, total trade flows, and the federal budget deficit. The partial equilibrium models are discussed first.

One set of analyses of various liberalization scenarios was reported in a modeling symposium organized by the American Agricultural Economics Association in the summer of 1988 (Center for Agricultural and Rural Development, 1988). The analysts were asked to use their individual models to evaluate similar scenarios so that the results conducted by various analysts and models could be compared. The three models reporting similar scenarios on the elimination of U.S. commodity programs unilaterally were POLYSIM of Oklahoma State University, the FAPRI modeling system of Iowa State University and the University of Missouri, and the NAC/BLS modeling system of Iowa State University and the International Institute for Applied Systems Analysis.

POLYSIM begins with a given baseline and a set of policy and behavioral parameters that represent supply, demand, and price responses to changes in external conditions or policies.

The FAPRI modeling system is a set of commodity models of crops and livestock for the U.S. and foreign countries and regional trade models for the major crop commodities in international trade. It is an econometric model with explicit policy parameters for the major trading countries. The model generates its own baseline and then generates deviations from this baseline arising from changes in external and policy variables (Devadoss et al., 1989).

The NAC/BLS modeling system is a global, applied general equilibrium model, which is more complete in its commodity coverage and more general in its interaction with the nonagricultural sector than either of the other two models (Fischer et al., 1988). Because it goes beyond the commodity models and allows resource adjustments as well as interaction with general economic sectors of the countries, it is more appropriate than the other models in looking at long-term effects (ten years or more) of policy changes.

QUANTITATIVE EFFECTS OF ELIMINATING U.S. PROGRAMS

Three modeling systems were used to estimate the effects of phasing out U.S. commodity programs during the 1989–1991 period (table A-1). The results of the models are consistent in terms of the directions of impact on acreage, prices, exports, receipts, and costs. Among the models the range of magnitudes in the effects is quite large. With all models, estimates of idled acreage is down, planted acreage is up, and the total acreage planted and idled declines by 1.7 million acres (POLYSIM) to 12 million acres (NAC/BLS). The largest estimated decline in total acreage (−12.0 million acres) is associated with the 1990–1992 period, when the amount of acreage idled under annual programs is estimated to be larger than that idled during the 1992–1996 period. The overall implication is that some of the total land resources (currently planted plus idled acreage) would move out of program commodities into forage production or nonagricultural use in the absence of the current programs.

The estimated increase in area planted to program crops in all cases reduces crop prices. The range of crop price reductions is from 1.8 to 11.3 percent. As feed prices fall, livestock production increases and livestock prices decline. The estimated range of livestock price reductions is 7.8 to 12.8 percent. Although the volume of all exports increases slightly, the decline in prices is large enough to reduce the value of exports in all of the model estimates.

Cash receipts from sales of crops and livestock decrease by $6.5 billion to $12 billion in the various estimates. Government payments decline by about $3 billion, except in the 1998–2000 period, when payments are assumed to be relatively insignificant to begin with. Production costs are also lower, but the net effect on income (receipts minus costs in table A-1) is a reduction of $8 billion to $10 billion. This reduction in income to producers is fairly consistent in all the model estimates.

An implication of the substantial decline in net returns is that the total government savings (payments plus indirect costs) of $4 billion to $5 billion

TABLE A-1. Comparison of Estimated Effects of Phasing Out All U.S. Commodity Programs During the 1989–1991 Period, Using Three Models

	FAPRI	POLYSIM	NAC/BLS	
	(1992–1996 average)	(1992–1996 average)	(1990–1992 average)	(1998–2000 average)
Item (change)				
Acreage idled (millions of acres)	−7.7	−6.9	−21.1	−5.1
Acreage planted (millions of acres)	1.5	5.2	9.1	2.4
Crop prices (percent change)	−4.3	−11.3	−1.8	−3.1
Livestock prices (percent change)	−7.8	−11.9	−10.9	−12.8
Value of U.S. exports (billions of dollars)	−0.7	−1.0	−1.2	−2.1
Cash receipts (billions of dollars)				
Crops	−4.1	−4.8	−0.5	−1.5
Livestock	−4.0	−3.1	−6.0	−9.7
Government payments (billions of dollars)	−2.8	−2.8	−3.1	−0.3
Production cost (billions of dollars)	−2.6	−3.3	−0.3	−1.8
Receipts minus costs (billions of dollars)	−8.3	−7.4	−9.3	−9.7
Total government cost (billions of dollars)	−4.6	n.e.	n.e.	n.e.

Notes: The models did not include any direct (decoupled) income assistance to farmers, but the Conservation Reserve Program was retained. All prices and values are in real terms.

n.e. means not estimated.

Source: From the modeling session of the American Agricultural Economics Association, Knoxville, Tenn., July 1988.

per year would not be sufficient to compensate producers for the loss in income resulting from the elimination of government programs. A primary reason for government savings being insufficient to compensate producers is that current programs keep a significant amount of productive land out of production, thus helping to increase program and nonprogram commodity market prices.

QUANTITATIVE EFFECTS OF MULTILATERAL TRADE LIBERALIZATION

Two of the models (FAPRI and NAC/BLS) were used to evaluate the phase-out of both U.S. and foreign commodity programs during the 1989–1991

TABLE A-2. Comparison of Estimated Effects of Phasing Out U.S. and Foreign Commodity Programs During the 1989–1991 Period, Using Two Models

	FAPRI[a] (1992–1996 average)	NAC/BLS[b] (1990–1992 average)	NAC/BLS[b] (1998–2000 average)
Item (change)			
Acreage idled (millions of acres)	−7.7	−21.1	−5.1
Acreage planted (millions of acres)	4.9	8.9	2.6
Crop prices (percent change)	5.2	17.6	1.4
Livestock prices (percent change)	2.2	5.2	−4.7
Value of exports (billions of dollars)	2.9	2.2	−1.5
Cash receipts[c]			
Crops	1.8	11.3	1.0
Livestock	0.2	5.5	−4.4
Government payments[c]	−2.9	−5.1	−1.0
Production cost[c]	1.3	2.1	0.8
Receipts minus costs[c]	−2.2	9.6	−5.2
Total government cost[c]	−4.6	n.e.	n.e.

Note: n.e. means not estimated. All prices and values are in real terms.

Source: From the modeling session of the American Agricultural Economics Association, Knoxville, Tenn., July 1988.

[a]FAPRI analysis eliminates programs in developed market economies, plus those in Argentina and Brazil.

[b]NAC/BLS analysis eliminates programs in all market economies.

[c]In billions of dollars.

period relative to the baseline (see table A-2). Compared to the unilateral case, the change in idled acres was the same, but planted acreage was unchanged or higher because of higher world commodity prices. The FAPRI and NAC/BLS models predicted that acreage planted would rise 4.9 to 8.9 percent in the short to medium term. Total planted and idled acreage still decline, but by a smaller amount in the medium- to long-term period.

Because of the removal of commodity programs outside the United States, foreign demand for commodities increases substantially and world prices are driven higher. In the short to medium term both crop and livestock prices increase by 2–18 percent. As shown in table A-2, the FAPRI and NAC/BLS estimates of world price increases are between 5.2 and 17.6 percent for crops and between 2.2 and 5.2 percent for livestock. These results are consistent with price effect, estimated in several earlier studies (shown in table A-4). The models also indicate that the value of U.S. exports would rise by $2.2

TABLE A-3. Comparison of the Estimated Effects of Phasing Out U.S. Commodity Programs Using Three Computable General Equilibrium Models
(in billions of dollars)

	Robinson et al. (for 1991)	Hertel et al. (long run)	Feltenstein (1986)
Gross national product	9.6	4.9	28
Budget deficit	−26.1	n.a.	−34
Trade deficit	−15.6	n.a.	−42

Note: n.a. means not applicable.
Source: Data presented at the conference on Agricultural Policies and the Non-Farm Economy, Washington, D.C., May 26–27, 1988 (cosponsored by the American Enterprise Institute, Washington, D.C., and the Centre for International Economics, Canberra, ACT, Australia).

billion to $2.9 billion as both world prices and the volume of U.S. production and exports increased. Since these price increases are export driven, both the volume and value of exports increase in the short and medium term. The longer-term (1998–2000) estimates show a decline in livestock prices because of a lagged production response to higher livestock prices in the earlier period.

Cash receipts are also estimated to increase in the short and medium term both for crops and livestock, but in the long run livestock receipts decline and crop receipts increase by only a small amount. Government payments decline, as occurred in the scenario of unilateral liberalization by the United States. Production costs increase because of higher commodity prices. While NAC/BLS shows a $9.6 billion increase in the short term for income, the FAPRI model shows that medium-term receipts minus costs would fall by $2.2 billion with multilateral liberalization. In the long term, decline in income is associated with the decline in livestock prices and cash receipts.

This case differs from the unilateral scenario in that net returns to producers either increase or at worst decline by a smaller amount than the government savings in program costs. Therefore, either producers are better off when both U.S. and foreign commodity programs are eliminated or the government has at least sufficient savings from the elimination of programs to compensate producers. With multilateral liberalization, however, the crop commodities currently receiving program supports would not be the ones suffering substantial losses and needing compensation. Losses would be experienced by nonprogram crops and livestock.

COMPUTABLE GENERAL EQUILIBRIUM MODELS

Computable general equilibrium (CGE) models estimate the effects of liberalization on the farm sector and the macroeconomy. We will report the macroeconomic effects estimated by three models here (see table A-3 for a summary of results).[3] The first study (Robinson, Adelman, and Kilkenny, 1988) estimated the effects of unilateral and multilateral elimination of subsidies on consumption, investment, the gross national product (GNP), and the budget and trade deficits. The authors found that unilateral elimination of subsidies in 1986 would, by 1991, raise the GNP by $9.6 billion and lower the budget and trade deficits by $26.1 billion and $15.6 billion, respectively. The second model (Hertel, Thompson, and Tsigas, 1988) estimated the effects of unilateral liberalization on domestic farm structure, farm income, and the GNP for both the short and long term. The authors found that while farm income would fall in both the short and long run, total economy GNP would increase by $2 billion in the short run and $4.9 billion in the long run.

The third study (Feltenstein, 1988) also estimated the economywide effects of unilateral elimination of U.S. agricultural subsidies. In 1982 dollars elimination of subsidies in 1986 would have increased the GNP by $28 billion, reduced the budget deficit by $34 billion, and reduced the trade deficit by $42 billion.

PREVIOUS STUDIES OF AGRICULTURAL TRADE LIBERALIZATION

A number of studies of trade liberalization were conducted during the 1980s. Although they covered various time periods and were conducted by many analysts using different models, the estimated impacts on world commodity prices have been surprisingly consistent. The more recent studies indicate liberalization would generally produce higher world price increases, mainly because current programs have a greater depressing effect on world prices than did programs during the late 1970s and early 1980s.

For wheat and feed grains world prices increased 1 to 13 percent (see table A-4). Two of the three studies that included soybeans found that soybean prices would fall; the other showed prices of oilseeds and products, in aggregate, rising. This drop in soybean prices suggests that current programs tend to stimulate soybean demand and restrict production (because of, for example, export taxes in Brazil and Argentina). There is less consistency among the results for rice, sugar, and dairy products, although generally prices of these commodities were estimated to increase by more than those for wheat and feed grains.

Another generalization that can be made from all these results is that market price increases are smaller than the differences between market price

[3]These CGE study results were originally presented at the conference on Agricultural Policies and the Non-Farm Economy, Washington, D.C., May 26-27, 1988. It was cosponsored by the American Enterprise Institute, Washington, D.C., and the Centre for International Economics, Canberra, ACT, Australia.

TABLE A-4. World Price Effects of Agricultural Trade Liberalization Scenarios
(percent change from base period)

Source/liberalization scenario	Time frame[a]	Wheat	Corn	Soybeans	Rice	Sugar	Dairy products	Beef	Poultry
USDA—Krissoff and Ballenger (1987)									
Complete multilateral agricultural liberalization	1984	12.0	8.6	−2.0	22.9	60.5	42.0	19.7	4.5
USDA—Roningen, Sullivan, and Wainio (1987)									
Liberalization by U.S., Canada, EEC, Australia, New Zealand, and Japan	1984	10.0[b]	7.0[c]	3.0[d]	—	—	27.0	10.0[e]	—
World Bank—Tyers and Anderson (1986)									
Complete multilateral liberalization	1985	8.8	4.2[f]	—	−7.9	7.6	66.9	15.9[g]	—
Liberalization by OECD countries	1985	2.0	1.0[f]	—	5.2	4.9	27.0	16.1[g]	—
World Bank—Zietz and Valdes (1986)									
Complete liberalization by OECD countries	1979–1981	12.7	11.7	—	—	16.7	—	18.5	—
FAPRI									
Complete liberalization by OECD, Argentina, and Brazil	1992–1996	12.7	11.8	−7.1	22.3	—	—	—	—
NAC/BLS									
Complete liberalization by all countries	1993–1995	12.7	14.0	—	34.5	—	11.8	23.4[g]	—

Note: Dashes mean not applicable. OECD, Organization for Economic Cooperation and Development; EEC, European Economic Community.

Sources: Adapted with permission from the Agricultural Policy Working Group, 1988, "Decoupling: A New Direction in Global Farm Policy," Washington, D.C., February, p. 31; modeling session of the American Agricultural Economics Association, Knoxville, Tenn., July 1988. (See also reference list for specific studies.)

[a]The period for which effects are estimated.

[b]Food crops.

[c]Feed crops.

[d]Oilseeds and products.

[e]Meat and eggs.

[f]Coarse grains.

[g]Beef and lamb.

and government-supported prices for most developed countries. Even in the United States, where government-supported prices are not as far above market prices as in many developed countries, these market price increases are not sufficient to compensate producers for the loss in income support programs.

REFERENCES TO APPENDIX

Center for Agricultural and Rural Development. 1988. "Policy Scenarios with the FAPRI Commodity Models." Working Paper No. 88-WP41 (Ames, Iowa State University), December.

Devadoss, S., P. Westhoff, M. Helmar, E. Grundmeier, K. Skold, S. R. Johnson, and W. H. Meyers. 1989. "The FAPRI Modeling System at CARD: A Documentation Summary," CARD Technical Report 89-TR13, Center for Agricultural and Rural Development (Ames, Iowa State University), December.

Feltenstein, A. 1988. "Agricultural Policies and the U.S. Federal Budget and the Trade Deficit." Paper presented at conference, Agricultural Policies and the Non-Farm Economy, Washington, D.C., May 26–27.

Fischer, G. K., K. Frohberg, M. A. Keyzer, and K. S. Parikh. 1988. *Linked National Models: A Tool for International Food Policy Analysis* (Dordrecht, The Netherlands, Kluwer).

Hertel, T., R. Thompson, and M. Tsigas. 1988. "Economy-Wide Effects of Unilateral Trade and Policy Liberalization." Paper presented at conference, Agricultural Policies and the Non-Farm Economy, Washington, D.C., May 26–27.

Krissoff, B., and N. Ballenger. 1987. *Effects of Protection and Exchange Rate Policies on Agricultural Trade: Implications for Argentina, Brazil, and Mexico.* Economic Research Service, Staff Report No. AGES 870825, September.

Organization for Economic Cooperation and Development. 1987. *National Policies and Agricultural Trade.* Paris.

Robinson, S., I. Adelman, and M. Kilkenny. 1988. "The Effect of Trade Liberalization in Agriculture on the U.S. Economy: Projections to 1991. Paper presented at conference, Agricultural Policies and the Non-Farm Economy, Washington, D.C., May 26–27.

Roningen, V., J. Sullivan, and J. Wainio. 1987. "The Impact of the Removal of Support to Agriculture in Developed Countries." Paper presented at the American Agricultural Economics Association Conference, East Lansing, Mich., August.

Tyers, R., and K. Anderson. 1986. "Distortions in World Food Markets: A Quantitative Assessment." Background Paper No. 22 (Washington, D.C., World Bank), January.

Zietz, J., and A. Valdes. 1986. "The Costs of Protectionism to Developing Countries: An Analysis for Selected Agricultural Products." Staff Working Paper No. 769 (Washington, D.C., World Bank).

TARGETING AND THE DISTRIBUTION OF PROGRAM BENEFITS

DANIEL A. SUMNER

The idea of targeting government programs has a long history and broad appeal. Who could be against targeting, when a lack of it suggests random scattering of the effects of policy? Obviously, one should design programs so that their impacts are aimed directly toward the objectives. But just as obviously, if targeting were so easy, we would be doing it instead of bothering to write and talk about it.

In the context of farm commodity programs in the United States, targeting usually has a more narrow connotation than simply aiming impacts toward objectives. In this context, targeting generally is taken to mean changing the distribution of benefits so that commodity programs confer benefits on certain farms rather than others. Operationally, targeting is usually proposed for direct government payments. Further, it is usually stated that the favored farms should be those that are relatively small, are only marginally viable, or are operated by families that are relatively poor.

Concern about the distribution of effects of commodity programs across farms is pervasive. The Roman Catholic bishops' recent statement urging policy changes to "protect and advance the interests of small or moderately sized farms" is indicative (Stepp, 1989). Such concern is expressed in the notion that farm programs exist to aid "needy" farms or families. This notion often is combined with the presumption that gross farm sales are a reliable indicator of the farm operator's personal income or the farm's financial security. Since most farm program payments do not go to small, vulnerable farms operated by poor people, the implication is that the existing distribution of direct payments is a result of poorly targeted programs or of pervasive abuse by farmers who find loopholes in payment rules.

The charge that farm programs have missed the target has come from several sides of the political arena. For example, in proposing major policy initiatives for farm price and income supports as a part of the president's budget for fiscal year (FY) 1990, the U.S. Office of Management and Budget (1989, pp. 71–72) notes that "most of this assistance has gone to large-scale farm operations rather than the small family farmers who need it the most. Only 22 percent of direct payments to farmers go to those in severe financial distress." Notice the non sequiturs in the implied claims that large-scale farms have less "need" for assistance and are less likely to be in "distress."

The budget document goes on to state that "while farms with annual sales in excess of $250,000 represented 4 percent of all U.S. farms in 1986, they received 25 percent of the total amount of direct Federal payments."

These comments are factual, but the tone suggests that the distribution of payments results from a fundamental problem in program design or operation. This opinion is not unique to the administration. For example, during 1987 the U.S. General Accounting Office (1987a–1987e) issued at least five reports that supported targeting farm programs and modifying payment limitations.

Broad public opinion is represented by a recent editorial from a rural North Carolina newspaper. The editorial comments specifically on the new Payment Eligibility Form 502, which requires all producers to demonstrate that their farm operations will not violate payment limitation rules, but the outlook is more general:

> As with all government programs, as soon as some official devises a new way to "help" farmers, some farmers with questionable scruples devise a way to cheat the system and derive greater "benefits" than the law intends. . . . Because of this abuse, ASCS [Agricultural Stabilization and Conservation Service] committees must now review the farm programs of each participant. . . . We never seem to learn that government "giveaways" never really work. While few citizens would object to helping farmers who are in need of help, most of us feel as if our tax money ends up in the trough of a greedy few determined to turn programs intended to relieve financial burdens of the needy into gravy trains. (*Kinston* (N.C.) *Daily Free Press,* 1988)

This editorial again makes the commonly argued point that farm programs are intended for the "needy" and that adjustments of a farm's organization to comply with program rules amounts to abuse. The editorial also suggests how difficult the targeting goal is to reach and how administrative complexity places burdens on both bureaucrats and farmers.

SOME BASIC ISSUES

Effective targeting of any government program requires a clear understanding of what the program is intended to do and how to achieve that objective. In the case of farm commodity programs in the United States, there seems to be no consensus on either of these conditions. Farm programs often are assumed to be aimed at many distinct objectives, most of which are inconsistent if not mutually exclusive (see Allen, in this volume). As a practical matter, targeting has meant imposing side conditions on the individuals or organizations that receive direct government payments under commodity programs or placing limits on the amounts of payments that can be received. To date, proposed side conditions or limits have been largely ineffective when applied or proposals have been politically unsupportable. The problems in developing effective and successful targeting may in fact may be inherent in the nature of farm programs.

In this chapter, targeting farm programs will mean adjusting programs to aim at three distributional objectives: (1) alleviating poverty among farm families; (2) securing the viability of financially vulnerable farms; and (3) encouraging maintenance of smaller or mid-sized farms relative to large farms or very small part-time farms. These distributional targets are distinct (different farms would be aided depending on which objectives were primary), but they are all included in the phrase "helping farms that need help."

Revising farm commodity programs to contribute to these distributional objectives raises two general issues that must be addressed in any evaluation of proposed program changes. First, several government programs such as farm credit subsidies, job training subsidies, and general antipoverty or rural development programs may contribute to distributional objectives more directly than would revised commodity programs. Second, changing commodity programs may interfere with their contribution to other major objectives, such as enhancing market stability, encouraging U.S. competitiveness, and creating more "equity" between the farm sector incomes and the rest of the economy. As Kristen Allen (in this volume) points out, farm commodity programs are already clumsy instruments for meeting these other objectives, and targeting to satisfy distributional aims may further loosen the correspondence with these other objectives.

The distinction between targeting the effects of farm commodity programs and targeting government payments should be discussed explicitly. The latter is easier to accomplish but is of little fundamental interest. We really know little about the overall distributional effects of programs, and this makes changing programs to change their distributional effects very difficult (see Sumner, 1985). Most of the ultimate benefit from farm programs has gone to owners of farmland or other resources in inelastic supply. Further, more than half of all farmers receive most of their net income from off the farm; thus, the relationship between farm income and family income is even more tenuous (Brooks and Reimund, 1989). In spite of these problems, almost all proposals for targeting are focused on targeting payments rather than actual net income or wealth effects.

Given this skeptical beginning, the rest of this chapter briefly covers the following ground: a discussion of how payments are generated; a description of recent distributions of farms by sales category and the recent distributions of direct government payments; an outline of the recent policies to limit direct payments and payment eligibility; an analysis of the effects of some representative targeting schemes; a discussion of the distributional effects for examples of commodity programs that do not rely on direct payments; and, finally, a summary and a few concluding observations.

THE CALCULATION OF DIRECT PAYMENTS

Direct payments have been paid to producers under a variety of long-term and ad hoc farm programs. Paid land diversions are sometimes used to encourage reduced output and, hence, higher market prices of program crops. Land rental payments have been used in the Conservation Reserve Program (and previous land bank programs). Subsidies are paid for participation in the Farmer-Owned Reserve (FOR) and to encourage added storage facilities. The payments under the Whole-Herd Dairy Buyout paid producers to liquidate their dairy herds and to agree to leave the dairy industry for five years. There are also disaster payments, such as those authorized in the summer of 1988, and more regular payouts under the government-subsidized crop insurance program. Most of these payments are such that larger producers get larger payments.

The majority of all payments are deficiency payments (or, since 1985, so-called Findley payments and marketing loans that are calculated in ways very similar to the regular deficiency payments). This section presents a few simple equations to illustrate how payments are determined for the deficiency payment program and related schemes. The government payment to producer i for crop j—G_{ij}—is calculated roughly as follows:

$$G_{ij} = (T_j - M_j)\, Qg_{ij} \tag{1}$$

where T_j is the legislated target price and M_j is the higher of an average market price Pm_j and relevant loan rate L_j:

$$M_j = \text{Max}\,(Pm_j,\, L_j).$$

For marketing loans and Findley payments the relevant loan rate for payment purposes, L_j, may be below the regular loan rate and may be defined such that it remains below the effective market price. Qg_{ij} is the program payment quantity defined as:

$$Qg_{ij} = [(Yg_{ij})\,(Ab_{ij})\,(1 - r_j)]$$

where Yg_{ij} is the program yield established in 1985 by the farm's own proven yield history or by the county average; Ab_{ij} is acreage base established by past production history of the specific farm; and r_j is the proportion of base acreage required to be left idle under the Acreage Reduction Program (ARP) to qualify for payment on the crop j. Participating producers may not produce on more than $(1 - r_j)Ab_{ij}$ acres for program crop j, so the programs tend to hold down output; this raises market price even for nonparticipants. When the ARP rate, r_j, is high or when the expected deficiency payment rate, $T_j - M_j$, is low, producers are more likely to maximize profits by foregoing payments in favor of producing more, so they will decline to participate in the program. Participation is also affected by anticipated future benefits because one-fifth of a farm's base is lost for each year of nonparticipation. Of course, there are also additional gains to participation (such as access to loans) as well as additional costs (such as cross compliances, paperwork, and payment limits), but the essence is captured by the deficiency payment rate and ARP rate.

In the standard case, the payment rate does not vary by farm. Recently, however, a significant proportion of direct payments for commodities programs have been made in kind through certificates redeemable for commodities. Recipients have had considerable flexibility as to how and when the certificates were used. The dollar value to a farmer of these generic certificates issued in the place of cash payments has varied by the time of year, location, and the recipient's success in anticipating price movements. In this sense, the effective deficiency payment rate has been different for different growers.

Across commodities the total direct payment per producer (G_i) is the sum of payments per program:

$$G_i = \sum_{j=1}^{J} G_{ij} \tag{2}$$

where J is the number of program crops and for most producers, the G_{ij} will be zero for many of these. The amount of direct payments, G_i, is the primary focus of most proposed targeting schemes.

In most years the bulk of all payments is accounted for by payments distributed along the lines of equation (2). There are, of course, many complications and details that have not been fully represented. For example, an individual may own several farms and receive payments on more than one. Further, many farms are operated as partnerships, with several farmers receiving the payments. As noted earlier, total direct payments per farmer also include those for some minor programs (such as wool) and ad hoc programs such as weather-related disaster programs or the dairy herd termination program. These recurrent program payments are also distributed roughly in proportion to the quantity of farm output of the specified commodities. Nonetheless, discussions of targeting often exclude these ad hoc or special objective payments, presumably because their stated rationales (such as reduced cow numbers or less soil erosion) are more specific and focused than those for standard commodity programs. As indicated in the section on current targeting efforts, however, some of these special objective payments are at least nominally affected by payment limits.

DISTRIBUTIONS OF FARM SIZE AND GOVERNMENT PAYMENTS

The distribution of direct payments has been used as evidence that current farm programs are not appropriately targeted. The empirical relationships between payments and farm size, farm incomes, or farm financial position motivate various schemes for targeting payments. This section presents some recent evidence on the farm size and payments distributions as a background to a discussion of efforts to modify them.

At the outset two basic points are important. First, the distributions of payments presented below are dominated by the characteristics of the commodity industries to which payments are made. It is important to recognize how commodity-oriented commodity programs are. A program is targeted to a particular commodity to affect its market and thereby its producers. It is an economic and political fact that even though most farmers produce several commodities, they exert much of their collective political influence along commodity lines. This results in a set of farm programs centered around particular commodities with cross-commodity issues handled through other policies. These other policies (such as farm credit subsidies, crop insurance, water subsidies, rural electrification, farm labor regulation, and agricultural research and extension) are not treated here.

Second, although the distributions discussed in this section are presented in terms of farms, payments are associated with "persons" and, as mentioned above, their real incidence is associated with asset ownership. If the payment associated with the production of a commodity is linked to a person's production history, then the benefit accrues to that person. On the other hand, a payment linked to any specific asset or instrument accrues to the owner of that asset or instrument. For example, deficiency payments are associated with acreage base in a program crop. Therefore, the benefits from these payments go to the landowner through higher rental or sale value of the land even if the direct payment goes initially to the producer who rents the land. A large producer who rents all the program cropland he farms may be almost unaffected by reduced payments because his land rental costs would fall by the (expected) amount of the reduced payments received on a per acre basis. In looking at distributions of payments, we should remember that the farmer who receives the payment is not necessarily the person who receives the benefit.

Before examining the distribution of payments, it is useful to consider the size distribution of farms. Table 1 shows the distribution of farms by sales group for 1987 (the most recent year for which data are available). The table shows that most farms (52.4 percent) are very small, producing less than $10,000 in gross sales. Most of these farms produce less than $5,000 worth of output, and their farm production clearly is not a significant contribution to the economic output of the sector. Even the 20.7 percent of all farms that produce between $10,000 and $40,000 in sales contribute only about 8 percent of gross farm income. Many of these farms are large enough to be significant factors in the economic lives of the farm families involved, but typically they are not the sole or even the major income source. Also, they do not contribute a significant share of total output in any but a few commodity industries (one notable exception is burley tobacco).

More than 85 percent of gross farm income in 1987 was accounted for by the 27 percent of farms that had gross sales of $40,000 or more. But half these farms, those with sales of less than $100,000, may not be large enough to

TABLE 1. Distribution of Farms by Sales Class, 1987

Sales class	Number of farms (thousands)	Percent of total number of farms	Percent of gross farm income
>$500,000	29	1.3	32.2
$250,000–$499,000	71	3.2	16.9
$100,000–$249,000	201	9.2	22.2
$40,000–$99,000	286	13.2	14.3
$10,000–$39,000	450	20.7	8.2
<$10,000	1,139	52.4	6.2

Source: U.S. Department of Agriculture, Economic Research Service, 1987, *Economic Indicators of the Farm Sector, National Financial Summary, 1987,* Washington, D.C. USDA data based on Census of Agriculture.

generate the full net incomes of their operator families. Many in this group also have significant off-farm earnings. The larger of the commercial farms, the 1.3 percent of all farms that produce more than $500,000 in sales, contribute almost one-third of all gross farm income. All except a handful of this group of farms, however, are not large by general business standards. Almost all the operations in the larger sales categories are family farms in the sense that they are owned and operated by a single family or a small group of related members of one extended family and not by unrelated shareholders and hired managers. Certainly the concentration of agricultural production does not raise the monopoly or antitrust concerns typical for manufacturing industries.

Table 2 uses data based on the Farm Cost and Returns Survey of the U.S. Department of Agriculture (USDA) to examine the distribution of direct payments by sales class. Note that small farms are underrepresented in

TABLE 2. Distribution of Direct Payments by Sales Class, 1987

Sales class	Percent of payments	Percent of program crop sales	Percent of farms with payments	Percent of all survey farms	Average payment per farm with payments ($1,000)	Payments as percent of gross cash income
>$500,000	10	15	3	2	69	3
$250,000–$499,000	21	24	8	4	46	10
$100,000–$249,000	36	34	24	12	27	13
$40,000–$99,000	22	19	27	15	14	12
$10,000–$39,000	9	7	25	23	6	9
<$10,000	2	1	12	44	2	6

Source: U.S. Department of Agriculture, Economic Research Service, 1988, *Agricultural Income and Finance, Situation and Outlook Report,* table 1, Washington, D.C., December. Based on data from USDA's Farm Cost and Returns Survey.

relation to the total population, constituting only 44 percent of the survey total (compared to 52.4 percent in table 1). Nevertheless, even in these survey data one basic message is clear: small farms are much less likely to get payments than larger farms, but their share of payments is twice their share of program crop sales. Comparing columns 4 and 5 in table 2 indicates that mid-sized farms, those with between $40,000 and $500,000 in sales, are more likely to receive payments than other farms. Comparing columns 2 and 3 indicates that mid-sized farms tend to get a higher proportion of total payments than they produce of total sales. For these farms, payments are more than 10 percent of gross cash income compared to 3 percent for the larger farms. Although payments are distributed roughly in proportion to program crop sales, they tend to be skewed toward mid-sized farms in relation to sales. Also, production of program crops tends to be relatively concentrated on farms in the middle sales classes.

Grains and cotton long have dominated federal payment programs, so most payments have been made in the Midwest, Plains, and Delta states. For example, although California, which is important in rice and cotton production, receives significant payments, it gets a much smaller share of payments than its share of total crop sales. The distribution of payments for selected states for 1987 (see table 3) shows that since more than half of all direct payments were distributed through feed grain programs, those states important in feed grains were important on the list of total payments. This pattern was more pronounced for 1987 than for earlier years in the decade, but the list of high payment states has changed little. The year 1987 also had relatively high outlays for a variety of miscellaneous and ad hoc programs. Conservation payments were $1.5 billion, and payments for storage programs, disaster programs, dairy termination, and other miscellaneous programs totaled almost $2 billion. However, the benefits under these programs were distributed across states in a way roughly similar to the payments for major program crops.

Direct payments vary by year because (1) changes are made in program parameters (e.g., the Food Security Act of 1985 reduced loan rates, thus increasing deficiency payments for 1987); (2) new one-time programs are introduced (e.g., the Whole-Herd Buyout involved large direct payments); (3) intermittent weather changes cause changes in payment levels (e.g., regular disaster payments and ad hoc programs, such as the Disaster Assistance Act of 1988, are weather sensitive); and (4) supply and demand conditions change (partly due to U.S. weather), thus changing the difference between the market price and target prices (e.g., low production in the Soviet Union helped raise world wheat prices in 1988 and 1989). For the most part, however, over the last decades the distributions of direct payments have not varied from year to year nearly as much as have the levels of direct payments.

Finally, it is useful to examine the relationship of payments to one often-mentioned program criterion—financial vulnerability. Table 4 shows the distribution of direct payment by indicators of financial vulnerability. In 1987 only 7 percent of all farms were vulnerable as defined by negative income flow and relatively high debt positions. (Most of the negative income farms

TABLE 3. Distribution of Direct Government Payments for Selected States and Commodities, 1987
(in millions of dollars)

	Feed grains	Wheat	Rice	Cotton	Total payments[a]	Total cash receipts
Arkansas	25	62	199(1)	70	398	3,143(16)
California	36	60	89(2)	168(2)	462	15,522(1)
Illinois	1,279(2)	56	—	—	1,478(2)	6,174(5)
Indiana	586(5)	32	—	—	670(8)	3,872(10)
Iowa	1,553(1)	1	—	—	1,988(1)	8,780(3)
Kansas	368	484(1)	—	—	966(6)	5,722(7)
Minnesota	732(4)	152	—	—	1,194(5)	5,809(6)
Nebraska	922(3)	108	—	—	1,275(4)	6,823(4)
North Dakota	196	409(2)	—	—	720(7)	2,308(24)
Texas	311	185(5)	65(4)	429(1)	1,441(3)	9,086(2)
Total U.S.	8,490	2,931	474	1,204	16,747	67,239

Notes: Only selected important agricultural states are included in this table. Numbers in parentheses are rank of the state among all states.

Dashes mean not applicable, zero, or negligible.

Source: U.S. Department of Agriculture, Economic Research Service, 1987, *Economic Indicators of the Farm Sector, National Financial Summary, 1987,* Washington, D.C. USDA data based on Census of Agriculture.

[a]Total includes payments for other crops and for wool, Conservation Reserve, Disaster, Storage, and Dairy Termination programs, among others.

were very small part-time operations with little or no debt and few or no prospects of contributing significantly to family income.) These "vulnerable" farms received 7 percent of payments and represented 7 percent of farms with payments (see also U.S. General Accounting Office, 1987d). However, payments were skewed slightly toward these farms in that payments represented a higher percentage of gross cash income for this group than for farms in the other financial categories. This fact is mostly an artifact of the severe stress placed on cotton and grain markets and on Midwest land values in the mid-1980s. The next round of general farm financial stress could hit the livestock, sugar, tobacco, or horticultural industries, instead of grain and cotton, in which case direct payments would tend to be skewed away from "vulnerable" farms. In the mid-1980s, financial stress happened to hit the crop industries that have direct payment programs, but the programs themselves were not targeted to this criterion.

This section has related direct payments to the distributions of gross sales and some financial characteristics of the farm operations. As noted earlier, however, these distributions may have little relation to farm family income. In the last thorough study of the effects of programs on income distribution, Gardner and Hoover (1975) found that elimination of commodity programs

TABLE 4. Distribution of Direct Payments by Financial Situation, 1987

Financial position	Percent of payments	Percent of program crop sales	Percent of farms with payments	Percent of all survey farms	Average payment per farm with payments ($1,000)	Payments as percent of gross cash income
Favorable	63	65	61	49	19	9
Marginal income	10	9	17	37	11	10
Marginal solvency	21	20	16	8	23	9
Vulnerable	7	6	7	7	17	13

Note: Categories of financial position are defined by the Economic Research Service, U.S. Department of Agriculture, as follows:

Favorable—positive net income and debt/asset ratio ≤0.4
Marginal income—negative net income and debt/asset ratio ≤0.4
Marginal solvency—positive net income and debt/asset ratio >0.4
Vulnerable—negative net income and debt/asset ratio >0.4

Source: U.S. Department of Agriculture, Economic Research Service, 1988, *Agricultural Income and Finance, Situation and Outlook Report,* table 1, Washington, D.C., December.

would significantly reduce inequality of farm family incomes. Their work analyzed the effects of programs on the full permanent income of farm families in 1969 by examining the returns to ownership of resources (especially land, allotments, quotas, and human capital). They show empirically that the farm's gross sales class is a poor indicator of the net income of the farm family.

Unfortunately, the confusion between gross sales and net income continues to plague public discussion of the distributional effects of farm programs. A recent *Wall Street Journal* editorial (June 7, 1989, p. A32) states that "two-thirds of direct government payments in 1985 went to the 314,000 farms with annual sales exceeding $100,000. This transfer from middle- and low-income taxpayers to well-to-do farmers also is highly inefficient." What is left out in the jump from sales class to income is any evidence about the net incomes of the farmers receiving payments or even any discussion of the incomes of those landowners who actually benefit from the payments.

The data on distributions of farm numbers by sales class and of government payments by various farm characteristics have been used to support the notion that commodity programs are poorly targeted. This chapter has indicated that reality is more complex. The case for targeting is certainly not obvious from these data without more information on specific aims and linkages from a policy proposal to those aims.

CURRENT TARGETING EFFORTS—PAYMENT LIMITATIONS AND ELIGIBILITY

Given all the discussion of the maldistribution of the benefits of farm programs, very little targeting to distributional objectives has ever been at-

TABLE 5. Summary of Payment Limitations Applied to Farm Programs

Payment limit[a] ($)	Applicability	Legislative source
50,000	Basic deficiency or diversion payments	FSA, 1985
100,000	Disaster payments	FSA, 1985
50,000	Conservation reserve payment	FSA, 1985
250,000	Overall maximum for total of *all* payments	Appropriations, 1986
250,000	Honey loans	Appropriations, 1986
100,000	No payments from more than three entities: $50,000 for first; $25,000 each for next two	Reconciliation, 1987
	No eligibility for nonresident aliens	Reconciliation, 1987
	Only persons "actively engaged" in farming activities eligible, payments commensurate with contribution	Reconciliation, 1987
50,000	Feed assistance benefits	Disaster, 1988
100,000	Crop disasters payments	Disaster, 1988
Normal income	Combined drought and insurance payments limited to income with normal yields	Disaster, 1988
	Maximum gross income for eligibility: $2.5 million for feed assistance; $2 million for crop payments	Disaster, 1988

Note: FSA refers to the Food Security Act of 1985 and "Disaster" refers to the Disaster Assistance Act of 1988.

[a]Per person, per entity.

tempted. The only significant efforts to affect distribution of payments directly have been in payment limits and eligibility rules. The Agriculture Act of 1970 introduced a $55,000 payment limit. Several changes occurred during the 1970s, including lowering and then raising again the basic limit and providing alternative (i.e., higher) limits for rice.

Table 5 summarizes recent and current limits on direct government payments. Considerable interest has focused in recent years on how farmers adjust to payment limits and on revising the payment eligibility and limitation rules. Each of the last four years has seen legislation imposing new restrictions on who qualifies for payments and how much they may receive.

The basic limitation applied to standard deficiency payments has been $50,000 per eligible producer since enactment of the Agriculture and Food Act of 1981. The Food Security Act of 1985 added additional limits on disaster payments and conservation reserve payments. The following year a limit of $250,000 per person was applied to all forms of direct payments. This new overall limit was designed to control the marketing loan payments that made up the difference between the market price and the standard loan rate for rice and cotton. In addition for grains Findley payments based on the difference between the regular and special lower Findley loan rates are exempt from the $50,000 limit (Glaser, 1986).

The determination of who is eligible for payments and how an individual may qualify on multiple farms and as multiple "persons" was addressed in the Reconciliation Act of 1987. Prior to the 1989 crop year, an individual could qualify for payments as an individual and as a member of any number of partnerships or corporations as long as the overall ownership of the organizations was different (U.S. General Accounting Office, 1985, 1987a). The 1987 act limits to three the total number of separate entities from which an individual can collect payments. Further, since an individual can have at most a 50 percent share in each separate entity for which he or she collects payments, the effective limit for basic payments is $50,000 for the solely owned entity and $25,000 shares of the overall limit for each of the half ownerships for a total of $100,000 per individual (Lederer and Pollack, 1988; U.S. General Accounting Office, 1987c, 1987e).

The 1987 act also restricts eligibility for direct payments. Landlords, partnership owners, and others are not eligible for direct payments unless they provide labor or management to the farm or accept risk. Further, nonresident aliens are excluded even if they would otherwise qualify as active participants in the farming enterprise.

The 1987 act and its implementation are exceedingly complex, requiring detailed interpretation to define precisely who is eligible and what legal arrangements are allowed (see, for example, U.S. Department of Agriculture, Agricultural Stabilization and Conservation Service, 1988). The law has not yet been in effect for a full payment year, so it is not clear how effective it will be in changing the payment distribution. Of course, even if the direct payment flows are changed, the distribution of economic returns may not change appreciably. For example, a cash landlord may not receive direct payment, but the rental rate per acre will tend to include the value of the direct payment received for the crop base associated with the land. By shifting from share to cash rent (at least on paper), a landowner may still be the "recipient" of the payments. Further, since expected future payments are capitalized into the land price, the current landowner will receive the capitalized value of expected future payments, even for land that is sold.

The 1988 disaster assistance legislation introduced new eligibility criteria that require comment. Producers with more than $2 million in "qualifying" gross revenue from any business enterprise, including any farming enterprise, are not eligible for crop disaster payments. (The criterion is $2.5 million for livestock payments.) If a producer receives more than half of his or her total gross revenue from farming, only the farm gross is counted; otherwise, gross revenue from all enterprises, farm and nonfarm, is counted toward the eligibility limit. The gross income measure applies even if the business is operated at a loss. Further, some businesses, such as retail stores, traditionally have large gross sales in relation to capital or earnings, but no allowance is made in the law for variations across enterprises. Under this criterion, for example, a producer with typical revenues from a 900-cow dairy herd or a farm equipment dealership would have been ineligible for disaster aid on crops lost during the 1988 drought. Although it is not clear what financial records were actually used in certifying eligibility in 1988 or if any grower was actually declared ineligible under this provision, the 1988 law represents the first time an income criterion has been used to exclude a grower from participating in a commodity program.

In the last four years payment eligibility and limits have become more stringent. It is not yet clear to what degree the new rules actually will affect the distribution of economic benefits from the commodity programs. In the past, eligibility rules and payment limits have been judged to be a bother to large producers, a benefit to rural lawyers, but mostly just irrelevant to the distribution of program effects. This overall lack of impact indicates either that creating rules to exclude those not favored is difficult or that eligibility rules and limits are purposely ineffective.

PROPOSALS TO TARGET PROGRAM PAYMENTS

This section discusses recent proposals to target farm program payments. The three basic options are (1) reducing payment limits; (2) determining eligibility based on size, vulnerability, or other farmer or farm characteristics; and (3) adjusting payment rates so that they depend on farm output or other measures of size. These and related options were reviewed intensively by the General Accounting Office in reports released during the summer of 1987 (see also Reinsel, Craven, and Allen, 1987; Debertin and Infanger, 1987).

Reducing Payment Limits

Payment limits take the form of simply placing a maximum on G_i as defined in equation (2). As mentioned earlier, however, the serious administrative issue concerning these limits is how to define the effective payment unit, i, relative to the farm operation as a business or legal entity and to the farmer as a person. The discussion below considers some economic effects of reducing limits but does not deal with administrative feasibility.

A binding reduction in the payment limitation would cause a series of complex reactions within the affected industries on nonprogram commodities and on other farm-related markets for inputs and marketing services. A lower limit would cause some relatively large and low-cost producers to leave the program but remain in the industry producing for sale at the market price without being required to hold land out of production. This group of producers would expand production of program crops. A second group of relatively large and low-cost producers would remain in the program and would reduce output because the market price rather than the higher target price would become the relevant supply price for production above the payment limit. For those that remain in the program, the relevant cost curve remains the

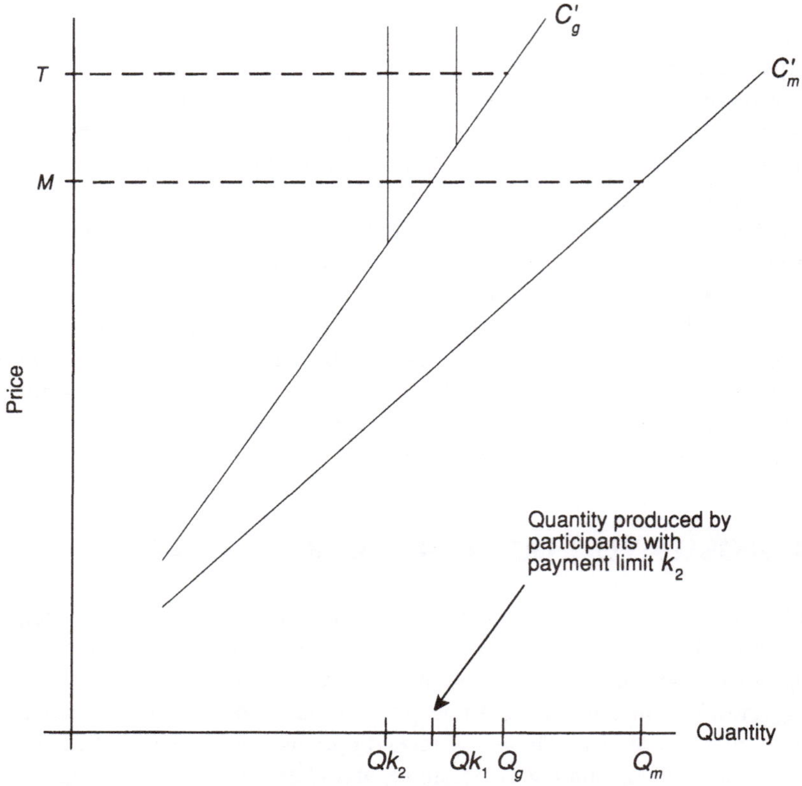

FIGURE 1. Responses to payment limits. (See discussion in text.)

higher cost curve applicable to the participants; therefore, the output of participants falls. (Here the target price is assumed to be the relevant supply price for participants, even though in reality it is significantly more complex.)

Figure 1 illustrates the reaction of an individual producer to payment limits in a very simple example. Marginal cost without participation is represented as C'_m and marginal cost with program participation and associated ARP requirements is represented by C'_g. Target price is T and market price is M. With no participation, the quantity produced is at the level Q_m where marginal cost is equal to the market price ($C'_m = M$). With participation and no payment limit, the quantity produced is Q_g, where marginal cost under participation equals the target price ($C'_g = T$). Total program payment amount is represented by the rectangle $(T - M)Q_g$. Now consider this farm's reaction to a payment limit. When the payment amount is limited and $(T - M)$ is held constant, one may solve for the implied limit quantity upon which payments are made. Figure 1 shows the implied limit quantity for a dollar limit K_1. This quantity is $Qk_1 = K_1/(T - M)$. In this case, the amount produced by the farm declines when the limit is imposed. Now consider the reaction to a more stringent limit of K_2, which translates into quantity $Qk_2 = K_2/(T - M)$. In this case the payment limit is low enough that participation in the program is no longer profitable, so actual output of the farm is Q_m. This decision is based on the finding that payments from participation, shown in figure 1 as the rectangle $(T - M)Qk_2 = K_2$, are less than profits foregone because of participation, as measured by the region between C'_m and C'_g up to price level M. The figure also shows that if the farm remained in the program when the payment limit was K_2, it would produce Qk_2 for payments but also continue to produce additional quantity up to the point at which $C'_g = M$. In this case, for participants, production would have been effectively decoupled at the margin.

Figure 1 indicates how individual farm responses to payment limits depend on the level of the limit. A similar illustration could be used to indicate that the reaction to a given limit depends on the farm's cost conditions. Different farms have different cost curves and so react differently. Aggregate response is more complicated because M and T are likely to be endogenous. Aggregate response depends only partly on summation across firms, some of which expand and some of which contract. In the aggregate, and given a few years for adjustment, the effects on the market price through the supply response, on the target price through political reactions, and on implicit and explicit land rental rates, which will help determine the entry or exit of resources, are also important.

The impact of payment limits on aggregate production, market price, and asset values is not obvious. Nonetheless, program costs will almost surely fall unless political forces cause the target price to go up. For program costs to rise, producers leaving the program must expand output enough so that the resulting lower prices cause sufficiently higher payment rates to the remaining program participants. (For a further discussion of payment limits on supply, see Foster and Babcock, 1987.)

One reaction often seen in the past in response to payment limits has been farm reorganization. A producer nominally cuts back production in reaction to the newly binding payment limit. But at the same time a new farm is established that uses resources released from the original cutback, and the new farm ownership and management are closely related to those of the original farm. The result is that no resources leave the program crop industry and government payments do not decline. U.S. General Accounting Office reports (1987a, 1987c) suggest that reorganizations added several thousand new eligible producers to payment roles during the mid-1980s and added several hundred million dollars per year to program costs. These added costs, while significant, account for only a small percentage of total payments.

If smaller payment limits reduced participation rates of farms producing a significant share of output, they would thwart Acreage Reduction Programs and thereby fail to raise prices to the whole industry, participants and non-participants alike. Further, if significant numbers of producers received lower payments (perhaps zero), the total farm sector income would fall and farm asset values, especially land values, also would fall. Finally, if targeting shifted policy-induced size distributions so that higher-cost producers of program commodities produced a larger share of the total, the competitive position of the United States in world markets would be reduced and costs to U.S. consumers would be raised. (See Knutson et al. [1987] for evidence that the cost of production impacts of effective payment limits may be significant, especially in cotton and rice.)

Redefining Program Eligibility

Defining eligibility for payments in terms of debt-equity ratios, overall sales class of the farm, or amount of off-farm work also has been discussed (U.S. General Accounting Office, 1987b, 1987e). This targeting option would simply set $G_i = 0$ for less favored farms. The previous section noted that the 1987 legislation introduced new criteria for direct payment eligibility. The rules limiting direct payments to active farm participants and disallowing payments to nonresident aliens are not directly related to farm size, farm financial stress, or income distribution. These limitations will affect few recipients and will mainly encourage some changes in the business organization of a few very large farms. However, they do represent an effective political force in restricting at least the initial distribution of direct payments if not the distribution of benefits to active domestic farmers.

An additional innovation in the Disaster Assistance Act of 1988 was to include the criterion of nonfarm sales for producers who received most of their gross revenue from nonfarm enterprises. Such a criterion, if applied broadly, would discourage participating producers from diversifying family income sources. If applied to net incomes, such a criterion could remove eligibility for small part-time farms as well and change the common pattern of combining farm and off-farm work in addition to the pattern of combining farm and nonfarm business investments. The restriction on total farm sales used as an eligibility criterion is very similar to basing payment rate on sales where the rate is set to zero for large producers. Therefore, this option will now be discussed as a special case of applying equation (3).

Basing Payment Rates on Sales or Output

Several methods of basing payment rates on farm production have been proposed as targeting schemes (U.S. General Accounting Office, 1987b). One example of basing the target price for farm i and for crop j on a measure of gross sales of all program crops may be written as follows:

$$T_{ij} = T_j \cdot \text{Min}[S_g/(\Sigma_j P_{ij} Q_{ij}), (1 + s)] \quad (3)$$

where T_j is the base target price for crop j, P_{ij} is the market price, and Q_{ij} is output, so that $\Sigma_j P_{ij} Q_{ij}$ is the total revenue from program crops. S_g can be defined as the gross sales of program crops per farm for a mid-sized farm and is set by legislation or administration. A farm of this size would get exactly T_j as a target price. The parameter s determines the maximum target price for small farms. For example, a farm with sales larger than S_g would get a target price less than T_j, and a farm with sales less than S_g could get at most $T_j \cdot (1 + s)$ per unit. At a cost of added administrative expense, this option could be written to include all gross sales rather than just program crops.

An alternative scheme is to base the target price for each program crop on the output of that crop relative to some government norm Q_{gj}. This may be written as:

$$T_{ij} = T_j \cdot \text{Min}[(Q_{gj}/Q_{ij}), (1 + s_j)] \quad (4)$$

where $T_j(1 + s_j)$ represents the government established maximum target price for crop j for small farms.

The broad implications of these options are similar, and in both cases even qualitative effects depend on how the option is applied. For example, if the lowered target prices affect farms producing a large share of sales or output, then both options would cause small part-time farms to have an advantage in the production of program crops.

Equation (3) presents a version of this option that uses an adjustable target price to make the deficiency payment rate a function of the gross value of output of program crops. Obviously, this is only one of many ways that size could be defined for payment purposes. It does have the advantage of depending only on program crop data that would be available in Agricultural Stabilization and Conservation Service (ASCS) records. One disadvantage of this formula is that program crop production is only loosely related to whole farm size. As defined, this option would encourage diversification across enterprises that is socially inefficient and also perhaps uneconomical vertical integration with crop and livestock enterprises on the same farm. Of course, any targeting option that causes a lower target price for farms with certain characteristics will cause farms to avoid those characteristics whenever feasible.

The implications of these options depend on how they are applied. If a significant proportion of full-time farmers faced lower target prices, there could be relatively large sector-wide effects, but significant effects may occur even if few farmers are affected directly. In 1985 only about 1.4 percent of wheat program participants produced more than 25,000 bushels (U.S. General Accounting Office, 1987b). This fact may seem to indicate that if a payment limit, eligibility criterion, or drop in target price were keyed to this output level, only a few of the largest farms would be affected. However, farms in this group produced almost 20 percent of the total wheat output of participants in 1985. Based on a deficiency payment of approximately $1.00 per bushel, those near the 25,000 bushel mark were well below current payment limits. Further, at $4.38 per bushel these growers had a total gross value of output of about $110,000—hardly enough for an extravagant family income once farm costs were paid. Similar figures for feed grains would illustrate the difficulty in designing a targeting program that affected a

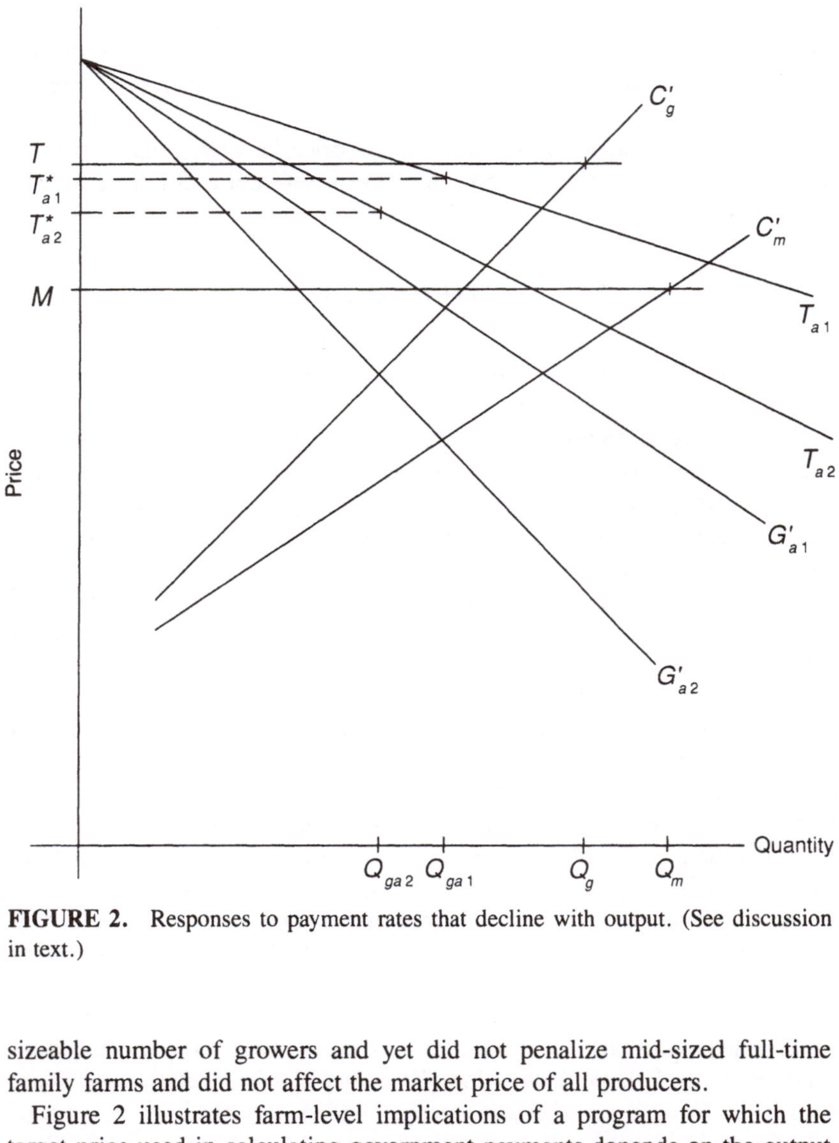

FIGURE 2. Responses to payment rates that decline with output. (See discussion in text.)

sizeable number of growers and yet did not penalize mid-sized full-time family farms and did not affect the market price of all producers.

Figure 2 illustrates farm-level implications of a program for which the target price used in calculating government payments depends on the output level of a particular crop for a single farm (as in equation (4)). Before instituting the program with target price dependent on output, the target price was T, the relevant marginal cost for the participating farm was C'_g, and the marginal cost for the nonparticipating farm was C'_m. This farm would choose to participate and produce output Q_g, the output level for which $C'_g = T$, rather than Q_m. At this quantity the benefit from payments is larger than the loss from having higher costs of production.

An alternative, downward sloping target price schedule is shown as T_{a1}. Farms with an output to the left of where T_{a1} intersects T would enjoy an increase in the deficiency payment rate. However, for the farm shown in figure 2, the implied target price is lower than T. For the target price schedule T_{a1}, the relevant marginal revenue schedule is indicated by G'_{a1}. Line G'_{a1} shows the marginal payment rate given that each additional unit of output causes the target price per unit to decline. The farm would choose output Q_{ga1} (at the intersection of G'_{a1} and C'_g) and receive a deficiency payment of $(T^*_{a1} - M)$ per unit. As the figure is drawn, the farm continues to participate. If, however, the alternative target price schedule were T_{a2}, the marginal revenue schedule would be G'_{a2}. In this case, the deficiency payment of $(T^*_{a2} - M)\ Q_{ga2}$ would not be enough to compensate for the cost of participating. So the farm would leave the program, and output for the farm would rise to Q_m.

As with payment limits, different farms would react differently, and analysis of aggregate response to this scheme is quite complex because of market price and political reactions.

Figure 3 shows the situation for a program option under which the farm gets target prices T_0 for all output up to Q_{t0}, then target price T_1 for output between Q_{t0} and Q_{t1} and target price T_2 for additional output. In this case, the farm would choose to produce Q_g, where marginal cost curve C'_g equals T_2. The total deficiency payment to this farm is $T_0Q_{t0} + T_1(Q_{t1} - Q_{t0}) + T_2(Q_g - Q_{t1})$. For the farm shown, because the higher target prices apply to inframarginal output, these target price levels and the quantities to which they apply only determine the level of a lump sum transfer and therefore do not affect the output of large participating producers.

Figure 3 also illustrates how a target price program can cause farms of different cost conditions to produce the same output. The farm represented by costs C'_{ga} and the farm represented by C'_{gb} will both produce at output level Q_{t1}, even though farm "*b*" has much lower marginal costs.

In September 1987 the U.S. General Accounting Office (1987a) reported on the simulated effects of several targeting scenarios. Among the options considered was a 50 percent increase in payments to favored farms and a 50 percent decrease in payments to other disfavored farms. Most farms were in the middle and not affected directly. The groups were defined as follows: The favored class had an equity of less than $500,000, off-farm income of less than $30,000, negative net farm income, sales greater than $40,000, and a debt-asset ratio of between 40 and 70 percent. This group represented relatively low-wealth farms that also had significant farm sales and not too much off-farm activity but that were in financial trouble in terms of income and debt. Disfavored farms, facing a decline in payments, were those with an equity above $500,000 *or* with off-farm income above $30,000, *or* with net farm income above $50,000. These farms had significant wealth or off-farm activity or were making good farm income. Neutral farms, facing no payment increase or decrease, were farms with positive net farm income of less than $50,000, *or* with either very low or very high debt relative to assets, *or* with sales of less than $40,000. These farms were not disfavored in terms of farm or off-farm income or equity but had very low sales, or had no debt problem or so much debt that they were not considered viable. The

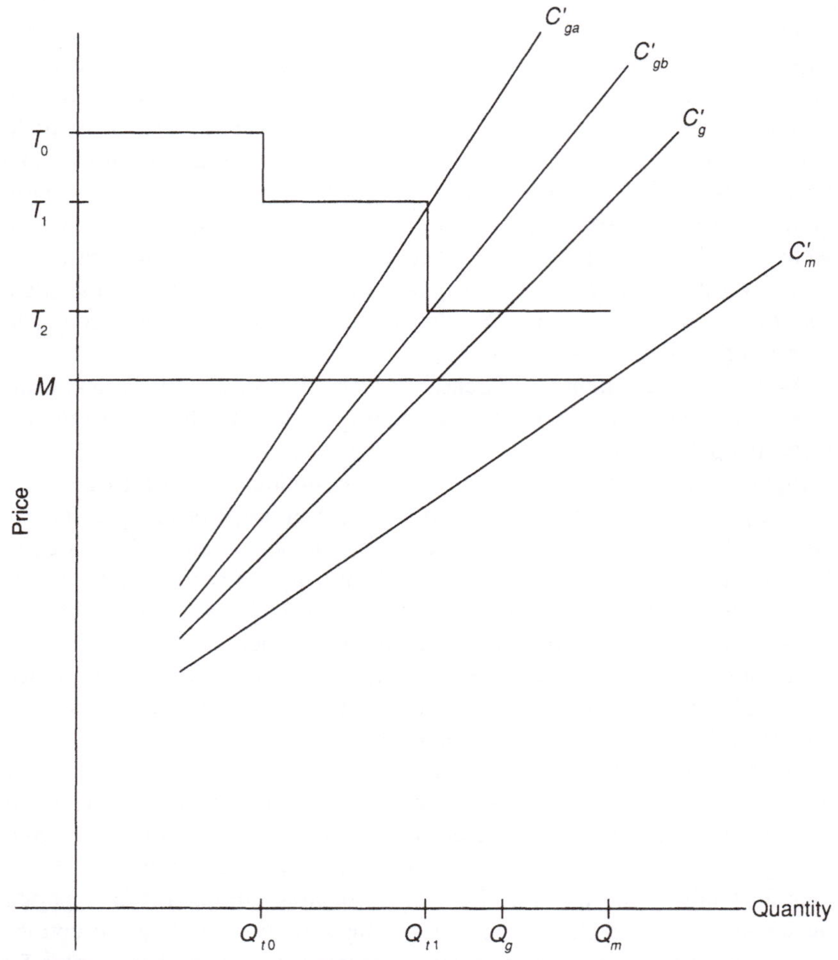

FIGURE 3. Responses to payments with inframarginal regions and discrete jumps. (See discussion in text.)

U.S. General Accounting Office (GAO) found about 24,000 farms in the favored category, 139,000 farms in the disfavored category, and most farms in the neutral group, based on the Farm Cost and Returns Survey data. In 1985 the favored farms as a group would have gotten an increase of $182 million while the disfavored farms would have lost $1.279 billion in deficiency payments.

The GAO also used the FAPSIM farm sector model of the USDA to examine the impact of this shift in payments. Using a 1985 baseline, acres of major crops would have declined by a small percentage, but output would have been unchanged (or slightly higher) because of the increased output of the farms that left the program. With prices slightly lower, exports would have increased, but net farm sector income would have been lower by several percentage points. These results are reported here in very broad terms because the simulations examined can give only very rough indications of how the farm sector might react to targeting. In particular, farm reorganizations to meet new criteria, changes in participation rates, and shifts among crops as reactions to the program changes are simply not well understood.

Overall reactions to various targeting schemes are very difficult to anticipate, particularly in the aggregate. There are many potential specific ways of modifying commodity programs to address targeting interests. This section has focused on only a few stylized prototypes. A major problem in developing general conclusions is that even the broad effects of targeting may depend on specific features of the program introduced.

Benefits from Programs Without Payments

Although attention has focused on direct payments, especially deficiency payments, farm programs affect markets and thus create winners and losers mainly through market forces. This point is general across all farm programs, but made most clear by the programs that involve no payments and may involve no net government outlays at all. (Nelson [1989] provides an analysis of benefits to nonparticipants in U.S. grains programs based on analysis of 1982 data. Bollman [1989] provides a Canadian counterpart.) The U.S. sugar and tobacco programs provide useful examples and will be considered briefly here. (The peanut program is another example that combines features of both sugar and tobacco; see Rucker and Thurman [1989].)

Direct government payments have never been an important feature of tobacco programs. Further, the net federal outlays for tobacco have been quite small relative to the income transfers involved, the value of output in the industry, and the outlays on other commodity programs. Since 1982 the so-called no-net-cost provisions have required that the industry itself be responsible for costs associated with the operation of the quota and loan program. A similar no-net-cost rule now operates in the U.S. sugar program. These two programs are also similar in that the essence of each program is that it restricts available supply so that the market price is increased. The programs differ in the method used to control supply and whose relevant supply is controlled.

The United States is both a major importer and a major exporter of leaf tobacco but has a large net export balance. Imports arrive with few barriers (most of the relatively low GATT-bound tariffs are refunded under a duty drawback system). Price enhancement comes from a domestic production quota, with a loan and storage program used to facilitate interyear and intergrade quota adjustments. U.S. tobacco and foreign tobacco are not perfect substitutes. Therefore, even with unrestricted imports, the price of U.S. leaf can be raised above free-market levels in both domestic and export markets. The key to the tobacco program is that the demand for U.S. tobacco is not perfectly elastic even in the long run, so restricting supply is profitable for many of those in the industry. The U.S. government's role is to facilitate a cartel that raises its price to domestic and international buyers, thereby increasing total net revenue. Benefits accrue mostly to owners of tobacco quota, but details of program operation allow some input suppliers, marketing firms, and others to gain in relation to a no-program option (see Sumner and Alston, 1985; Alston and Sumner, 1988).

In the case of sugar, the United States is a large net importer, and U.S. growers face no production restriction. Price enhancement comes from a strict import quota. Sugar imports are perfect substitutes in demand for domestic output, and the USDA limits imports to keep the domestic price above the legislated loan rate. The law requires that no domestic sugar be taken under loan so that net government outlays remain near zero. With no quota limiting production, the beneficiaries of the program have been growers and others who own land or other resources that are relatively well suited to sugar production. Without the program, much of the land now devoted to sugar would shift to other crops or out of agriculture. Such a shift would entail a loss of implicit or explicit land rents and an associated loss of the capital value of land and specialized equipment, facilities, and human capital (Clauson and Hoff, 1988). In some areas, a major loss would be the capitalized value of sugar-refining facilities, some of which are owned by growers through cooperatives. The other major group of farmer beneficiaries of the sugar program are corn producers. High sugar prices have allowed expansion of the high fructose corn syrup industry. Corn producers face a higher demand for their product because of the increased price of a substitute in the integrated sweetener industry. One should note, however, that this spillover effect is small relative to the benefit to sugar producers or the transfers to corn growers from the corn program itself (Lopez, 1989).

A further difference between the tobacco and sugar industries is the size distribution of farms and enterprises. The tobacco industry is made up of about 100,000 farms and tens of thousands of additional quota owners who rent their land and production allotment or lease their quota to producers. Two tobacco types account for almost all U.S. production: The burley and flue-cured tobacco industries produce distinct products that are imperfect substitutes in final demand. The industries are located separately and have different size distributions of farms (U.S. Department of Agriculture, *Tobacco Situation and Outlook Report,* various issues). Of the two major tobacco types, quota for burley tobacco may not be sold but may be leased to growers. That industry is dominated by growers producing less than $40,000 worth of tobacco. A significant proportion of burley output (and benefits from the program) comes from farms producing less than $10,000 worth of tobacco on fewer than five acres. Relatively few quota owners or growers and even a relatively small proportion of output are associated with tobacco enterprises of more than forty acres or $100,000 of production per year. The burley tobacco industry is dominated by quota owners and growers for whom tobacco is distinctly a part-time supplement to family income. Further, the major production regions are relatively low-income, low-wage areas, and many of the beneficiaries are relatively low-income families, including many low-income elderly. The flue-cured tobacco industry has experienced considerable farm consolidation in the last few decades; changes in the 1980s—the sale of quota was legalized in 1982 and the lease and transfer of quota was outlawed in 1986—has resulted in greater consolidation of quota ownership into the hands of growers. However, the flue-cured tobacco industry remains one with almost all its output produced by farms with tobacco sales of less than $200,000, although very small enterprises are far less common than in the burley industry (Grise and Griffin, 1988). Recent quota lease rates or land rental rates indicate a price increase caused by the tobacco programs in the range of about $0.25 per pound. These revenue gains total about $400 million and are distributed among more than 100,000 producers and quota owners for an average of $4,000 per year per person (Alston and Sumner, 1988).

The U.S. sugar industry at the farm level is made up of two distinct industries: the cane sugar industry and the beet sugar industry. The beet sugar industry is scattered in particular areas across the Midwest and West, with major production areas at the Minnesota-North Dakota border, in California, at the Oregon-Idaho border, at the Colorado-Wyoming-Nebraska border, in Montana, and in Michigan. The cane sugar industry is located mainly in Florida, Louisiana, and Hawaii, with a little output in Texas. Sugar cane industry production is dominated by large farms that sell $1 million or more of sugar. There are fewer than 1,000 cane producers in the whole industry. The beet industry has more small to mid-sized farms but fewer than 9,000 producers, and more than 20 percent of its producers are in the $250,000 sales class or above (Clauson and Hoff, 1988).

In the last few years the U.S. price of sugar has been held at between two and four times the world price of sugar, suggesting that direct transfers are more than half of total revenue for domestic sugar producers. For total sales in the range of about $0.95 billion for beets and $0.8 billion for cane, transfers from consumers amount to more than $1 billion total and well over $100,000 per producer per year. Because at world prices U.S. production would probably be much below current levels, the net benefit of the high domestic sugar price is less than the price impact times current output. Further, it is the landowners and others controlling inelastic resources who really capture the long-term program benefits (Lopez, 1989).

The essential point of these examples is to show not the specific distribution of gains but rather that income transfers from farm programs do arise without any direct outlays by the government. Both the tobacco and sugar programs can be viewed domestically as a specific tax on demand, with the receipts being distributed to growers in proportion to output. In both industries the incidence of benefits depends on who controls inelastic resources used in production. This latter point is well illustrated by the tobacco case, in which clear beneficiaries are quota owners and not necessarily growers or even landowners.

It may be of interest to speculate about how one could apply a targeting scheme to industries such as sugar and tobacco that do not have direct government payments. It might be possible to arrange a sliding scale of loan rates for the output of firms with less favored characteristics. Such a change in the programs would cause adjustment in the industries as farms reorganized to meet the criteria of targeted firms. Alternatively, the government might require that a certain share of the market be reserved for farms that satisfy the targeted criterion, as is now done with affirmative action quotas and small business set-aside programs. How such a farm "affirmative action" policy could be monitored and enforced is, of course, problematical.

CONCLUSIONS

Targeting commodity programs has been proposed to help poor people, to keep vulnerable farms in business, and to aid mid-sized farms. This chapter has raised questions about the potential effectiveness of targeting for any of these objectives.

There is absolutely no doubt that farm commodity programs are lousy welfare programs and deliberately so. Some people seem to justify farm programs as an aid to the poor, but in fact the programs transfer income to relatively wealthy families. About this fact there is no dispute. Owners of farmland tend to have much more wealth than the average American. This chapter has developed arguments and evidence that suggest that commodity programs simply cannot be reasonable welfare programs even with various targeting adjustments. Obviously, the distribution of benefits can be changed. It is not clear, however, that commodity programs could ever be used effectively to provide substantial aid to poor rural people without overwhelming leakage to the nonpoor or unbearable administrative complexity and cost.

Sometimes an expanded farm welfare system is proposed as an alternative to targeting. In the early 1970s a three-year experiment with a rural negative income tax program—the Rural Income Maintenance Experiment (RIME)—was conducted by the U.S. Department of Health, Education, and Welfare (HEW). In the experiment farm families with low net incomes received negative income tax payments from the federal government. These experiments mainly demonstrated how difficult such a program would be to administer.

In reviewing some of the results of the experiments, Finis Welch noted the unique burdens that had been placed on American farmers (1978). In addition to providing food and fiber for a nation and the world, for decades they had also regularly provided lessons to officials at the U.S. Department of Agriculture in the principles of supply and demand analysis. For example, if the price of sugar is held at triple the world market price, more sugar will be produced, and if government programs require that wheat acreage be reduced, then land of unusually low productivity will be idled and higher yields will occur on the land left in production. The negative income tax experiments caused the participating farmers a further burden—to demonstrate to HEW officials some additional economic principles. These poorly educated, low-income farmers showed how, given the proper incentives, one could shift expenses into current periods and receipts into later periods to reduce one's taxable income and increase one's payments.

The behavior that the temporary negative income tax project encouraged was perfectly legal and clearly within the rules of the policy, but HEW and the program evaluators were nonetheless caught unprepared. Its seems hard for economists and bureaucrats to conceive of (or perhaps to concede) the creativity of farmers and other entrepreneurs.

The general propositions demonstrated by the results of the negative income tax experiment apply directly to most attempts to target farm programs. The consequences of government programs are hard to anticipate fully and farmers often demonstrate a surprising degree of ingenuity and flexibility. Further, complex programs often yield even more scope for ingenuity. Three implications of these facts should be noted. First, analysts are prone to overestimate the effectiveness of government programs to regulate behavior of farmers in ways legislators might wish. As a result, programs tend to cost more and achieve less than predicted. Second, the effort that goes into reacting to farm programs uses real resources. Complex programs cause complex reactions, and the effort and activity devoted to finding ways around programs would be more productively spent at real economic activity if the program was not there. Third, by reacting in ways not fully anticipated, farmers also subvert some of the negative consequences of programs. For example, by idling the least productive acres, the nation wastes fewer resources under an Acreage Reduction Program.

These principles apply fully to attempts to limit program payments or otherwise target the effects of commodity programs. For example, with a $50,000 limitation on direct payments, farm operators have been creative in the variety of business organizations that maximize the number of "persons" associated with a single farm. More complicated rules would effectively limit benefits to some farmers while missing others. However, farmers can be expected to be innovative and do the unexpected, especially when farm programs are as lucrative as they have been in recent years. It has often been noted that programs for research and technology together with innovative farmers motivated by profits can cause two blades of grass to grow where only one has grown before. Government commodity programs, together with innovative farmers motivated by profits, can cause two farms to grow where only one has grown before, even though productivity declines.

If the objective of targeting is simply to reduce direct payments that now go to a few thousand large producers, this can be achieved by some sort of targeting scheme at the cost of some added administrative complexity, some reduced economic efficiency in farming, and some legitimate claims of arbitrary distinctions between economically similar operations. The budget savings from such targeting would likely be minor, and the basic truth that the programs shift income to the relatively wealthy would remain unchanged, but some embarrassing headlines might be avoided.

There are several major problems with trying to use targeting to lower financial distress. First, paying more to those with financial losses rewards farms for poor management or excessive risk taking. Allowing higher payments to farms with more debt or negative cash flow would effectively penalize otherwise similar farms that are more efficient and prudent. Second, using relatively arbitrary debt, income, or similar criteria for targeting will cause farms to adjust their income statements or balance sheets to be included in the favored status. Third, it is exceedingly difficult to devise criteria that would provide significant help to keep favored farms solvent without inadvertently missing other equally deserving farms or inadvertently including farms that were not meant to be targeted. A related point is that financial distress varies over time and across farm commodity industries. Targeting based on financial distress was recently proposed during a period of severe problems for Midwest grain farms. However, since that time, grain markets have improved, and the distress in these industries and regions has lessened. The next agricultural recession may instead hit industries that do not have direct payment programs; thus, targeting would be of little help. Fourth, government policy includes many programs aimed at helping relieve financial problems for farmers. The Farmers Home Administration acts as a lender for troubled farm borrowers. Nonrecourse loan repayment schedules and disaster programs have been used to aid farmers during troubled times, and nonfarm training programs and poverty programs provide aid for transition out of farming. These sorts of programs all seem more suited to providing assistance to farms with troubled financial situations.

Finally, the major problem with using targeting to affect the size distribution of farms or related structural characteristics is that very little is now known about how farm programs relate to the size distribution. Although there are views on all sides of the question, a reasonable assessment is that commodity programs are roughly neutral with respect to the farm size distributions, both within and across commodity industries. The current observed size distributions are not due to government programs that favor large rather than small or mid-sized farms. Targeting could shift the size distribution, but the reactions to any specific program would be complex and hard to fully anticipate. Among the reactions are changes in program participation, output levels in program and nonprogram commodities, and prices of crops, livestock, and assets, especially land. Finally, if relatively large, low-cost farms were placed at significant competitive disadvantage, the overall efficiency of the farm sector would decline.

This chapter has expressed considerable skepticism about trying to adapt farm commodity programs to serve objectives related to income distribution, farm financial distress, or size distributions of farms. It may be possible to move toward the objectives with targeted commodity programs, but the programs are not efficient, other policy tools are better suited, and other purposes of commodity programs are likely to be sacrificed. Using commodity programs as distributional tools is like using a pipe wrench to drive in a nail. It is hard to get the nail in, a hammer is a much better tool for that job, and afterwards the wrench won't be much good for its own job.

ACKNOWLEDGMENTS

For their comments and suggestions the author would like to thank Laura Martin, Kristen Allen, Ross Korves, James Oehmke, Don Reeves, Randal Rucker, and other participants in the Food and Agriculture Committee meetings, February 3 and April 28, 1989, in Washington, D.C., and in the Western Economics Association meeting, June 20, 1989, in Lake Tahoe, Nevada.

REFERENCES

Alston, Julian, and Daniel A. Sumner. 1988. "A New Perspective on the Farm Program for U.S. Tobacco." Working Paper no. 88–19. Department of Agricultural Economics, University of California, Davis.

Bollman, Ray D. 1989. "Who Receives Farm Government Payments?" Presidential address, Canadian Agricultural Economics and Farm Management Society, Montreal, July.

Brooks, Nora L., and Donn A. Reimund. 1989. "Where Do Farm Households Earn Their Incomes?" Agricultural Information Bulletin no. 560 (Washington, D.C., U.S. Department of Agriculture, Economic Research Service, February).

Clauson, Annette L., and Frederic L. Hoff. 1988. *Structural and Financial Characteristics of U.S. Sugar Beet Farms.* Agricultural Economic Report no. 584 (Washington, D.C., U.S. Department of Agriculture).

Debertin, David L., and Craig L. Infanger. 1987. "Targeting Federal Farm Programs: Options and Consequences." Staff Paper no. 229, Department of Agricultural Economics, University of Kentucky, Lexington.

Foster, William E., and Bruce Babcock. 1987. "Commodity Supply Response with Program Payment Limits." Unpublished draft, University of California, Berkeley, June.

Gardner, Bruce L., and Dale M. Hoover. 1975. *U.S. Farm Commodity Programs and the Inequality of Farm Household Income, 1969.* Economic Research Report no. 35, Department of Economics and Business, North Carolina State University, Raleigh.

Glaser, Lewrene. 1986. *Provisions of the Food Security Act of 1985* (Washington, D.C., U.S. Department of Agriculture).

Grise, Verner, and K. F. Griffin. 1988. *The U.S. Tobacco Industry.* Agricultural Economic Report no. 589 (Washington, D.C., U.S. Department of Agriculture).

Kinston (N.C.) *Daily Free Press.* 1988. Reprinted in *Raleigh* (N.C.) *News and Observer.* "Abuse Increases Farmers' Paperwork" (editorial, December 25, A24).

Knutson, Ronald D., James W. Richardson, Edward G. Smith, Edward M. Rister, Warren R. Grant, Lawrence A. Lippke, and Craig L. Israelson. 1987. "Economic Impacts of Payment Limits." Unpublished paper, Texas A&M University, College Station.

Lederer, Thomas H., and Susan L. Pollack. 1988. "1987 Budget Acts: Description of Changes in Agricultural Policies and Programs." Staff Report no. AGE5880114 (Washington, D.C., U.S. Department of Agriculture).

Lopez, Rigoberto A. 1989. "Political Economy of the United States Sugar Policies," *American Journal of Agricultural Economics* vol. 71, no. 1 (February), pp. 20–31.

Nelson, Frederick J. 1989. "Profile of Farms Benefiting From the 1982 Farm Commodity Programs." Staff Report no. AGES 89-32 (Washington, D.C., U.S. Department of Agriculture).

Reinsel, Robert, John Craven, and Joyce Allen. 1987. "Redistributing Farm Program Benefits." Agricultural Information Bulletin no. 522 (Washington, D.C., U.S. Department of Agriculture).

Rucker, Randal R., and Walter N. Thurman. 1989. "The Economic Effects of Supply Controls: The Simple Analytics of the U.S. Peanut Program." Unpublished paper, North Carolina State University, Raleigh, May.

Stepp, Laura S. 1989. "Catholic Bishops Pledge Support for Family Farm." *Washington Post,* November 9, p. A20.

Sumner, Daniel A. 1985. "Farm Programs and Structural Issues," in Bruce L. Gardner, ed., *U.S. Agricultural Policy: The 1985 Farm Legislation* (Washington, D.C., American Enterprise Institute).

Sumner, Daniel A., and Julian M. Alston. 1985. "The Removal of Price Supports and Supply Controls for Tobacco: An Economic Analysis of the Impact" (Washington, D.C., Food and Agriculture Committee of the National Planning Association, November).

U.S. Department of Agriculture. Various issues. *Sugar and Sweetener Situation and Outlook Report* (Washington, D.C.).

———. Various issues. *Tobacco Situation and Outlook Report* (Washington, D.C.).

U.S. Department of Agriculture, Agricultural Stabilization and Conservation Service. 1988. *ASCS Handbook: Payment Limitations.* Short Reference 1-PL, Amendment 1 with Amendment 2 incorporated (Washington, D.C., Government Printing Office).

U.S. General Accounting Office. 1985. *Examples of USDA's Application of the $50,000 Payment Limitation.* Fact sheet for the Chairman, Committee on Agriculture, Nutrition, and Forestry, U.S. Senate, RCED-86-29FS (Washington, D.C., Government Printing Office).

———. 1987a. *Farm Payments: Farm Reorganization and Their Impact on USDA Program Costs.* Report to Congressional Requesters, RCED-87-120BR (Washington, D.C., Government Printing Office, April).

———. 1987b. *Farm Payments: Implications of Targeting Farm Income Support.* Report to the Honorable Jesse Helms, U.S. Senate, RCED-87-99 (Washington, D.C., Government Printing Office, June).

———. 1987c. *Farm Payments: Basic Changes Needed to Avoid Abuse of the $50,000 Payment Limit.* Report to Congressional Requesters, RCED-87-176 (Washington, D.C., Government Printing Office, July).

———. 1987d. *Farm Programs: Analysis of Options for Targeting Payments and Crop Loans.* Report to the Chairman, Committee on the Budget, U.S. Senate, RCED-87-144 (Washington, D.C., Government Printing Office, September).

———. 1987e. *Farm Payments: Analysis of Proposals to Amend the $50,000 Payment Limit.* Briefing report to the Honorable Silvio O. Conte, House of Representatives, RCED-88-42BR (Washington, D.C., Government Printing Office, October).

U.S. Office of Management and Budget. 1989. *The 1990 Budget of the United States* (Washington, D.C., January, pp. 71–72).

Welch, Finis. 1978. "The Labor Supply of Farmers," in John Logan Palmer and Joseph A. Pechman, eds., *Welfare in Rural Areas: The North Carolina Iowa Income Maintenance Experiment* (Washington, D.C., The Brookings Institution).

TEN TRUTHS ABOUT SUPPLY CONTROL

THOMAS W. HERTEL

Supply controls as we know them today were first introduced in the 1930s. Between 1929 and 1932 farm prices and net farm income declined precipitously, putting the Roosevelt administration under pressure to do something for the farm sector. It responded with the Agricultural Adjustment Administration, which implemented the basic set of farm programs that we now have. Since then we have experienced a wide range of supply controls used on a broad cross-section of crops. Acreage Reduction Programs for grains and cotton and the conservation reserve are the most important, but marketing quotas have also been employed (e.g., for tobacco and peanuts). Marketing orders, too, serve to limit supplies of various perishable products (e.g., fluid milk, fruits, vegetables, and nuts). In short, the United States has actively employed supply controls as an instrument of farm policy over the past five decades.

There is a fairly predictable policy cycle for supply controls. It is the mirror image of movements in farm prices. These movements are determined by a variety of factors, including weather, macroeconomics, and farm policies, both at home and abroad. Apart from these cyclical variations, rapid technical change coupled with income- and price-inelastic demands guarantees that the overall, long-term trend in real prices will be downward. As prices fall, inefficient (often smaller-sized) producers are forced out of business, which helps solidify broader public support for price-enhancing measures such as supply control.

The implementation of public policies designed to retain resources in agriculture effectively contributes to the problem of excess capacity in the agricultural sector, which in turn lowers the rate of return on farm labor and assets (in the absence of government transfers). Thus, the effectiveness of such an approach for supporting farm incomes is questionable in the longer run. Any effort to raise prices by restricting supplies is at best a countercyclical policy.

The most recent cyclical price trough occurred in the mid-1980s and was accompanied by extraordinarily high levels of farm debt as well as high real interest rates. Important factors contributing to the depressed market for farm commodities were the agricultural policies of the United States, the European Community, and other developed market economies that were increasingly insulating producers from world price signals. The resulting surplus commodities were dumped on the international marketplace, causing further declines in world prices. It was in this environment that Senator Tom Harkin (D-Iowa) and others proposed legislation in 1986 (S. 2869) to restrict the supplies of farm products (Harkin, 1987). Their objective was to raise prices of grains, oilseeds, cotton, and dairy products dramatically by implementing marketing quotas. However, the combination of a strong Republican administration, rebounding export markets (in part due to the Food Security Act of 1985), and, finally, the drought of 1988 undermined any chances that these proposals might have had.

Despite the failure of these proposals, however, it cannot be said that the United States weathered this cyclical trough without significant supply control measures. In an effort to deal with grain surpluses, almost 80 million acres of farmland were taken out of production under the Payment-in-Kind (PIK) program of 1983. Combined with a drought, this program caused a temporary rebound in the marketplace. However, by the mid-1980s the grains programs were once again idling large amounts of land under the annual Acreage Reduction Programs (ARPs). Also, introduction of the Conservation Reserve Program under the Food Security Act of 1985 mandated long-term retirement of an added 40 million to 45 million acres by 1990.[1] In short, supply control has remained an important policy instrument in the 1980s.

The purpose of this chapter is to provide a thorough discussion of agricultural supply controls as an instrument of public policy in U.S. agriculture. Two alternative means of implementing supply controls are considered: (1) acreage controls and (2) output quantity controls. Acreage controls reduce farm supplies indirectly by limiting the use of an important factor of production, namely, land. Output controls limit supplies directly and may be implemented by the allocation of quotas, which limit the amount that may be produced (i.e., sold). I will assume that such quotas are tradeable, and thus they are similar to the tobacco marketing quotas. However, unlike the tobacco quotas, I will assume that they are not tied to a specific location (e.g., a county).

The ten main conclusions, presented throughout the chapter, are stated as "truths." While some might argue with the wording or tone of one or more of these points, I believe there is now sufficient empirical evidence (cited throughout, as appropriate) to state them as truths, not hypotheses.

The next section discusses the basics of supply control. The "truths" stated in this section are common to any policy that attempts to raise farm prices by restricting supply, either directly or indirectly. A comparison of direct (output) and indirect (acreage) controls is then presented, followed by a discussion of some of the practical problems associated with the implementation of supply controls.

[1]It should be noted that this Conservation Reserve Program had important environmental as well as supply control objectives. In particular, the program was designed to target highly erodible land.

THE BASICS OF SUPPLY CONTROL

1. Agricultural supply controls transfer purchasing power from nonfarm households to those households currently controlling the production in agriculture.

It is no secret that someone must pay the bill for higher farm prices. It may be consumers, taxpayers, or foreigners, but the money must come from somewhere. Supply controls are often viewed as a mechanism for reducing government expenditures on farm subsidies. By restricting the amount of production permitted, subsidy levels fall, even if per unit subsidies are unchanged. If producers face a constant target price for their output, then restricting output and raising market prices will also reduce per unit subsidy levels. This is why Senator Harkin projected large budgetary savings under his "Save the Family Farm" bill. Of course, higher market prices for farm products translate directly into higher food prices, and so the costs are effectively transferred from taxpayers to consumers.

But aren't most consumers also taxpayers? To evaluate accurately the distributional implications of supply control, we need to examine how households spend their money and how they earn it, taking into account the taxes they must pay. Table 1 provides this type of complete accounting for a hypothetical supply control program for grains, oilseeds, and cotton. The entries represent changes in real income attributable to a 10 percent increase in market prices for these crops (over 1984 levels). These results refer to the medium run (four years) and assume complete adjustment to this new policy. Data are presented for four types of households: farm and nonfarm private domestic households, the public (government) household, and foreign households. (The public household's "real income" is a proxy for the welfare derived from its provision of public goods and services.)

Each household group in table 1 wears a number of hats. Farmers, for example, are also consumers. As such they will be affected by higher prices for consumption items (see the first row of table 1). Farmers are also producers; that is, they own factors of production employed on (and perhaps off) the farm.[2] Changes in real income attributable to changes in the returns to these factors of production are shown in the second row. Finally, farmers are taxpayers (row 3) and potential recipients of quota rents (row 4). All these dimensions of farm household well-being must be considered in order to arrive at the net change (the row labeled "total") in this group's purchasing power as a result of supply control.

Consider first the case of output controls. As shown in table 1, virtually all households are worse off as consumers because they must now pay higher

[2]In the study cited here, Hertel and Tsigas (1989) make the simplifying assumption that farm households solely comprise those individuals supplying land, labor, and capital to agriculture. In practice, it is increasingly difficult to distinguish between farm and nonfarm households, as nonfarm income becomes more and more significant for a large number of farm families.

TABLE 1. Distributional Consequences of Supply Controls: Implications of a 10 Percent Increase over 1984 Market Prices of Grains, Oilseeds, and Cotton

(millions of 1987 dollars)

	Output controls				Voluntary acreage controls			
	Private domestic households				Private domestic households			
Dimensions of each household	Farm	Nonfarm	Government households[a]	Foreign households	Farm	Nonfarm	Government households[a]	Foreign households
As consumers	−213	−7,094	−1,628	−2,704	−236	−7,893	−1,694	−2,769
As producers	−5,119	1,736	0	678	7,180	1,833	0	859
As taxpayers	52	1,679	0	0	4	115	0	0
As quota rent recipients	13,252	0	0	0	—	—	—	—
Total[b]	7,972	−3,679	−1,628	−2,026	6,948	−5,945	−1,694	−1,910

Note: Figures reported represent equivalent variations, i.e., the money that households could give up after the change and still be left as well off as before.

Source: Hertel and Tsigas (1989), table 5.

[a]The change in government household welfare is a proxy for the change in welfare derived from the provision of public goods and services. Thus, this decline in government household welfare reflects a decline in the provision of such services. This is a consequence of a fixed budget in the face of increasing prices for products purchased by the government household.

[b]The entries in the total row do not sum to zero because of changes in the efficiency with which resources are utilized as a result of supply control. The output control option increases this efficiency by moving resources from the farm to the nonfarm sectors. The acreage control option results in a net loss to society, largely because of the idling of productive land. These points are developed more fully in this chapter in the context of the eighth truth.

prices. Perhaps more surprising is the fact that farmers as producers are worse off. This is because I have assumed that the quota rents are divorced from land ownership. If this is so, then, following the supply reduction, diminished demand for farmer-owned assets forces their market prices to fall. Of course, adding the $13 billion in quota rents more than compensates *current* farm households for their other losses.[3]

It should be borne in mind that quota rents for current farm owners translate into increased production costs for future farm sector entrants. Thus, output controls are not effective in raising the incomes of future generations of farmers. This is nowhere more evident than in the tobacco program, where an active market exists for rental and resale of quotas (Gardner, 1981, pp. 77–78; Hoover and Sumner, 1984). Thus, anyone interested in expanding production must compensate the initial quota owner for the windfall gains that were realized when the program was initiated. Since initial quotas are typically assigned on the basis of land ownership, the net effect of implementing (or strengthening) this form of supply control is to widen the distribution of income within the farm sector (Gardner and Hoover, 1984).

Finally, there is the question of budgetary savings. This is a complicated issue, because, as the entire pattern of production and consumption in the economy adjusts, so too does the tax base. Table 1 shows the net budgetary position of the Treasury improving by $52 million + $1,679 million = $1,731 million. In this experiment, the budgetary surplus is distributed back to the private households in proportion to their initial incomes, and shows up as two positive entries in the "taxpayer" row of table 1. The net effect of output controls is to leave nonfarm households $3,679 million worse off. By contrast, real incomes of the farm households have risen by about $7,972 million, with most of the difference accounted for by the income losses of public and foreign households.[4]

Turning to acreage controls, we see from table 1 that the pattern of consumer losses is similar. However, the mechanism for implementing this form of supply control is quite different, resulting in a very different pattern of producer and taxpayer effects. In particular, it is assumed that a type of "bid ARP" is used, whereby idled acreage is rented by the government. Thus, taxpayer costs change very little, since deficiency payment savings are expended on land rental payments. Note that the per unit rental cost of land increases rapidly as additional acres are bid out of production and market prices rise. This rise in land rents benefits farm owners so that the aggregate farm household is almost as well off as under output controls ($6,948 mil-

[3]If the quota rents were tied to the land, then farmers as producers would be better off. A rough approximation of the amount of gain may be obtained by summing the quota rent and producer rows in the first column of table 1.

[4]Because of changes in the efficiency with which resources are utilized, as a result of supply control, entries in the "total" row in table 1 do not total to zero. This point is discussed in detail in a later section of the chapter.

lion). However, the drop in real income of nonfarm households is much larger (−$5,945 million) than in the case of output controls, since they no longer enjoy the large taxpayer savings.

If the acreage controls were mandatory and producers were not compensated for the idled land, then the farmers' gains would diminish. This is because the reduction in planted acreage (about 11 percent) partially offsets the increased rental rate on land. Since farmers are not compensated for their idled acreage in the mandatory case, taxpayer costs are diminished and the nonfarm household is not as badly hurt.

In sum, supply control, in the form of either voluntary acreage controls or mandatory output controls, is an effective means of increasing real incomes of farm households—at the expense of nonfarm households. By restricting availability, the price of agricultural products is raised and farm incomes go up. Truths 2 and 3 address more precisely the mechanisms for this transfer and the magnitude of requisite supply reductions.

2. The single most important parameter in determining the impact of a given supply reduction is the farm-level demand elasticity.

The farm-level demand elasticity tells us how much aggregate farm supplies must be cut if farm prices are to rise by 1 percent. Thus, if the farm-level demand elasticity equals −0.5, then a 0.5 percent reduction in farm supplies will cause farm prices to rise by 1 percent. Ideally this parameter captures all of the adjustment, both domestic and foreign, to higher U.S. farm prices. The greater the ease with which other countries adjust to changes in the U.S. supply of commodities, the larger the farm-level demand elasticity and, hence, the greater the production cut needed to attain a given price target.

Brandow (1961) devoted the majority of his 1961 study on supply control to an analysis of farm-level demand elasticities. He estimated that the aggregate farm-level demand elasticity for U.S. agriculture, based on domestic uses, was equal to −0.23. (At the time of his analysis, export markets played a small role in determining the price responsiveness of farm-level demands.) This means that a 2.3 percent cut in aggregate farm output at that time would have translated into a 10 percent price increase.

Since the time of Brandow's study, the price responsiveness of production and consumption in export markets has come to play a much more important role in determining the price elasticity of demand facing U.S. farmers. A more recent study places the value of this parameter at −0.53 (Hertel et al., 1989). As a result, supply control today is less than half as effective as it was at the time of Brandow's study; that is, an equiproportional price hike now requires aggregate output to be cut back by more than twice as much.

In practice, supply control is never applied to all farm products simultaneously. Rather, it is employed on a crop-by-crop basis. Because there is more potential for substitution away from individual farm products than for substitution away from the farm aggregate, we can logically expect product-

TABLE 2. Product-Specific Farm-Level Demand Elasticities

Product	Own-price elasticity	Aggregate price elasticity[a]
Dairy	−.513	−.512
Poultry	−.205	−.136
Red meats	−.407	−.398
Cotton	−.997	−.966
Food grains	−.956	−.253
Feed grains	−1.021	−.917
Sugar crops	−.044	−.026
Oilseeds	−.793	−.514
Other crops	−.350	−.388
All farm products	—	−.526

Notes: Elasticities refer to a four-year time horizon. Dash means not applicable.

Source: Hertel, Ball, Huang, and Tsigas (1989), table 8. Reprinted with permission.

[a]Aggregate price elasticity reports quantity adjustment when all farm prices are simultaneously increased by 1 percent.

specific farm-level demand elasticities to be larger than their aggregate counterpart. This gives rise to truth no. 3.

3. Supply controls are most effective when they are applied to multiple commodities—in particular, those that are close substitutes in consumption.

Table 2 documents truth no. 3. The first column reports product-specific elasticities. In the case of export crops, these own-price elasticities of farm-level demand are relatively large—in the neighborhood of −1—reflecting the fact that export markets are inherently more price-responsive than the domestic market. There are several reasons for this. First of all, the demand for food is generally more price-responsive in the rest of the world than in the United States. Second, the export demand elasticity facing the United States includes a supply response component as well as a demand component; that is, foreigners can adjust to an increase in the price of products by either (1) consuming less food or (2) producing it themselves. This is summarized in truth no. 4.

4. As U.S. agriculture comes to rely more heavily on export markets, the farm-level demand elasticity will continue to rise, making supply control increasingly costly.

The second column of table 2 reports the total farm-level demand elasticity for each product. These figures measure the percentage change in quantity demanded following a simultaneous 1 percent increase in all farm prices. When several products are strong farm-level substitutes, their total elasticity will be considerably smaller (in absolute value) than their own-price elasticities. For example, red meats and dairy products are significant farm-level substitutes for poultry. As a result, the aggregate poultry demand elasticity is one-third smaller than its product-specific counterpart.

The difference between the first and second columns in table 2 is even more dramatic for food grains. Empirical evidence indicates that an increase in the U.S. feed grains price has a significant effect on the rest of the world's demand for U.S. food grains (Seeley, 1985). (Foreigners produce less and utilize more wheat in the feed mix.) As a consequence, when both food and feed grain prices go up by 1 percent, the farm-level demand falls by much less than when only food grain prices rise.

The theme of substitution in both production and consumption is central to any analysis of supply control. By definition, supply control involves creating artificial scarcity in the farm product market (and, in the case of acreage controls, in the land market). Economists describe consumer and producer responses to scarcity in terms of product substitution. To the extent that individuals can substitute away from farm products, they can partially avoid the scarcity effects of supply control, which leads to truth no. 5.

5. Substitution, both in production and in consumption, is the foremost enemy of those who advocate supply control.

When I have asked crop farmers whether or not they alter the amount of fertilizer applied in response to relative price changes, I often receive a clear no for an answer. Similarly, hog farmers often claim that they do not alter their feed mix in response to changing ingredient prices. Consumers, too, are not immune to this nonsubstitution mindset. Many people will claim that their purchases of food are not responsive to price. Yet when one observes the data on fertilizer use, feed mix, and food demand, all are significantly responsive to price (other factors constant).

In addressing this apparent discrepancy of statement and fact, it is important to distinguish between intrafirm-substitution and interfirm-substitution effects. Even if no individual farmers altered their feed mix in response to relative price changes (zero intrafirm substitution), the mix of ingredients sold to hog producers could change as a result of interfirm substitution. In particular, as corn prices rise, those producers who happen to use less corn in their feed mix become relatively more profitable. Since more profitable firms tend to expand, these firms play an increasingly important role in the aggregate demand for feed. As a result, the proportion of corn in the aggregate feed mix falls, and it appears that farmers have substituted away from corn!

Substitution in response to relative price changes is in fact pervasive in the economy. Some livestock farmers do alter their feed mix in response to relative prices. Also, consumers do engage in some product substitution. All it takes is a subgroup of consumers who are price-sensitive in order to alter the observed mix of food consumed in response to relative price changes.

Of course, the longer supply controls remain in place, the greater the degree of accommodation (i.e., substitution) in response to the altered rela-

TABLE 3. Implications of Increasing Price Responsiveness in the World

	Indices of supply control (2 years = −1.00)		
Time horizon	Food grains	Feed grains	Oilseeds
2 years	−1.00	−1.00	−1.00
4 years	−1.40	−1.24	−1.14
6 years	−1.43	−1.43	−1.16
8 years	−1.77	−1.53	−1.22
10 years	−2.13	−1.56	−1.22

Note: Assumptions: Domestic use farm-level demand elasticities are −0.12, −0.67, and −0.18, for food grains, feed grains, and oilseeds, respectively (based on results in Hertel et al., 1989). They are treated as being invariant over time. Domestic use shares are 0.647, 0.783, and 0.619 (1987 values). Export demand elasticities are taken from Seeley (1985) and increase over time as a result of adjustment in the rest of the world to higher U.S. prices.

tive prices. In the long run firms that refuse to alter their input mix tend to go out of business. Efficient firms expand, and new technology evolves. The U.S. sugar program provides an excellent example of how persistent artificial scarcity induces product innovation and investment designed to permit substitution away from the scarce commodity. In 1975 the vast majority of domestic sweetener demand was satisfied by sugar. Today, high-fructose corn syrup supplies almost half of domestic sweetener demand (U.S. Department of Agriculture, Economic Research Service, 1989). It has virtually taken over the soft drink market and is now threatening the market for table sugar. These phenomena would likely not have occurred in the absence of the sugar program.

This type of substitution is most dramatic in the case of export demands for U.S. farm products. As noted earlier, foreigners can adjust to higher U.S. farm prices both by restricting consumption and by increasing production. The latter leads to the most dramatic loss of U.S. export markets in the long run. By way of illustration, consider the results in table 3, which show the implications for U.S. supply control of gradual accommodation in the rest of the world to higher U.S. prices for food grains, feed grains, and oilseeds. The numbers are indices indicating the proportionate reductions in output quotas required to maintain a given domestic price target in the face of increasingly responsive export demands. For convenience, all indices are set equal to 1.00 for the two-year time horizon. Thus, a food grains price target that requires a 10 percent reduction in quantity over a two-year horizon will require a 21.3 percent reduction in quantity over a ten-year horizon! The differences between two- and ten-year responses are smaller for feed grains and oilseeds but still substantial.

The indices in table 3 are based on responses from the rest of the world to U.S. price changes as simulated by the International Institute for Applied Systems Analysis model of world agricultural trade (Seeley, 1985; Fisher et al., 1988). They incorporate the fact that it takes time for other countries to adjust their allocation of resources between farm and nonfarm sectors. They also reflect the fact that some countries attempt to insulate producers and/or consumers from world price changes. Over time, however, virtually every country in the world transmits world price signals, at least in part, to its domestic economy.

Of course some countries may only become competitors in certain crops after a time period sufficiently long to permit adaptive research and perhaps major infrastructural investments. For example, cotton production may be economical, but only after the development of irrigation facilities. For the long-term view of this phenomenon, consider the following quote and numerical example provided by Don Paarlberg to illustrate the damaging effects of the U.S. cotton price support program from 1928 to 1977 (Paarlberg, 1980, p. 35):

> Through the cotton program we held our cotton production essentially constant and conceded market growth to our overseas rivals who, under our price umbrella, quadrupled their output.

	1928–30	1976–77
United States production of cotton (millions of bales)	14.4	12.6
Foreign production of cotton (millions of bales)	12.2	49.2

While, in the absence of the cotton program, comparative advantage may have dictated some of the overseas shift in production, there can be little doubt that supply controls and high support prices contributed to this reconfiguration of world cotton production. (This also explains why U.S. competitors are such enthusiastic supporters of U.S. supply control.)

Because producers and consumers tend to adjust to scarcity over time, the farm-level demand elasticity increases and the artificial price stimulus achieved by a given production quota wears off. Supply reductions beget further supply reductions in order to maintain high farm prices. Thus, truth no. 6 about supply control.

6. Supply control is like an addictive drug—the longer it is used, the larger the dose needed to get high (high prices, that is).

Truth no. 6 has a corollary. Like a drug habit, long-standing supply controls to which the economy has accommodated itself are the most difficult to change. Once the quota rents and/or increased land values have been transferred to succeeding generations of farmers, they become very much like any other cost of production. New entrants pay the "fair market value" for the right to produce for the controlled markets. At this point, any change in the rules of the game threatens the viability of these younger farmers, regardless of how efficient they might be. Meanwhile, the real beneficiaries of supply control, namely those who were farming at the time of implementation, may have long since retired!

This problem of capitalization of farm program benefits is pervasive, and extends beyond the commodity market in which the supply control measures are introduced. As noted earlier, by introducing artificial scarcity into certain markets, supply controls send a signal to consumers and other producers to substitute away from this product. This has important consequences that contribute to the difficulty of reversing a long-standing supply control habit. Consider the interests of individuals in related markets who have made investments based on this artificial scarcity. They soon become proponents of the supply control regime. If these controls were eliminated, they too would experience a capital loss, since their investment was based on inflated prices for the controlled products. A classic example of this phenomenon is, again, the sugar program. It now enjoys strong support from a relatively new constituency, namely the producers of high-fructose corn syrup. They support a continual tightening of the import quota in order to maintain high prices for the sugar product with which they are in direct competition.

If carried to their logical extreme, supply controls can completely eliminate the market for a commodity. Initial controls, followed by tighter controls, lead to yet more substitution in production, and the system feeds on itself. High prices beget substitution, which in turn requires additional controls. Eventually one passes the point of no return, at which point elimination of the controls becomes irrelevant, since the market has evaporated.

OUTPUT AND ACREAGE CONTROLS COMPARED

Since both output and acreage controls share the objective of raising farm prices by restricting supply, the truths discussed above apply to both. However, the choice between acreage and output controls is quite important for other reasons. Consider, in particular, truth no. 7.

7. Acreage and output controls have dramatically different effects on land rents and the pattern of input use and innovation in agriculture.

As shown in table 1, acreage and output controls transfer income to farmers in very different ways. In the case of output controls, the scarce factor of production becomes the output quota itself, and initial recipients of these quotas are the beneficiaries of the higher food prices. Assuming that the quotas are divorced from land ownership, land rents fall, as does the employment of nonland crop inputs (see the first column of table 4). The biggest drop (−7.7 percent) comes in the use of fertilizer and chemical inputs. Since land is now more abundant, relative to the amount of farm output supplied, purchased input usage falls, resulting in lower yields than would otherwise have been realized. Thus, output controls are compatible with "lower variable input" agriculture.

By contrast, acreage controls restrict supply by making the land input more scarce. As a result, land rents rise by 26.8 percent (table 4, second column),

TABLE 4. Factor Market Effects of Output and Acreage Controls
(percentage change)

	Output controls[a]	Acreage controls[a]
Land rents	−19.3	26.8
Fertilizer and chemicals	−7.7	0.2
Farm labor	−1.2	−2.7
Crop capital	−5.3	−5.1
Livestock capital	0.6	0.6

Note: Results refer to a four-year time horizon.
Source: Hertel and Tsigas (1989), table 3.
[a]Both experiments raise market prices for cotton, grains, and oilseeds by 10 percent over their 1984 levels.

and there is an incentive to raise yields by using purchased inputs more intensively. Note that total fertilizer and chemical applications actually rise (marginally), despite the fact that planted acreage falls. Thus, the intensification effect slightly outweighs the effect of restricting output.

The results reported in table 4 refer to the medium run, i.e., a four-year period of adjustment. As such, they reflect anticipated substitution based on existing technology. In the longer run, however, relative prices can be expected to influence the development of agricultural technology (Hayami and Ruttan, 1970). Thus, a program of acreage controls such as that employed in the United States might be expected to encourage new technology to be more land-saving in nature than would otherwise be the case. Indeed, recent econometric evidence for the post–World War II period indicates that this is the case (Offut and Shoemaker, 1988).

The differential effects of output and acreage controls on the intensity of cultivation and the pattern of agricultural innovation are documented in a recent study by Foster and Babcock (1989). They studied the growth rate in tobacco yields over the 1940–1986 period. Statistical analysis leads them to conclude that the switch from acreage to marketing quotas in 1965 had a dramatic effect on tobacco yields. They found an immediate drop in yields in 1965, followed by a much slower annual growth rate—0.8 percent per year under the marketing quota versus 3.8 percent per year under acreage controls.

To summarize, the choice of acreage controls is the choice of a high-yield/high-input system of agricultural production. This has potential implications for non-point source chemical pollution and groundwater quality. The concentration of larger amounts of chemicals on fewer planted acres as a result of acreage controls can exacerbate environmental pollution.[5]

[5]The implications of current Acreage Reduction Programs for soil erosion vary depending on the program in question. The Conservation Reserve Program targets highly erodible land and thus has beneficial effects insofar as it reduces soil loss. If such a program were also targeted at the acreages most prone to chemical runoff and leaching, then it might have additional environmental benefits.

In addition to this potential for negative externalities, acreage controls distort market signals by idling productive land. One way of evaluating the net social cost of this policy intervention is to sum the total real income changes for the four households in table 1 (see Hertel and Tsigas [1989] for further details). This equals −$2,600 million, indicating that the gainers could by no means compensate the losers without the gainers themselves being made worse off. Hence truth no. 8.

8. From a societal perspective, idling acreage represents a notably costly form of supply control.

By contrast with acreage controls, the estimated change in societal welfare following output controls equals a positive $647 million. Does this mean that output controls of this sort are a good thing? Before jumping to such a conclusion, it is important to point out that in 1984 the pattern of resource utilization in the U.S. economy was already highly distorted (Hertel, Chattin, and Tsigas, 1989). The U.S. Department of Agriculture's Economic Research Service (1987b) estimated the total (direct and indirect) support to agriculture at about 22 percent of the value of farm output during the 1982–1984 period. This pattern of support served to deter the movement of resources out of agriculture, reinforcing instead a pattern of policy-induced excess capacity. By directly limiting the amount of crop production, output controls encourage the movement of some farm resources into other sectors of the economy. This reallocation of productive capacity to the nonfarm economy creates the marginal gains for society, relative to the observed situation in 1984.[6]

Of course, the preceding analysis has been based on the presumption that the basic structure of farm and food programs will remain in place and that output controls on grains, cotton, and oilseeds are simply added to the mix of 1984 farm and food policies. I would reach a very different conclusion if output controls were compared to the alternative of *no* policy intervention, which would generate much larger welfare gains for society. Unless accompanied by "decoupled" payments, however, the unilateral elimination of market interventions entails substantial income transfers from farm to nonfarm households (Hertel, Thompson, and Tsigas, 1989).

Thus far I have focused on the distinction between farm and nonfarm households. However, there is a large group of farm-related businesses with a direct interest in supply control. Consider the case of the fertilizer industry. This group has been a vocal opponent of supply control in general and of acreage controls in particular.[7] This opposition is perhaps somewhat surpris-

[6]Hertel and Tsigas (1989) develop this point in greater detail. They also decompose by sources these aggregate efficiency gains.

[7]For example, in 1987 the Fertilizer Institute paid to republish and widely distribute the results of a USDA study documenting the costs of supply control (Harrington, Schluter, and O'Brien, 1986).

ing, given the results reported in table 4, which indicate that total fertilizer use would slightly *increase* as a result of additional acreage controls. These are medium-run results, however, and there is little doubt that the immediate, short-run impact of planting fewer acres would be a decline in fertilizer sales. Only after farmers begin to seek heightened yields in response to the increased scarcity of land will total fertilizer use begin to rebound. And, in the very long run, eroding export markets may again lead to a decline in total fertilizer use.

The fertilizer example provides a good illustration of the conflicting forces at work in determining how agribusinesses are affected by supply control.[8] On the one hand, reduced farm output tends to hurt these firms; on the other hand, the relative price changes that accompany supply control tend to induce substitution away from the scarce products. The net effect on fertilizer sales depends on which is larger, the output effect or the substitution effect induced by the intensification of production. Of course, many other agribusinesses are unambiguously hurt by the reduced volume of agricultural production. Certainly over the long run, tight output quotas that price U.S. commodities out of export markets cannot be in the interest of most of the U.S. agribusiness community.

ISSUES OF IMPLEMENTATION

The issue of supply control implementation is perhaps best introduced with the following quote from Don Paarlberg, which I will label truth no. 9.

9. "The tendency of those who defend the [supply control] programs is to compare individual decision-making at its worst with centralized decision-making at its best." (Paarlberg, 1980, p. 31)

In order to implement the current system of acreage controls, the U.S. government requires an elaborate system of county Agricultural Stabilization and Conservation Service (ASCS) offices to monitor compliance. Despite the large expenditure required to implement this system, much slippage occurs. For example, in 1985 the established wheat base was 94 million acres, yet even at the height of the export boom in 1981, with no acreage in set-aside, the wheat crop never exceeded 88 million planted acres and 80 million harvested acres (U.S. Department of Agriculture, Economic Research Service, 1987a, table 2). This type of "phantom acre" slippage is further compounded by farmers' tendency to set aside their least productive land. Thus, extremely wet areas as well as nonirrigated "corners" of square fields with circular (center-pivot) irrigation systems tend to find their way into the set-aside acreage.

[8]For further discussion of the effects of farm policy on agribusiness and vice versa, see Alston, Carter, and Wohlgenant (1989).

Despite such problems, acreage controls are easier to monitor than production controls would be, particularly in the case of feed grains. Given the high market prices prevailing under supply control, there is a strong incentive for producers to err on the side of overproduction and so to be assured of fulfilling their quota. Thus, the question immediately arises: What will happen to the surplus production? How can farmers be prevented from feeding this surplus to their livestock? If such a precedent for surplus disposal were to be established, there would be a strong incentive for feed grain producers who would not otherwise raise livestock to do so. This would result in an inefficient pattern of managerial and physical resource use in agriculture—solely to take advantage of a loophole in the implementation of output controls.

Another likely outlet for surplus production in excess of an output quota is the export market. In fact, the 1981 farm bill altered the peanut quota program to permit unlimited export sales at world market prices (Hoover and Sumner, 1984). Since export demand is relatively price-responsive, this had a stimulative effect on exports and total production. Between 1982 and 1986 shelled peanut exports increased from 2.7 billion to 3.6 billion pounds, and harvested acres increased from 1.37 million to 1.54 million acres (U.S. Department of Agriculture, Economic Research Service, 1987a, tables 154 and 158). This increase in peanut exports is particularly notable since it occurred at a time when other farm exports were falling. For example, grain exports fell from 97.2 million metric tons in 1982 to 75.4 million metric tons in 1986 (U.S. Department of Agriculture, Economic Research Service, 1987a, table 2).

By diverting production in excess of the quota onto the world market, the United States is in effect forcing domestic consumers to bear the entire burden of supply control. Furthermore, supply control becomes much less effective; that is, the cut in production, as well as the accompanying increase in farm household income, is diminished. Hertel and Tsigas (1989) provide a complete analysis of this type of two-price scheme for grains, oilseeds, and cotton. Assuming the same domestic price targets as used in table 1 (10 percent increases over 1984 levels), they show that permitting surplus disposal in the export market amounts to roughly a $2 billion transfer from domestic to foreign households. This point is summarized in truth no. 10.

10. By permitting surplus production that exceeds a quota to be disposed of on the world market, we effectively transfer dollars out of U.S. pockets, into those of foreign households.

SUMMARY AND CONCLUSIONS

In conclusion, the predominant method of supply control in U.S. agriculture—voluntary acreage controls—is an extremely costly method of raising prices and supporting farm income. By idling productive resources, distorting factor choices, and altering the pattern of agricultural innovation, the United States forgoes large amounts of real income. Furthermore, without environmental targeting these programs likely exacerbate the chemical pollution of streams and groundwater. By contrast, output controls are compatible with lower levels of purchased input use in agriculture. Also, in the context of current farm programs, they do not appear to cost society additional resources in the medium run; they simply transfer money from nonfarm to farm households.

The big problem with any form of supply controls is that prolonged use diminishes their effectiveness. The rest of the world has proven itself very capable of adjusting to artificially induced scarcity in the United States. In the course of that adjustment, this country loses export markets and fails to give farmers the incentives necessary to keep them competitive. Once the initial boost from supply control has worn off, the United States is left in the unenviable position of having either to (1) restructure dramatically the agricultural sector (in the absence of massive export subsidies) or (2) abdicate its role as an agricultural exporter altogether. These are hardly tolerable alternatives with which to enter the 1990s, a decade that promises tight federal budgets and an intensely competitive global marketplace.

ACKNOWLEDGMENTS

The author would like to thank Don Paarlberg, Marinos Tsigas, John Antle, Ronald Landis, Kristen Allen, and members of the Food and Agriculture Committee of the National Planning Association for valuable comments on this paper. Dr. Tsigas also assisted with the computations.

REFERENCES

Alston, J. M., C. A. Carter, and M. K. Wohlgenant. 1989. "Who Determines Farm Programs? Agribusiness and the Making of U.S. Farm Policy." International Agricultural Trade Research Consortium Working Paper no. 89–1, University of California at Davis.

Brandow, G. E. 1961. *Interrelations Among Demands for Farm Products and Implications for Control of Market Supply*. Pennsylvania Agricultural Experiment Station Bulletin no. 680, Pennsylvania State University, University Park.

Fisher, G., K. Frohberg, M. A. Keyzer, and K. S. Parikh. 1988. *Linked National Models: A Tool For International Food Policy Analysis* (Dordrecht, The Netherlands, Kluwer).

Foster, W. E., and B. A. Babcock. 1989. "Government Commodity Policy and Technological Change: Tobacco Yields, 1940 to 1986." Paper presented at the American Agricultural Economics Association meeting, Baton Rouge, La., 1989.

Gardner, B. L. 1981. *The Governing of Agriculture* (Lawrence, University of Kansas Press).

Gardner, B. L., and D. M. Hoover. 1975. *U.S. Farm Programs and the Inequality of Farm Household Income*. Economic Research Report no. 35 (Raleigh, North Carolina State University).

Harkin, T. 1987. "Time for New Directions," *Choices* vol. 2, pp. 5, 10–13.

Harrington, D., G. Schluter, and P. O'Brien. 1986. *Agriculture's Links to the National Economy*. Agricultural Information Bulletin no. 504 (Washington, D.C., U.S. Department of Agriculture, Economic Research Service).

Hayami, Y., and V. W. Ruttan. 1970. "Factor Prices and Technical Change in Agricultural Development: The United States and Japan, 1880–1960," *Journal of Political Economy* vol. 78, pp. 1115-1141.

Hertel, T. W., and M. E. Tsigas. 1989. "General Equilibrium Analysis of Supply Control in U.S. Agriculture." Unpublished manuscript, Department of Agricultural Economics, Purdue University, West Lafayette, Ind.

Hertel, T. W., E. Ball, K. S. Huang, and M. E. Tsigas. 1989. "Computing General Equilibrium, Farm Level Demand Elasticities for Agricultural Commodities." AES Research Bulletin no. 988, Purdue University, West Lafayette, Ind.

Hertel, T. W., B. Chattin, and M. E. Tsigas. 1989. "General Equilibrium Analysis of Producer Subsidies for Agriculture." Staff paper no. 89-6, Department of Agricultural Economics, Purdue University, West Lafayette, Ind.

Hertel, T. W., R. L. Thompson, and M. E. Tsigas. 1989. "Economywide Effects of Unilateral Trade and Policy Liberalization in U.S. Agriculture," in A. B. Stoeckel, D. Vincent, and S. Cuthbertson, eds., *Macroeconomic Consequences of Farm Support Policies* (Durham, N.C., Duke University Press).

Hoover, D. M., and D. A. Sumner. 1984. "Tobacco and Peanuts," in G. G. Rausser and K. R. Farrell, eds., *Alternative Agricultural and Food Policies and the 1985 Farm Bill* (San Leandro, Calif., Blaco Printers).

Offut, S., and R. Shoemaker. 1988. *Farm Programs Slow Technology-Induced Decline in Land's Importance*. USDA/ERS Technical Bulletin no. 1745 (Washington, D.C., U.S. Department of Agriculture, Economic Research Service, May).

Paarlberg, D. 1980. *Farm and Food Policy: Issues of the Eighties* (Lincoln, University of Nebraska Press).

Seeley, R. 1985. *Price Elasticities from the IIASA World Agriculture Model*. Economic Research Service Staff Report no. AGES 850418 (Washington, D.C., U.S. Department of Agriculture).

U.S. Department of Agriculture, Economic Research Service. 1987a. *Agricultural Statistics* (Washington, D.C., Government Printing Office).

———. 1987b. *Government Intervention in Agriculture: Measurement, Evaluation, and Implications for Trade Negotiations*. Staff Report no. AGES 861216 (Washington, D.C.).

———. 1989. "Sugar and Sweetener Situation and Outlook Report" (Washington, D.C., January).

PART 3

THE DYNAMIC FORCES SHAPING THE 1990 U.S. AGRICULTURAL LEGISLATION

STABILIZATION IN AGRICULTURE
An Uncertain Quest

JAMES P. HOUCK

Normality is a fiction of economic textbooks.

—Joan Robinson[1]

The quest for stability pervades the food and agricultural policies of every nation on this planet. Every national or international organization claiming even the remotest link to food and agriculture wishes to secure and promote stability. Little agreement exists on what to stabilize or how to stabilize it, but the desire for stability is ubiquitous. It finds expression in the language of policy pronouncements, in legislation, and in program mechanics affecting production, distribution, storage, consumption, and pricing of food and fiber products.

The quest for stability is perhaps the oldest, most venerable goal of agricultural and food policy. The Chinese economic historian Hu Jichuang (1988) quotes the ancient statesman Fan Li as follows:

> The selling price of grain will cause harm to the farmers if it falls to twenty and cause harm to traders if it rises to ninety. Grain prices fluctuating between thirty and eighty will benefit both farming and trade. To take measures for the stabilization of grain prices so as in turn to stabilize the prices of all goods and make marketplaces free from commodity shortage is the proper way of ruling a state. (p. 38)

This passage was written 2,500 years ago! Yet, it is astonishingly similar to price stabilization proposals being put forward today.

The central objective of this chapter is to take a nontechnical look at the economics of stabilization policy for food and agricultural markets. The problems and mechanics of market stabilization are described and analyzed. The main emphasis is on the U.S. farm and food economy, including its crucial international linkages. The scope of this chapter includes stabilization schemes and activities directly aimed at the prices, quantities, and returns

[1]Jakman, Michael (ed.), 1984, *The Macmillan Book of Economic and Business Quotations* (New York, Macmillan Publishing Co.), p. 23.

from specific commodities or product groups. Indirect or macroeconomic stabilizing efforts, such as progressive income taxation or countercyclical fiscal policy are not discussed. The philosophical tone of this discussion is essentially neutral: stabilization as a specific policy goal pursued actively by government is neither advocated nor attacked.

A BRIEF LOOK AT AGRICULTURAL INSTABILITY

It is beyond the scope of this chapter to present a great deal of empirical evidence showing the extent or severity of fluctuations in agricultural or food markets. This section does, however, provide some information that may be useful in connection with later arguments and assertions. For example, the graphs in figures 1 and 2 show the extent of instability in three important agricultural price series and in U.S. farm income.

The data in these graphs are percent changes from year to year. They show that episodes of instability tend to break out from time to time and then dissipate. More price instability seems to have occurred since the 1970s than the late 1950s to around 1970, although it is hard to argue that it has increased greatly since the chaotic changes of the 1970s subsided. Notice the very unstable behavior of net farm income in the United States in figure 2 (this series is net of direct government payments). Here, it seems reasonable to conclude that instability in net farm income has increased even though U.S. gross farm income does not appear to be particularly unstable.

Table 1 gives instability indexes for world prices of primary agricultural products, calculated by the World Bank over two historical periods. Indexes for wheat, corn, and soybeans appear around the middle of this listing, so one can infer something about what the instability graphs of other primary products might look like.

Figures 3 through 8 provide a long-run perspective on the nature of price and production instability from year to year for U.S. wheat, corn, and soybeans. These graphs, based on data from the U.S. Department of Agriculture (USDA), reinforce the notion that price and, possibly, production instability have increased since the early 1970s, at least in comparison to the immediately preceding period.

THE STABILITY "PROBLEM"

The economic problems in agricultural and primary commodity markets around the world can be divided into two broad categories. The first is the long-run tendency for these markets to display falling or stagnant relative prices and earnings when compared to other major sectors like industry and services. The second is the presence of chronic and excessive fluctuation in prices and returns from period to period. This chapter focuses on the chronic instability in commodity markets.

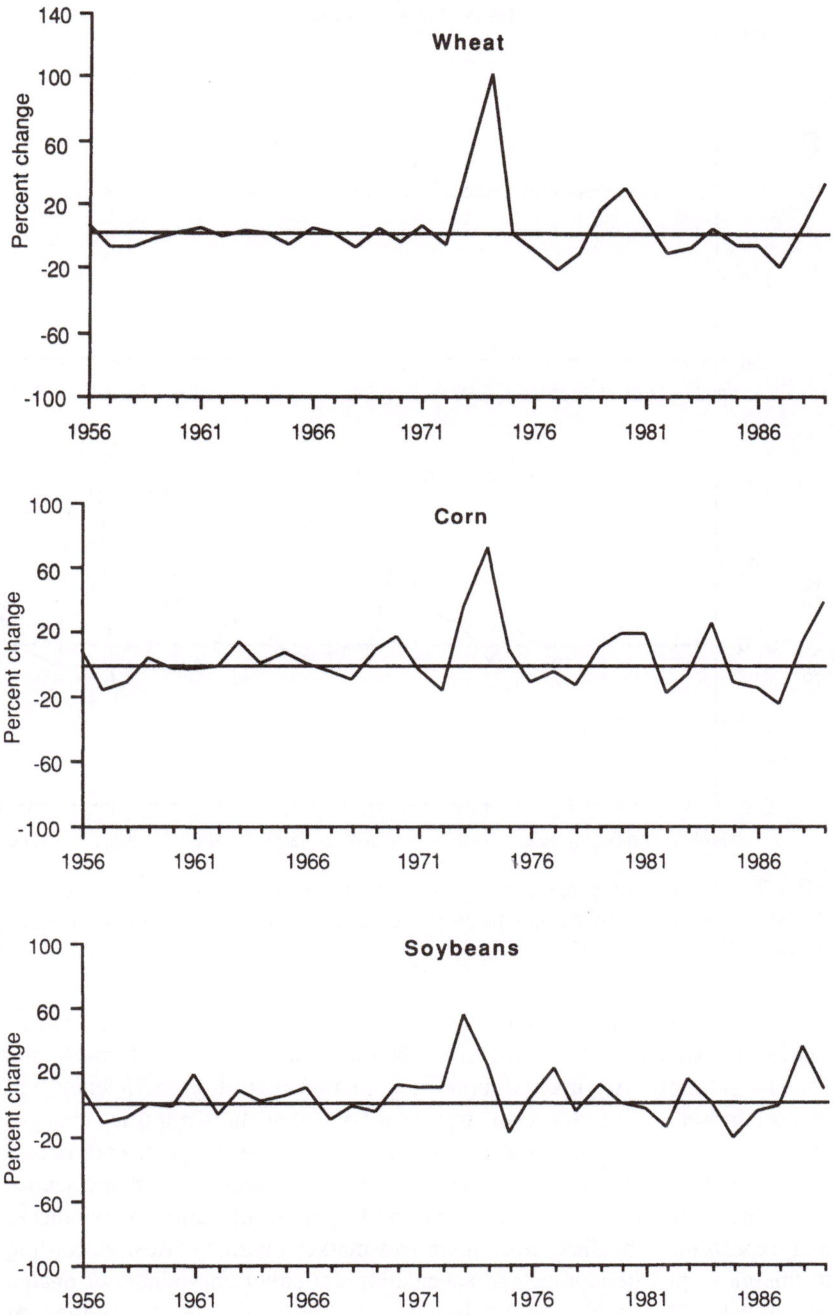

FIGURE 1. Annual percent change in world prices for wheat, corn, and soybeans, 1956–1989. *Sources:* Data from U.N. Food and Agriculture Organization (various issues) and USDA, Foreign Agricultural Service (various issues).

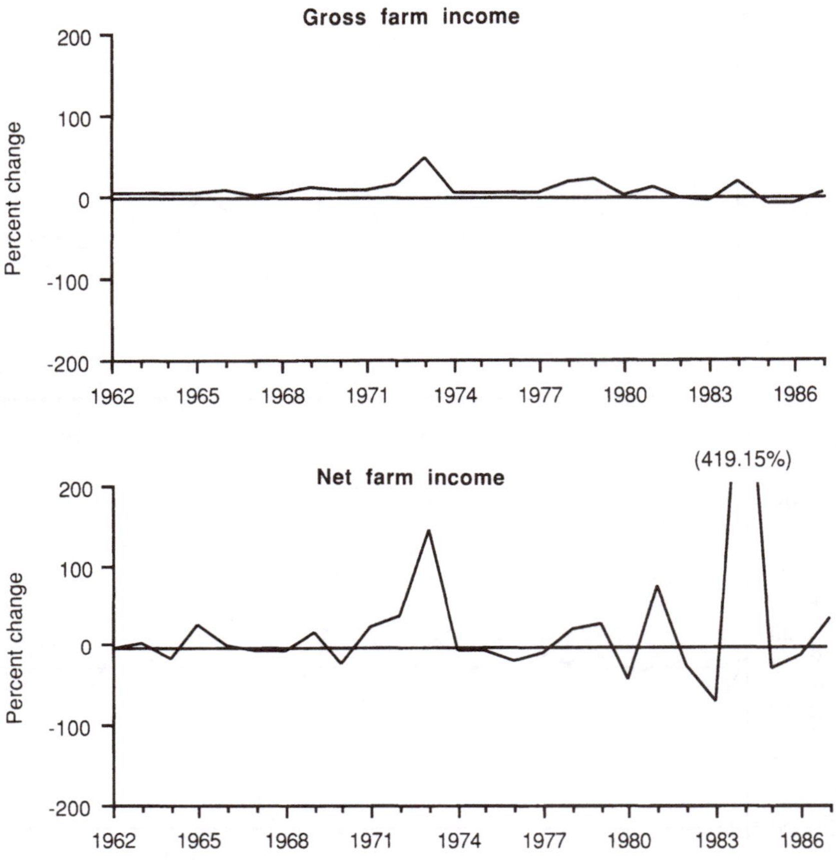

FIGURE 2. Annual percent change in U.S. farm income, 1962–1987. *Source:* Author's calculations based on data from records of the USDA, Economic Research Service, Washington, D.C.

There is widespread, but by no means full, agreement that agricultural market instability is excessive and, therefore, a problem to farmers and society at large. Arguments supporting this view rest on efficiency and welfare grounds. The efficiency argument holds that short-run price changes in many farm commodity markets are larger and more frequent than necessary to guide the efficient allocation of productive resources from one season to the next and to signal efficient period-to-period allocation of purchases and expenditures by final consumers and market intermediaries. According to this view, private storage and speculation are either unavailable or unable to mitigate undesirable market fluctuations. Instability in the volume of products moving through the system is also seen as creating inefficiencies in the use of transportation, storage, and processing facilities. Another important efficiency argument is that because instability increases the riskiness of agricultural production, cost-reducing specialization is partially discouraged, making total output lower at all price levels than it otherwise would be.

TABLE 1. International Price Instability Indexes, 1964–1984 and 1974–1984

	Price instability index	
Commodity	1964–1984	1974–1984
Sugar	90.8	51.5
Cocoa	37.3	34.1
Rice	33.0	21.9
Coffee	32.0	37.7
Palm kernels	27.5	32.5
Wheat	24.3	16.9
Tea	21.7	23.6
Jute	21.2	26.8
Soybeans	20.8	9.9
Beef	16.7	11.3
Corn	16.6	15.6
Rubber	16.1	14.0
Sorghum	15.6	13.6
Cotton	14.3	10.7

Note: Computed as the standard deviation of deflated year-to-year price differences relative to exponential trend price values and expressed as the percent of the trend value.

Source: World Bank, 1986, *World Development Report: 1986,* Washington, D.C., p. 86. Reprinted with permission.

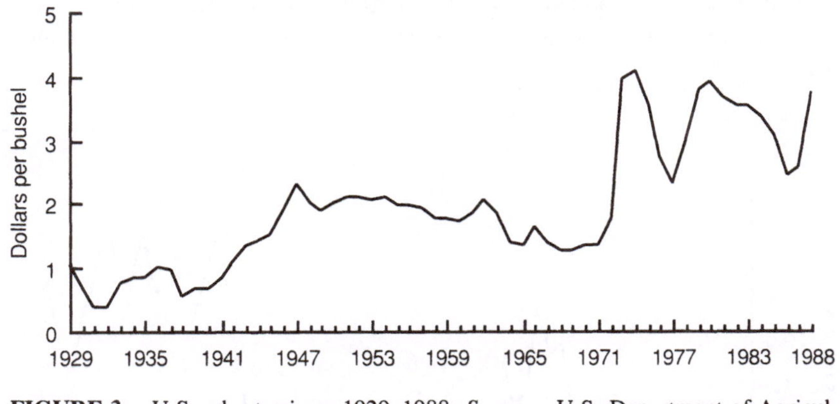

FIGURE 3. U.S. wheat prices, 1929–1988. *Sources:* U.S. Department of Agriculture (various years); USDA, Agricultural Stabilization and Conservation Service (1989).

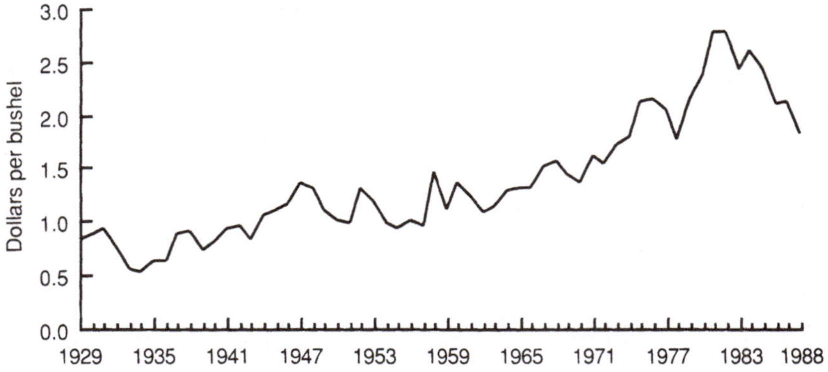

FIGURE 4. U.S. wheat production, 1929–1988. *Sources:* U.S. Department of Agriculture (various years); USDA, Agricultural Stabilization and Conservation Service (1989).

The welfare argument suggests that income uncertainty for farmers and rural economies from period to period poses several problems, particularly in specialized production areas. These include excessive difficulty in planning long-run investments and meeting regular debt repayment and other cash obligations. Widely fluctuating incomes may undermine the rational planning of household living expenses and may substantially increase the cash and asset base required for viable enterprises to exist. Hence, the overall size of existing farms and ranches may be larger than would otherwise be the case.

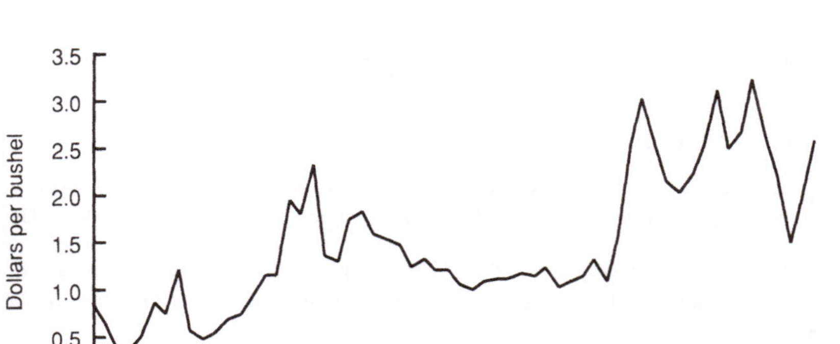

FIGURE 5. U.S. corn prices, 1929–1988. *Sources*: U.S. Department of Agriculture (various years); USDA, Agricultural Stabilization and Conservation Service (1989).

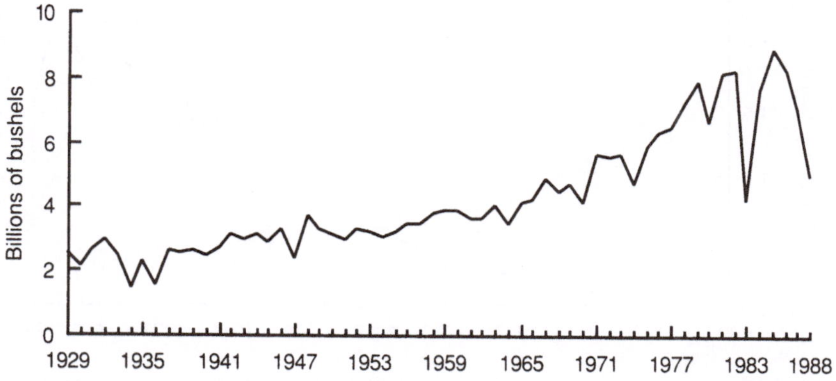

FIGURE 6. U.S. corn production, 1929–1988. *Sources:* U.S. Department of Agriculture (various years); USDA, Agricultural Stabilization and Conservation Service (1989).

Commodity market instability may induce fluctuations in income in the input and service industries linked to affected markets. Price and market instability may also create problems for consumers and other commodity users, including difficulties in planning, budgeting, storage, and finance.

The welfare consequences of market instability at the extremes should also be underscored. When food prices leap upward, low-income people in both rich and poor nations face rapid deterioration in real income and purchasing power. Many could face hunger or even starvation. When farm prices tumble dramatically, large and small farmers alike face serious income and cash

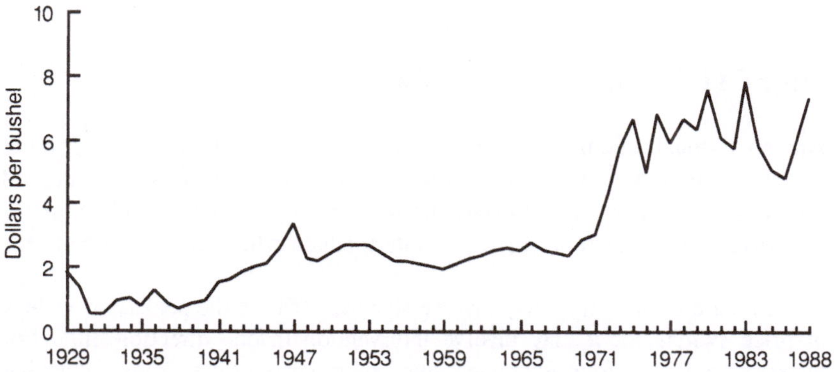

FIGURE 7. U.S. soybean prices, 1929–1988. *Sources*: U.S. Department of Agriculture (various years); USDA, Agricultural Stabilization and Conservation Service (1989).

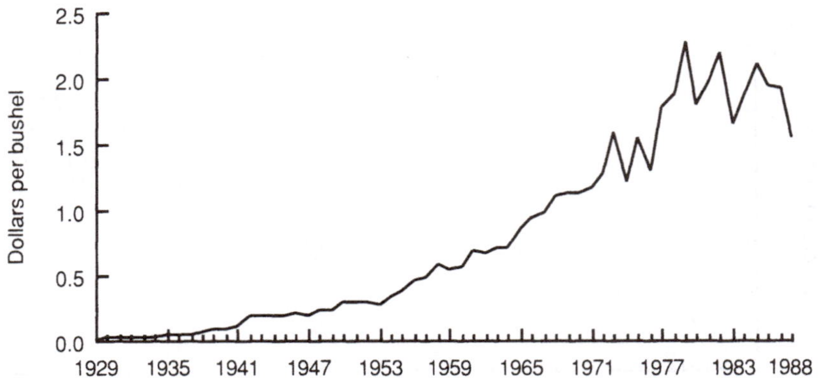

FIGURE 8. U.S. soybean production, 1929–1988. *Sources*: U.S. Department of Agriculture (various years); USDA, Agricultural Stabilization and Conservation Service (1989).

flow damage. Many otherwise well-managed and efficiently sized farms may be quickly forced into bankruptcy.

This nontechnical statement of the stability "problem" summarizes much of the conventional wisdom about food and agricultural instability that guides and shapes policies in virtually every nation on the globe. Each of the points that have been covered has long been subject to debate and disagreement on political, historical, and economic grounds. Yet, there is no doubt that a strong, generalized wish for agricultural stability exists almost everywhere. Stability is equated with normality and rationality. Instability implies confusion, even chaos. Each and every major piece of food and agricultural legislation ever passed by Congress has somehow identified stability as a specific objective. It is a fair bet that the same is true of virtually all similar laws, edicts, and directives elsewhere in the world.

OBJECTIVES OF STABILIZATION

Any operational scheme to promote stability must naturally alter one or more things that are otherwise unstable. First and foremost in agricultural and food policy is price. Here the stabilization objective might be farm-level commodity prices, consumer food prices, or intermediate wholesale or international prices.

Deliberate stabilization may involve shaving off both the peaks and valleys of price swings, or it may involve intervention in one direction only. For instance, a price floor guarantee may keep farm prices from collapsing without exerting control over upswings. On the other hand, administrative price controls might be used to keep food prices from skyrocketing, but have no ability to prevent a dramatic drop.

Another objective might be stabilization of physical quantities of a product that are subject to unwanted fluctuation from period to period, such as farm marketings, exports, imports, processing throughput, and final consumption. Stabilization of actual volumes implies either physical manipulation of inventories by the central authority (such as the U.S. Department of Agriculture) or administrative controls such as import or marketing quotas. Sometimes the core of an instability problem is seen as wide fluctuation in returns (price times sales) or in expenditures (price times purchases). In these cases, the stabilization objective becomes the value (price times quantity) itself rather than either component of it. Deliberate action to steady the flow of income or costs over time becomes the focus of such stabilization schemes.

As discussed in the section entitled "Unavoidable Trade-offs in Stabilization," these various stabilization objectives may be incompatible with each other in real-world contexts. Stabilizing price or quantity may destabilize income. Stabilizing income could cause additional fluctuations in prices. Stabilizing domestic marketings may destabilize exports or imports and vice versa. Trade-offs are unavoidable; their nature hinges mainly on the underlying source of any given episode of instability and the price elasticities of relevant demand and supply functions, which include the behavior of any private inventory holders.

THE FORMAL RATIONALE FOR STABILIZATION

The question of whether or not deliberate stabilization is beneficial to market participants and society as a whole has intrigued economists for generations. Formal theoretical inquiries date back at least to the 1930s. A flood of theoretical and empirical research on this topic engulfed the economics profession in general and agricultural economics in particular during the turbulent 1970s. A steady torrent of work, mainly within the regime of neoclassical welfare economics, continues to appear in professional journals and monographs. A brief list of suggested readings appears at the end of this chapter.

As this formal work continues to grow in scope, generality, and sophistication, fewer concrete assertions about either the advisability or folly of stabilization as a social objective can be made, at least from the viewpoint of rigorous economic theory. Conclusions are almost always contingent on underlying assumptions about the basic demand and supply structure, private inventory behavior, and how destabilizing disturbances come to pass. Most available theoretical and empirical analyses focus on the effects of price stabilization as the central policy goal.

Despite the tenuous hold on these questions that formal economic inquiry has produced, a few propositions about price stability and stabilization can be defended. These can be divided into two somewhat distinct categories—risk reduction gains and potential distributive gains.

Risk reduction ideas hinge on the notion or assumption that most producers are risk-averse with respect to their main business enterprise. That is, other things being equal, they prefer less business risk to more. For such producers, a reduction in price risk from period to period by some stabilization scheme will induce more input use and, hence, more output on the average at all price levels than would otherwise be the case. In a more stable environment, input demand and output supply functions will increase across the board as individual enterprises produce more and as more enterprises remain in business.

These changes on the supply side will cause average output prices to fall because more will be produced and sold at all supply prices. Consumers will gain in welfare from these lower prices. Some producers may gain and some may lose in terms of net returns to their labor and fixed capital. However, strict welfare valuations of economic theory show that society as a whole unambiguously gains from this adjustment, because the gainers could fully recompense the losers and still have some gains left over.

Now consider potential distributive effects. These are solely the result of stabilization of market prices around an average or trend value through intervention by the central authority. Some important theoretical results emerge if demand and supply relations are approximately linear within the relevant range and if destabilizing disturbances cause these relations to shift in parallel fashion. The specific results can be summarized as follows:

1. If the instability arises on the supply (production) side, producers gain and consumers lose from price stabilization.
2. If the instability arises on the demand (consumption) side, consumers gain and producers lose from price stabilization.
3. In either case, the gainers could recompense the losers and still have some gains left over.

These propositions and others in the same vein dealing with stabilized volumes give some theoretical underpinning to the yearning for agricultural market stability that is evident, especially among concerned but impartial observers and policymakers. Yet, each of these ideas is subject to criticism in the professional economics literature. One important criticism is that, in some instances, stabilizing prices can lead to the destabilization of aggregate producer income or consumer expenditures, a source of welfare loss not directly accounted for. Another important criticism is that a stabilization scheme in one market, say the domestic corn market, might lead to instability in another market, say, the international corn market or the market for a substitute product such as sorghum. Some critics also question the evidence about risk aversion as a characteristic of farmers.

INSTABILITY VERSUS LONG-RUN TRENDS

With policies and schemes designed to generate "pure," although perhaps not complete, stabilization, long-run trends in the demand for and supply of

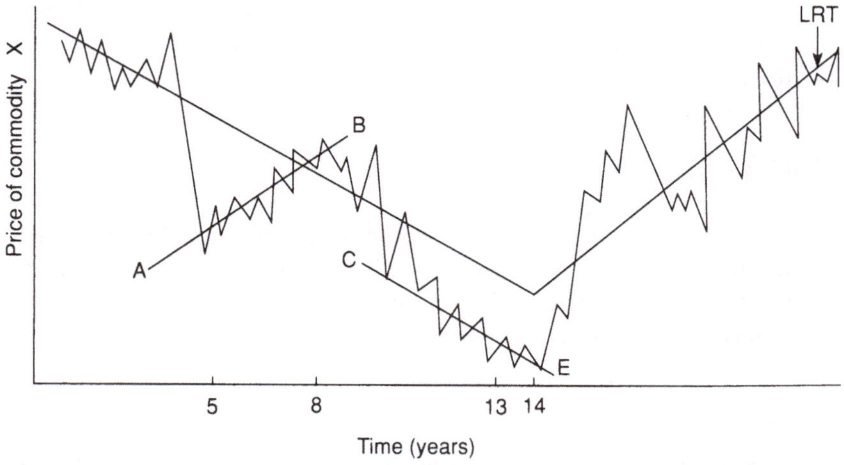

FIGURE 9. Short-term and long-term trends in the prices of hypothetical commodity X.

the products in question are allowed to function. The products to be stabilized are not subject to either net taxation or net subsidization over the long run, at least not by the stabilization authority. Pure stabilization is easy to describe but difficult to accomplish. Even apart from the politics of establishing and maintaining a scheme whose only purpose is stabilization, operating such a scheme with complete success from day to day or year to year is virtually impossible. Also, the very existence of an active stabilization program might alter the long-run trends themselves, by changing the total investment and technological structure of the sector in question.

Consider figure 9, which is a chart of prices for a hypothetical commodity X, plotted over time. Imagine that *we* know that the line LRT is the true long-run trend of these prices around which true stabilization should occur, but that the managers of the stabilization scheme do not possess this knowledge. As they experience each month and year, they do not know that market prices will trend down until years 13 and 14 and then begin a long upward sweep. So, in order to operate the scheme, they must somehow guess from period to period what the LRT really is. During the period from year 5 to year 8, for example, it might be very difficult to realize that the long-run downtrend was still in force. Steps might be taken, mistakenly, to stabilize around the dashed line AB.

Similarly, it would be difficult for the managers to know during years 13 and 14 that the long-run trend was shifting upward and not continuing along CE. The tendency of management would be to continue stabilizing around CE in year 14 and perhaps beyond.

Of course, perfect economic analysis and prediction would preclude such problems, but analytical perfection is itself rare, and agreement that perfection has been achieved is even rarer. Hence, stabilization programs always encounter operational difficulties when departures from past experiences actually occur or seem to be occurring. These problems can be minimized with good economic analysis, flexibility of operations, and sufficient resources, all of which are scarce.

CAUSES OF INSTABILITY

Why are agricultural and primary commodity markets unstable? One way to approach this question is to consider first the general preconditions for instability that such markets display and, second, the inherent forces that frequently set off episodes of instability.

Preconditions for Instability

The fundamental preconditions for instability in agricultural markets have to do with the severe price inelasticity of both demand and supply in the short run. In this context, inelasticity means that large changes in price induce only small changes in the quantities demanded or supplied. Agricultural product markets are price inelastic in the short run because of one or more of the following generic characteristics:

1. Farm price is usually a relatively small share of the final product price.
2. Few good substitutes exist in the short run for the basic commodity to be used in the final product.
3. The final products themselves are rather price and income inelastic.
4. The products are necessities.

Each of these conditions makes it difficult or unprofitable for users to alter the amounts demanded in response to price change. The more of these conditions that apply to any particular product, the more price inelastic will be the demand. As time passes, however, these demand functions will become less inelastic.

With inelastic short-run supply functions, large changes in current prices induce only small changes in amounts supplied for sale. Many agricultural short-run supply functions are severely price inelastic because of one or more of the following conditions:

1. Biological production requirements for both crops and livestock create a time lag between the decision to produce and the actual production. This relation is most pronounced in perennial crops such as coffee and apples.
2. Fixed production costs make up a large share of total costs.
3. The raw or semiprocessed products are perishable, or storable only at high cost relative to price.

Each of these conditions makes it difficult for producers to alter the amounts they offer for sale in the short run in response to market price signals. The more of these conditions that are reflected in any given commodity, the more price inelastic will be the supply function. However, these supply functions will become less inelastic over time.

Inherent Forces Setting Off Instability Episodes

The existence of highly inelastic demand and supply relations does not create instability by itself. More is needed. In particular, abrupt change must occur in the location of either demand or supply so that any uneasy equilibrium that might exist is upset. When supply shifts against a very inelastic demand function, or when demand shifts against an inelastic supply, prices and earnings gyrate in the same direction and short-term instability is present. In agricultural markets, both demand and supply are notoriously subject to sudden changes.

Supply functions can shift back and forth for some or all of the following reasons:

1. Changes in the environmental conditions of production, such as drought and flood, heat and cold, insect infestations, plant and animal diseases, and other phenomena of nature;
2. Natural production cycles;
3. Political instability or war;
4. Abrupt policy or administrative changes by governments concerning marketing, trade, or storage.

Demand functions typically do not change as quickly or dramatically as short-run supply relations, but they too are subject to destabilizing swings. Among the reasons for this are—

1. Rapid changes in incomes across an economy;
2. Violent changes in prices of substitutes, if these exist;
3. Political instability or war;
4. Dramatic changes in consumer behavior caused by reaction to publicity about an individual product's hazards or benefits;
5. Abrupt policy or administrative changes by governments concerning procurement, trade, or storage.

PATTERNS OF INSTABILITY

As a supply function shifts against an inelastic demand, or as a demand function shifts against an inelastic supply, price and earnings instability episodes are created. The more inelastic the relationships, or the more severe the shift, the more violent is the ensuing episode. If simultaneous supply and demand shifts occur, the resulting instability, up or down, can be even more extreme than otherwise. Some examples follow.

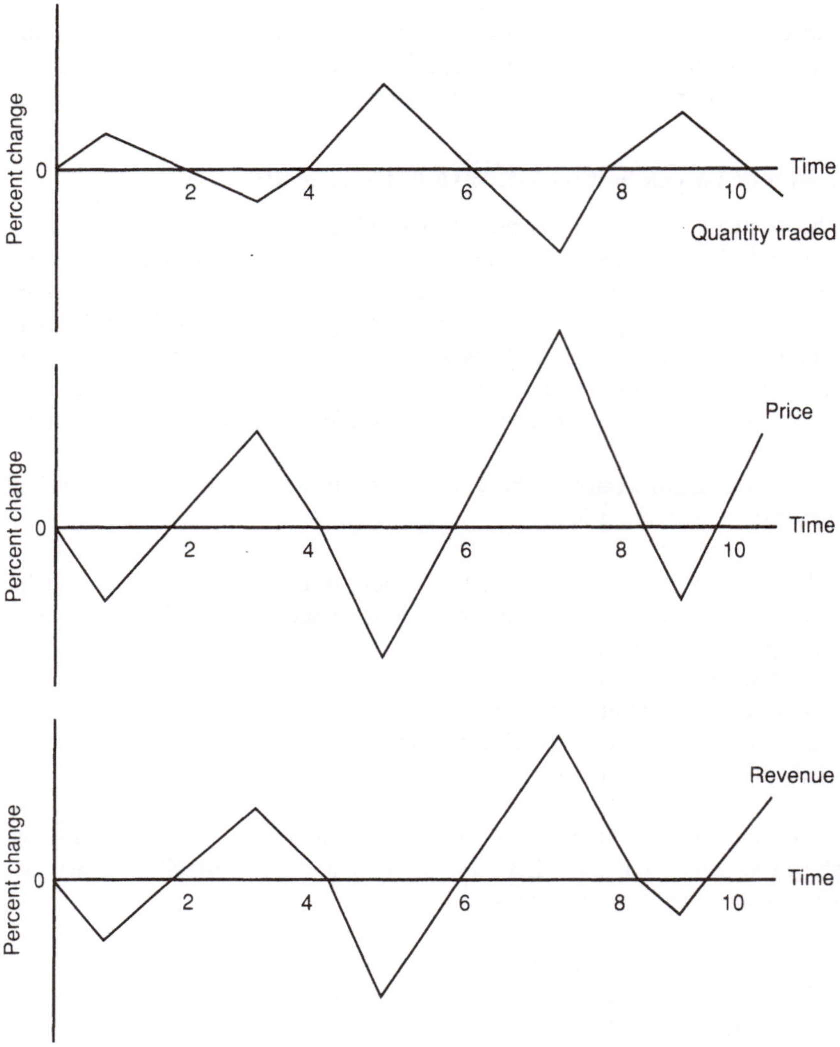

FIGURE 10. Time series when supply shifts against inelastic demand function.

Supply Shifting Against Stable Demand

The shifting of supply against stable demand is a very common situation in agricultural and other primary commodity markets—short-run swings in supply interact with a reasonably steady but inelastic demand. Short supplies, caused perhaps by a partial crop failure, send prices and earnings skyrocketing upward. Heavy supplies, caused perhaps by a bumper crop, send prices and earnings plunging downward. Figure 10 suggests how, in this particular hypothetical case, small changes in quantity traded are inversely related to severe price fluctuations and less extreme revenue or expenditures behavior.

As quantity fluctuates from period to period, price fluctuates in the opposite direction and with more amplitude. In addition, revenue follows price in its period-to-period behavior but less violently. Notice that in periods 4–8, the proportional changes in quantity are shown to be twice as large as in the periods before or after. These larger shifts are associated with even larger, but opposite, changes in price and revenue.

The regular changes shown in this diagram are, of course, hypothetical. Any real situation would show irregular patterns of change over time, but if quantity and price are inversely related, with revenue tending to follow price, it can be concluded that the instability being observed is the result of an unstable supply shifting against a relatively stable demand.

Demand Shifting Against Stable Supply

The pattern of demand shifting against stable supply is less common in agricultural markets, but when it does appear, extreme short-run instability can occur. Price swings caused by outbreaks of war or economic depression are notable examples. A strong surge in demand against a steady, inelastic short-run supply will cause prices and earnings to move quickly upward. A collapse in demand in the presence of steady supplies will create a corresponding collapse in prices and earnings.

Figure 11 illustrates that, for this case, period-to-period changes in quantities traded are directly related to large changes in prices and even larger changes in revenues. This diagram is also hypothetical, since changes over time are not likely to be so regular, but when quantities, prices, and revenues all tend to fluctuate in the same direction from period to period, the underlying cause is an unstable demand shifting against a relatively stable supply. Also notice that between periods 4 and 8, the quantity is shown to double in its proportional fluctuation. This in turn is associated with even larger fluctuations in the same direction in prices and revenues, the consequence of an inelastic supply function.

Simultaneous Shifts

Simultaneous shifts in demand and supply can create greater instability than would occur with movement in only one or the other. But it is also possible that such simultaneous shifts can be stabilizing. Imagine first that the supply falls (the curve shifts to the left). If demand increases at the same time, then the jump in prices and earnings will be even more acute than if only supply had fallen. However, if demand falls as supply decreases, prices and earnings will fluctuate less than otherwise. Similar reasoning can be used with a hypothetical increase in supply.

Now imagine that the demand for a commodity decreases. If the short-run supply increases at the same time, prices and earnings will fall even more than otherwise. But if instead supply decreases, the potential fall in prices and earnings will be mitigated and perhaps offset altogether. Similar reason-

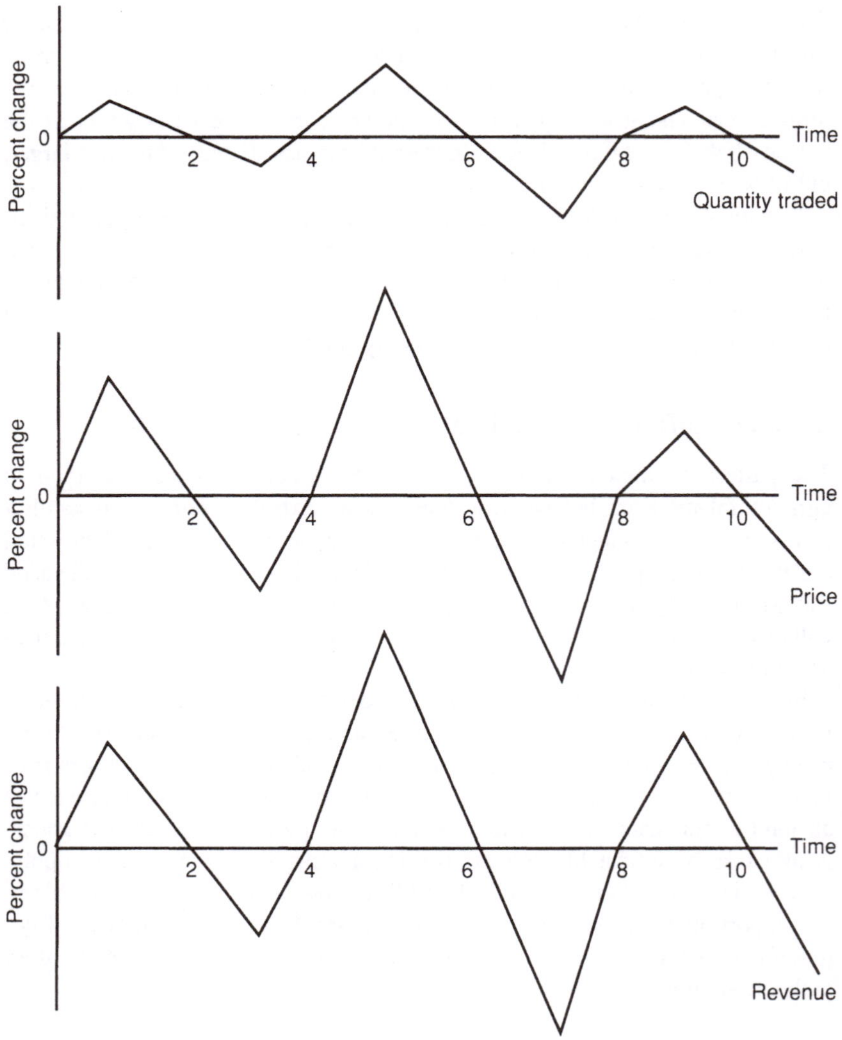

FIGURE 11. Time series when demand shifts against inelastic supply function.

ing can be applied to a hypothetical increase in demand. A graph of time series data with no apparent systematic pattern among quantities, prices, and revenues likely is reflecting simultaneous shifts in both demand and supply.

UNAVOIDABLE TRADE-OFFS IN STABILIZATION

The central authority can intervene to stabilize some aspects of a market, but typically it has neither the ability nor the resources to stabilize everything. Some things may become more unstable than before. Much depends on the underlying cause for any given episode of instability and the policy tools available to the authority.

Trade-offs with Price Stabilization

Most real-world stabilization schemes focus on price stability. The major objective is to keep prices in one or more central markets from moving outside some previously defined zone, which may be wide or narrow. The wider the zone, the less aggressive will be the stabilization scheme. Successful price stabilization within any zone may destabilize producer sales volume, market volume, earnings of producers, or consumption expenditures. A few examples will illustrate the main trade-offs. Notice that when prices are stabilized, the variables that become either more or less stable can switch places, depending on the instability source and the stabilizing device used.

Buffer Stocks. If price stability is achieved by purchases and sales of actual commodity stocks by the authorities in the face of shifting supply and a relatively constant inelastic demand, then sales quantities by producers are likely to be destabilized to some extent, but total volume flowing through the market and buyer expenditures will be stabilized. Producers' earnings in total may or may not be more stable than otherwise, but the pattern of instability will be reversed from that in figure 10—earnings will now follow output rather than price.

If the instability is caused by a shifting inelastic demand against a relatively steady supply, then total market volume will be destabilized as price stability is pursued with a buffer stock. However, producer earnings (and buyer expenditures) will tend to be stabilized, as will producer sales volumes.

Buffer Funds. If price stability to farmers is achieved by a neutral system of per-unit taxes or subsidies applied explicitly by a fiscal agency of government or implicitly through an institution like a marketing board, then somewhat different trade-offs will arise. With a steady inelastic demand and fluctuating supply, market volumes and producer sales will tend to be destabilized by a buffer fund. Producer receipts after taxes or subsidies may or may not be more stable than otherwise, but the pattern of instability will be reversed—earnings will rise and fall with output rather than with price.

A buffer fund for price stability operated with a steady supply and an unstable demand will tend to produce more stability than otherwise in market volume, producer sales volume, and producer earnings after taxes and subsidies. In general, buffer funding devices used for farm price stability also tend to generate wider swings than otherwise in wholesale and retail prices and expenditures, unless the board also manages a buffer stocking scheme to offset the underlying supply or demand fluctuations.

Administrative Controls on Quantities. An interesting and widespread form of domestic price-stabilizing intervention is regulation by the central authority of imports or exports so as to stabilize the internal price of an agricultural product. Import quotas, export controls, import tariffs and subsidies, and export taxes and subsidies are the primary means of controlling supplies on the domestic market. These measures may also be used to increase or decrease the long-run average level of domestic prices in relation to international prices.

When the link between unstable international prices and the domestic market is weakened or broken by systematic intervention, then the domestic market price usually is more stable than otherwise. The results across the internal market are much like those of a buffer stocking scheme, as described above. In addition, this kind of domestic stabilization policy usually creates instability in the volume of imports or exports and generates further instability in world prices if the trading nation is large in relation to the international market.

The most effective (and most notorious) international commodity trade policy to use when stability is a central objective involves variable levies on imports and variable subsidies on exports. These are simply calculated and applied around previously determined target prices.

Trade-offs with Stabilized Volumes

A stabilizing authority could focus on ironing out wide fluctuations in total volume flowing through the marketing or processing system. This is a plausible objective for an essential food staple in a low-income nation. It might also be sensible when volume fluctuations create substantial technical inefficiencies in transportation and processing. External additions to or subtractions from the market flow would involve buffer stock operations or centrally controlled imports and exports. When market volumes are directly stabilized, highly variable results emerge that depend on the basic source of the instability.

If the source is a fluctuating supply, then stabilization of the total volume in the system will also stabilize market prices and consumer expenditures. However, producer sales will tend to be more unstable than otherwise. Producer earnings may be more or less stable than before, but the pattern will be reversed, following supply fluctuations rather than price fluctuations.

If the destabilizing force is variable demand, then stabilization of the total market volume with appropriate additions and subtractions will tend to destabilize virtually everything else, including prices, producer sales and earnings, and consumer expenditures.

Trade-offs with Stabilized Income

A buffer stocking authority could also seek to stabilize the total income flow to producers, rather than either prices or quantities. The management of such a scheme through the marketplace is more complex than price or quantity stabilization alone, but it is possible if the authority can reasonably predict the amount supplied by producers and private inventory holders in advance of disposal or acquisition of buffer stocks.

If the instability in total income is created by supply fluctuating against a relatively constant inelastic demand, then stabilizing the income flow with a buffer stock will promote some stability in both prices and total market volume. If there is any price elasticity in the short-run supply function, then producer sales volume will be destabilized.

If the income instability is caused by a fluctuating demand function about a relatively constant supply, then, as income is stabilized with a buffer stock, prices and producer sales volume will also be more stable than otherwise. However, the total volume in the system will be more unstable because of the buffer stocking agency's activity.

Trade-offs in General

The trade-offs given in the previous subsections were described for specific sources of instability. In any real situation, the sources of instability may change over time for any given commodity, and simultaneous destabilizing shifts in demand and supply can surely occur. Therefore, the potential trade-offs between items stabilized and destabilized can change from period to period. For example, if wheat price stability is achieved over some time span, the package of related items—such as wheat marketings, farm income from wheat, or expenditures for wheat—that is simultaneously stabilized and the package that is destabilized can change markedly.

POLITICAL AND OPERATIONAL ASPECTS

There are a number of political and operational considerations that are likely to affect any potential stabilization effort, no matter what the commodity, country, or international organization involved, and no matter what mechanism is used. A few are presented here as brief assertions, even though pages could be written on each.

1. Resources are required for success. Income, expenditures, or prices at one level of marketing can be stabilized with money only, but to stabilize an entire market both money and commodity stocks are needed. Good and timely information is very important, especially for production, utilization, trade, and private stocks in all positions.
2. The narrower the zone within which the market is allowed to operate freely, the larger will be the financial and other resources needed for stabilization.
3. The political temptation to tinker with a stabilization system from year to year is always intense. Much of this pressure exists because the distinction between pure stabilization and support (or protection) is seldom clearly defined or understood. In practice, there is likely to be a tendency for

stabilization schemes to drift toward producer support. In some contexts, a large, momentarily idle balance of public funds in a stabilization account may be irresistibly tempting to politicians to use for other purposes.

4. It is virtually impossible for any agency to know the correct current level around which to stabilize. Only in retrospect can that be known. So, ongoing mistakes are unavoidable. The best that can be done to minimize errors is to provide the managers of the agency with as good an economic analysis as possible. In addition, it is also important to provide the stabilizing agency with sufficient financial resources and flexibility of operation to prevent episodes of uncertainty about long-run trends from paralyzing its work.

5. In most instances, it is essential to keep the stabilizing agency and the private market separate. Of course, the objectives of the agency should be clearly and openly stated and, naturally, the market will sooner or later discern what the agency is doing to achieve stability within any period. But the specific moves to be made in the short run should be either automatic, as with a fixed and known formula, or conducted without prior information being leaked from the agency to the trade. The less time that elapses between decision making and implementation the better.

STABILIZATION IN AN INTERNATIONAL SETTING

Much of the previous discussion is based on the notion of a would-be stabilizing agency facing a closed market. Yet, the reality of international markets and trade adds a dimension that profoundly alters what is possible and feasible. For an individual nation wishing to stabilize some aspects of its own internal market, the international nexus can make that task either easier or more difficult. The result depends mainly on the trade policy stance the nation wishes to adopt. Stabilization is easier if the nation is prepared to disconnect its internal markets from the sweep and tumble of international price-making forces. On the other hand, stabilization is vastly more difficult if the nation wishes its internal market to remain integrated with the world market.

Disconnecting

Internal price and market stabilization becomes relatively simple when price formation on the domestic and the international markets is disconnected by policy interventions. Free trade or even partially open markets for imports, exports, or both, however, are not possible. Traded volumes must be restricted to those amounts that will exactly fill the gap between quantities supplied and demanded domestically at the stabilized internal price. World prices for the products in question are largely irrelevant, except as they influence the tariffs to be collected on imports or the subsidies to be paid on exports.

To pursue such an approach for domestic stability (or other goals) is to abandon deliberately an open market, free trade stance in the international

TABLE 2. U.S. Share of World Production and Trade for Selected Commodities, Average of 1987 and 1988
(percent)

	U.S. share of world—	
Commodity	Production	Trade
Wheat	10	33
Coarse grains	25	57
Rice	2	19
Total grains	15	43
Soybeans	48	68
Soy oil	37	19
Soy meal	36	21
Total oilseeds	13	52

Source: U.S. Department of Agriculture records.

policy arena. Trade policy retaliation can be expected from the nation's trading partners, including its customers, suppliers, and competitors. Nations and economic blocs that systematically disconnect internal and external agricultural markets usually do so primarily to facilitate either domestic farm price and income supports or taxation of the agricultural sector. Price stabilization is typically a secondary, but readily achieved, objective. An important side effect when a large trading nation isolates its domestic market from the influences of world prices is that the world prices themselves become more unstable. All internal fluctuations are transmitted fully to the international market, but no world market fluctuations are absorbed by the insulating nation.

Remaining Integrated

Next consider the problem of stabilizing prices or other variables in an open, international market. By choosing not to isolate its internal market from world trade, the potential stabilizer must be in a position to exert influence over the whole world market in some important sense. It must bring sufficient resources of stocks, money, or administrative control over production to bear so as continually to nudge world prices or trade volumes into previously established target ranges. This task would be formidable for any single country in virtually any major farm or food commodity. To be effective, the stabilizing nation would have to achieve a large enough share of world sales so that its decisions about inventory acquisitions and disposals, for instance, would be sufficient to overcome any opposite reactions by other traders acting alone or together. In short, it would have to be able and willing to assert at least some monopoly power on its markets.

The data in table 2 show the recent U.S. share of the world's production and trade for several major agricultural products. Most experts agree that even the United States, acting alone, could not stabilize the world markets for these products without committing resources far beyond the amounts likely to be available for such purposes.

Recognizing the inability of any one nation to impose stability on international commodity markets, proponents of international stabilization have long sought effective, cooperative commodity agreements among producing and consuming nations. Both stability and income-increasing motives have been pursued over the years through such agreements but without much long-term success. Numerous international organizations, principally the Food and Agriculture Organization of the United Nations and the United Nations Conference on Trade and Development, have attempted to foster such agreements by offering advice, analysis, and organizational assistance to commodity groups.

MECHANISMS FOR STABILIZATION

This section contains a brief, selective catalog of the major classes of stabilizing interventions that economists, bureaucrats, political leaders, and others have invented and, for the most part, put into practice in recent times. The central purpose here is not to analyze and critique these schemes, either in general or against the particular circumstances to which they were applied, but to illustrate the broad range of existing stabilizing instruments in food and agricultural markets. The following stabilizing schemes are grouped into two broad categories—those that systematically alter actual market operations toward more stable prices, quantities, or incomes and those that function outside the market to stabilize income, earnings, or other financial flows by means of payments without directly intervening in purchases or sales.

Stabilization Via the Market

When market operations are altered toward stability, the government, acting through some agency of its own creation, intervenes on a regular basis either as a market participant or administrator.

Buffer Stocks. Buying and selling actual stocks to stabilize a market within an agreed zone is the classic mode of intervention. The 1930s concept of an "Ever-Normal Granary" remains a venerable ideal for proponents of stability. The buffer stocking components of various international commodity agreements adopted over the years have reflected this approach. International agreements that officially contain such provisions are those for cocoa, rubber, sugar, and tin. The history of these agreements is not encouraging for many reasons, and none have been successful in mitigating world price swings. Chronic organizational, financial, and operational problems have been endemic.

In the United States, the Farmer-Owned Reserve (FOR) program has operated to some extent as a buffer stock for corn and wheat, but without the kind of central control that would have really fostered price stability. In this program, farmers may designate some of their inventories as part of the FOR when prices are low and the reserve is opened. In return for this grain, producers receive nonrecourse commodity loans and storage payments. They may not sell FOR grain without penalty when prices are below the "trigger" level, but when prices reach the trigger, they may sell as they wish. When prices stay above the trigger for a specified period, the government, acting through the USDA, begins charging for storage and using other methods to bring the grain out of storage and onto the market.

A part of the International Coffee Agreement operates somewhat like a buffer stock, even though the core of that agreement is designed to be a price-enhancing export quota system. The stabilizing part is the so-called Indicator Price Mechanism. When international prices of selected coffees move above a preestablished level, export quotas of member producing nations are expanded slightly, allowing more coffee onto the market. When these indicator prices fall below a preestablished level, export quotas are tightened slightly, reducing supplies authorized to move onto international markets. In this scheme, the authority to control export supplies and sales belongs to the governments of the member nations.

Direct Payments or Buffer Funds. This category of stabilizing schemes involves financial payments for stability, which affects market operations directly. Direct payments are often associated with programs designed to raise prices and incomes, not merely stabilize them. In the context of the U.S. farm program, deficiency payments and marketing loan payments come under this category. They are per-unit payments made to qualifying farmers that have the effect of stabilizing the price received for at least some portion of the historical output. Since these payments are commodity specific and are based on established target or support prices, they tend to affect output decisions directly.

When prices to farmers and ranchers are disconnected from market fluctuations by statutory marketing agencies or boards, they are typically stabilized from period to period by financial payments often called buffer funds. These buffered prices may be higher or lower (usually lower) than an open market would deliver on average, but surely more stable. Marketing boards of this kind are common in many places around the world, especially Africa, Asia, and the centrally planned nations. The long-established wheat boards of Canada and Australia also operate in this vein. Whatever else marketing boards do, they usually add stability to prices received by farmers.

Another way that statutory marketing boards and similar agencies add stability to growers' incomes is by market discrimination. Quantities flowing to price inelastic market segments are controlled and stabilized. These segments typically include domestic consumption of internationally traded grains, milk for fluid use, and fresh fruits and vegetables. The balance of the output is sold for whatever it will bring on the more price elastic markets, which are typically the world market for grains and the processed product markets for milk, fruits, and vegetables. Such market discrimination, if conducted competently, can both raise and stabilize producers' incomes.

Variable Trade Controls. These stabilizing interventions in international trade flows were mentioned earlier in this chapter. Perhaps the preeminent example is the combination of the variable import levies and export subsidies used by the European Community. While holding prices of many agricultural goods above world levels, these interventions also promote price stability internally while aggravating instability abroad. It is probably fair to say that although price stability is a side-effect of Community price support decisions, it is welcome and, to some extent, celebrated. Numerous other countries, including Japan, Switzerland, and the Scandinavian nations, use variable levies and selective trade controls, many of arcane complexity, to enhance domestic price stability in their food and agricultural sectors.

One-Way Stabilization. Schemes that inhibit market price movements in one direction only are very common in agriculture. They are typically the operational features of price support policies. They prevent prices from falling below specified levels, but do little or nothing to keep prices from rising. In the United States, most of the machinery of nonrecourse loans, marketing quotas, commodity purchases, acreage diversion, market development, and the like, is designed to support market prices above levels that might otherwise occur. A side-effect of this effort is one-way price stability.

The other side of this coin is the machinery of administrative price controls that often comes into play during periods of heavy general inflation or rapid price jumps in critical commodities. When price controls are combined with an effective rationing scheme and the suppression of black markets, one-way price stability is achieved.

Stabilization Outside the Market

Aside from crop insurance, which is not discussed here, there are only a few actual examples of interventions put in place that deliberately stabilize important financial flows, mainly total income or earnings. Much has been written about this topic, and many programs have been proposed, but very few have been attempted. Two income-stabilizing efforts—the Canadian Western Grains Stabilization Program and the foreign exchange stabilization schemes operated on behalf of trading nations by the International Monetary Fund and the European Community—are considered here. These schemes are primarily one-way stabilization programs since they seek mainly to offset downward fluctuations in earnings without a corresponding requirement to skim off larger-than-usual incomes in good periods.

Canadian Western Grains Stabilization Program. This program, which has operated since 1976, is designed to stabilize net cash flows to crop producers in the four Canadian Prairie provinces. It covers cash receipts, net of cash costs, for the aggregate of seven major cash crops grown in the area. When aggregate cash receipts for these commodities in the four provinces fall below the average of the previous five years, stabilization payments are made to participating producers in proportion to levies they have paid individually into the stabilizing fund. The payments in total cover most of the net cash shortfall for the seven crops, but not necessarily that of any single farmer.

The program is not a true stabilizing fund for two reasons. First, supporting levies are paid in annually by farmers at the current flat rate of 1.5 percent of eligible crop sales receipts. Second, the Canadian government itself pays in an annual contribution equal to 3.5 percent of eligible crop sales receipts. Consequently, this scheme is a subsidized income insurance program except that individuals are not indemnified against income loss; rather, it is the seven-crop sector at large that is protected.

Foreign Exchange Earnings Protection. There are two schemes currently in force that attempt to protect trading nations from extreme shortfalls in national foreign exchange earnings that arise because of factors beyond the control of the victims. One is the Compensatory Finance Facility (CFF) of the International Monetary Fund. It has been available to fund members since 1963. The other is the European Community's 1975 STABEX scheme, which is available only to sixty-six developing countries, mainly former colonies, in Africa, the Caribbean, and the Pacific. These so-called ACP countries are formally associated with the European Community through the Lomé Convention of 1963.

Both schemes use a moving average computation as the reference point for calculating compensation rights. The CFF program is more of a true stabilization scheme because drawings are officially subject to full repayment within three to five years after a stabilizing loan is made. Foreign exchange drawings by ACP countries under STABEX are subject to much less stringent repayment terms than those of the CFF. For the lowest-income ACP nations, STABEX payments are essentially grants.

SUMMARY AND CONCLUDING COMMENTS

The quest for stabilization in the food and agricultural economy is at the core of policymaking everywhere. Large and unpredictable fluctuations in commodity prices and output create problems for buyers and sellers alike. However, it is probably fair to say that protection against downward fluctuations in agricultural prices and incomes has dominated both the rhetoric and the policy action over the years. Yet, in times of rapidly escalating food and fiber prices, attention quickly turns toward ways to contain these increases. True, unbiased stabilization around long-run trends remains an ideal seldom approached by real-world policies and programs.

Notorious inelasticities of agricultural demand and supply create the potential for instability, which fluctuations in natural and economic conditions then set in motion. Economic theorists are divided in their views about the advantages and disadvantages of stabilization as public policy. Except for the idea that more output is forthcoming in a stabilized environment, very few firm conclusions can be advanced about the social welfare results of stabilization, despite the widespread wish for economic stability that seems to be present worldwide.

A limited number of empirical studies are available on the effects of market stabilization. As with the theoretical analyses of this topic, comparable results and solid conclusions from study to study do not emerge. However, the overall social welfare effects, to the extent that they can be defined and measured, seem rather modest. One reason for this is that the benefits of price or income stabilization are sharply concentrated in periods when wide fluctuations otherwise would have occurred while the costs, which are easier to calculate, are more evenly distributed over the life of the program.

Any program to stabilize one or more aspects of a market will inevitably face trade-offs. Since not everything can be stabilized at once, difficult choices about targets must be made. In addition, stabilizing one thing will almost always destabilize something else. Deciding on the appropriate levels of prices, income, or other variables around which to stabilize is a formidable task, and the temptation is extreme for political leaders and bureaucrats to tinker with schemes once in place.

The fact that many highly unstable markets are international in scope adds complexity to stabilization ideas. For some trade policy regimes, a world market makes domestic stabilization easier; for others, it makes it much more difficult, if not impossible, without international cooperation. To date, international agreements to control instability in selected commodity markets have not been very successful or sustained despite bursts of interest in them over the years by numerous trading nations and commodity organizations.

It seems sensible to predict that interest in stabilization as a policy imperative will continue. Much of this interest will really be for one-way stabilization—producers want protection from extreme downward fluctuations and consumers want protection from extreme upward fluctuations. Perhaps separate, but well-designed schemes to meet the worst of each of these contingencies is the most that can be accomplished in the years just ahead.

ACKNOWLEDGMENT

The author gratefully acknowledges the assistance of Charles Rossman, graduate assistant at the University of Minnesota, in the preparation of this paper.

REFERENCES

Hu Jichuang. 1988. *A Concise History of Chinese Economic Thought* (Beijing, Foreign Languages Press).

United Nations Food and Agriculture Organization. Various issues. *Production Yearbook.* Rome.

U.S. Department of Agriculture. Various years. *Agricultural Statistics* (Washington, D.C., Government Printing Office).

U.S. Department of Agriculture, Agricultural Stabilization and Conservation Service. 1989. Supply and Use Tables. Unpublished mimeograph. July.

U.S. Department of Agriculture, Foreign Agricultural Service. Various issues. *Circular Series.* Washington, D.C.

SUGGESTED READINGS

Bigman, David. 1982. *Coping With Hunger: Toward a System of Food Security and Price Stabilization* (Cambridge, Mass., Ballinger Publishing Co.).

Gardner, Bruce L. 1987. *The Economics of Agricultural Policies* (New York, Macmillan Publishing Co.).

Hazell, Peter R. 1988. "Changing Patterns of Variability in Cereal Prices and Production," pp. 27–54 in *Agricultural Price Policy for Developing Countries,* edited by John W. Mellor and Raisuddin Ahmed (Baltimore, Md., The Johns Hopkins University Press).

Hu Jichuang. 1988. *A Concise History of Chinese Economic Thought* (Beijing, Foreign Languages Press).

Newbery, David M., and Joseph E. Stiglitz. 1981. *The Theory of Commodity Price Stabilization* (Oxford, Clarendon Press).

Robison, Lindon J., and Peter J. Barry. 1987. *The Competitive Firm's Response to Risk* (New York, Macmillan Publishing Co.).

Sumner, Daniel A. (ed.). 1988. *Agricultural Stability and Farm Programs* (Boulder, Colo., Westview Press).

World Bank. 1986. *World Development Report: 1986* (New York, Oxford University Press).

ENVIRONMENTAL PROTECTION AND AGRICULTURAL SUPPORT
Are Trade-offs Necessary?

KATHERINE REICHELDERFER

In theory an efficient, productive agricultural sector is entirely compatible with a clean, safe, and diverse natural environment. And in the minds of many Americans, this basic compatibility is apparently expected. In 1986, in the midst of the recent U.S. farm crisis, a national opinion poll by the Roper Organization found that 84 percent of respondents agreed that "special programs should be devised to insure the survival of the small farmer," while at the same time 56 percent believed that "we are spending too little on improving and protecting the environment" (*Public Opinion*, 1987). The intersection of these groups suggests that at least 40 percent of respondents place high enough values on both the survival of the family farm and the maintenance of environmental quality that they would support government intervention and expenditure to ensure that both objectives are met. One can only wonder how many of them realize that traditional means for supporting farm income tend to exacerbate environmental problems and that many forms of environmental regulation can adversely affect the financial position of small (and large) farmers. For, under current programs, agricultural protection and environmental protection manifest themselves as conflicting societal preferences.

There is strong hesitancy to acknowledge conflicts between agriculture and environmental quality. American farmers have long been viewed as stewards of the land and protectors of the natural environment. And since they would suffer disproportionately from degradation of the soil and water resources upon which they rely as factors of production, it makes sense that farmers would emphasize maintaining resource quality.

Yet, there is evidence which strongly suggests that many agricultural practices do have a negative effect on environmental quality. Cropland is the source of sediment causing annual economic losses in excess of $7 billion from siltation of navigation waterways, water storage facilities, drainage ditches, and irrigation canals, and causing interference with water-based recreational opportunities (Ribaudo, 1986). Nitrogen fertilizer and residues from agricultural pesticides have been widely detected in groundwater, at levels alarming to those who rely on contaminated sources for drinking water

(Batie, 1988). Agricultural runoff contributes one-fourth of the total nutrients and sediment interfering with the productivity of major estuaries, the source of many U.S. commercial marine fisheries (Crutchfield, 1987). Toxic salts have accumulated over a wide area of irrigated cropland (Young and Horner, 1986). Public reaction to these environmental problems suggests that the nonfarm population believes that its right to a clean, safe, and healthy environment may be infringed upon by farming activity.

In addition, the nonfarm community exhibits increasing demand for environmental amenities that challenge farmers' rights to decide how to use their own land. Demand for the preservation of endangered species, conservation of wildlife habitat, and the visual amenities associated with rural farmland is rising as average disposable income rises (Crosson, 1989). Farming can also affect these amenities. For example, conversion of land to agricultural uses was responsible for 87 percent of U.S. wetlands lost between the mid-1950s and the mid-1970s (Tiner, 1984). Increasingly scarce wetlands continue to be converted to agricultural uses.

It is not as if technology is unavailable to prevent agriculturally related environmental problems. A range of structural and crop cultural alternatives are available to restrict soil loss to levels sustainable through natural regeneration. Modern irrigation technology includes systems that dramatically reduce water application rates without interfering with crop development. And a variety of production practices classified as "low input sustainable agriculture" have proved in case studies to offer potential for profitable production with far lower than average use of pesticides and commercial fertilizers.

Why, then, does agriculture continue to generate adverse environmental effects? A partial explanation is that the expressed, long-run environmental goals of the American public (which includes farmers as a group) conflict with the short-run profit objectives of individual farmers. Adoption of many available technologies requires capital investment, high adjustment costs, increased operating costs, or some forfeiture of yield, any of which can reduce a farmer's income, at least in the short run. There also remains, in many states and locales, the legal question of whether farmers' rights to use land in ways that may result in environmental degradation are superior or inferior to the broader public's felt right to enjoy an undegraded environment. Decisions that result, through regulation or liability rules, in changing property rights also lower the income, wealth, or utility of the group that loses a former right.

Thus, Americans face a difficult problem: how to reconcile the apparent public interests in protecting both the environment and farm income. The traditional response has been a concurrent devotion of public funds, on the one hand, to subsidize the development and adoption of environmentally benign production technologies and, on the other hand, to support farm income. Unfortunately, the two approaches have not always worked in concert. In fact, the goals and means of traditional agricultural and environmental policies have often worked at cross-purposes, each offsetting, to some extent, the returns on public investment in the other.

It is not altogether surprising that agricultural and environmental policies are not entirely consistent with one another. They have largely evolved independently. The institutions charged with developing and implementing agricultural policies often are not those responsible for environmental policies affecting agriculture. Most major farm programs are federal, while many environmental programs affecting agriculture are legislated and/or implemented at state and local levels.

But, as we enter the 1990s, the implications of policy inconsistency take on new meaning in the context of the federal budget and farm bill processes. Given the current budget situation, it is unlikely that the funds available for implementation of either agricultural or environmental programs will be significantly increased. Efforts must instead focus on increasing the cost-effectiveness with which the current or a reduced level of funding is applied to meet policy objectives. An increasingly dim view is likely to be taken of expenditures that counteract one another. Thus, policy consistency takes on a new degree of importance, and programs for which internal inconsistency is suspected will likely receive closer scrutiny in policy decision making. The recent success of conservation and agricultural interest coalitions in integrating soil conservation, farm income, and production objectives of the Food Security Act of 1985 has encouraged environmental interest groups to view the farm bill as a vehicle for improving consistency between environmental and agricultural objectives. Some popular predictions of the outcome of the 1990 farm bill process go so far as to characterize it as an environmental bill. Whether or not environmental issues form the primary focal point for upcoming farm legislative debate, current social, political, and economic conditions clearly point to their playing a role (Reichelderfer and Hinkle, 1989). That role will include continued investigation of options for increasing the compatibility between agricultural and environmental programs.

This chapter addresses the sources of conflict between current agricultural and environmental policies and explores their implications as bases for proposing alternative approaches. The effects of agricultural policy on environmental quality are described and contrasted with the effects of environmental policy on the performance of the agricultural sector. Inconsistent and counteracting effects are identified, options for overcoming such effects are proposed, and, finally, conclusions are drawn with respect to the general pros and cons of alternative approaches for achieving policy consistency.

FARM PROGRAMS AND ENVIRONMENTAL QUALITY

For more than fifty years programs that support commodity prices and transfer income to the farm sector have formed the core of U.S. agricultural policy. These farm programs were not originally developed to affect environmental quality in any way. Until recently, they included no explicit conservation objectives. Conservation has not been their historical purpose.

The main objectives of agricultural policy are to support farm income and ensure a stable and reasonably priced food and fiber supply. These multiple objectives are achieved through a fairly complicated system of farm programs that insulate the farm sector from the market economy by artificially supporting certain commodities' prices and managing those commodities' supplies.

Each mechanism employed to achieve simultaneously commodity price support, farm income support, and supply control goals has unique, secondary, and unintentional effects on environmental quality. Furthermore, their combined and consistent use during more than five decades has tailored the evolution of American crop production toward systems that coincidentally pose particular environmental problems. The major means by which each of various components and the totality of commodity programs affect environmental quality are reviewed below.

Commodity Price Support

Increasing or sustaining a high price for a particular commodity sends two strong signals to farmers. First, it says to produce more, because the price supported commodity[1] becomes a more profitable candidate for production, relative to nonsupported commodities. Second, it says to use more inputs, because the value of a unit of output contributed by a unit of input has risen. Thus, two consequences of price support are that (1) the mix of crops produced shifts toward a greater proportion of supported commodities and (2) the average per acre use of agricultural inputs in the production of those commodities rises.

As long as production of the supported commodities presents no special environmental threat, these consequences of themselves pose no particular problem. But, in fact, the mix of commodities receiving price support happens to include crops that are among the most soil erosive and chemically dependent of all agricultural land uses (see table 1). By contrast, no production activities among the least erosive with low fertilizer and pesticide requirements receive price support.

The upshot is that the U.S. government is, in effect, subsidizing the production of precisely those crops most prone, by virtue of their agronomic

[1]Programs that support particular commodities' prices, independent of farm income support, include nonrecourse loans, the Farmer-Owned Reserve, government commodity purchases, marketing orders, import quotas, and the Export Enhancement Program. Although each of these programs operates by a unique mechanism, they all (1) raise a particular commodity's or commodities' prices above the levels they would have been in the absence of the program; (2) involve no direct payments to farmers other than those for the purchase of goods; and (3) provide benefits that are completely independent of the amount of land a beneficiary has in production.

TABLE 1. Agricultural Land Uses, Erosion Potential, Chemical Intensity, and Eligibility for Major Direct Farm Program Benefits

Land use/ production activity	Relative erosiveness	Relative fertilizer requirements[a]	Relative pesticide requirements[a]	Eligibility for direct commodity program benefits[b]
Cotton	Most erosive	High	High	Full
Soybeans	Most erosive	Low	Moderate	Partial
Corn	Moderately erosive	Highest	High	Full
Grain sorghum	Moderately erosive	High	Moderate	Full
Wheat	Less erosive	Moderate	Low	Full
Barley	Less erosive	Moderate	Low	Full
Oats	Less erosive	Moderate	Low	Full
Rice	Less erosive	Moderate	High	Full
Grassland	Least erosive	Lowest	Lowest	None
Hayland	Least erosive	Lowest	Lowest	None
Range and pasture	Least erosive	Lowest	Lowest	None
Forest and tree crops	Least erosive	Low	Variable	None

[a]Based on 1987 average, per acre applications of fertilizers (nitrogen and phosphorus) and pesticides (insecticides, herbicides, and fungicides).

[b]Includes loan rate and target price setting with associated deficiency payment programs.

characteristics, to lead to soil erosion and coincidental runoff or leaching of fertilizers and pesticides into water systems. Furthermore, while the production of crops such as cotton and corn requires relatively high use of fertilizers and pesticides even under normal circumstances, price support provides encouragement for further increases in agricultural chemical use rates in their production. These environmental effects are exacerbated by the system used to ensure that commodity price support translates into farm income support.

Deficiency Payments, the Base Acreage System, and Annual Acreage Reduction

Because farm income cannot be sustained through commodity price support if the resultant rise in commodity production leads to price-deflating surpluses, an elaborate system has been devised to (1) apply the brakes on production incentives provided through price support and (2) compensate participating farmers if the resulting level of production still exceeds the level at which target prices are reached through the market. Both the deficiency payment scheme and the base acreage accounting system on which it rests have implications for environmental quality.

The guaranteed price offered through deficiency payments for certain commodities reinforces the incentives provided by price support policies applied to the same set of commodities. In other words, commodity programs encourage expansion of production of those commodities that, while more profitable to produce, also include relatively soil-erosive and chemical-requiring activities (see table 1).

At the present time, deficiency payments do not provide direct incentives for increased per acre use of agricultural chemicals, but they could, if program yields are unfrozen.[2] The area upon which deficiency payments are based is fixed in any one year. So, too, have the yields involved in payment calculation been fixed since 1986, when program yields were frozen at 90 percent of recorded 1985 levels. Prior to 1986, program yields, and the deficiency payments based on them, were adjusted periodically to account for productivity changes. This practice provided strong financial incentives for increasing per acre yields by employing high rates of variable inputs—water, fertilizers, and pesticides. In fact, observed declines in total use and per acre application of agricultural chemicals since 1985 may be attributable, in part, to the 1986 freeze in base yields. If base yields are unfrozen, or if farmers expect they may be allowed to vary again, agricultural chemical use and average per acre yields would likely rise in direct response. In any case, the historical offer of deficiency payments brings more program commodity base acreage into production.

The greater one's total recorded base acreage, the greater the absolute number of acres and associated deficiency payments will be when acreage reductions are mandated. Thus, the system provides a strong incentive for maintaining as large a base acreage as possible. The need to maintain base acreage has special implications for environmental quality in that it prevents an extensive or permanent switch from production of program commodities to less erosive or less chemical-intensive commodities. In so doing, it also discourages the diversity of on-farm land use patterns that best sustain wildlife populations and are essential to the employment of low input sustainable agriculture systems.[3]

Annual Acreage Reduction. During the years that acreage reductions are required to manage supply of price-supported commodities, the aggregate consequences of a sustained, intensive production system are temporarily mitigated. Because less acreage is in production, decreased total quantities of potentially damaging inputs are used nationwide. A similar mitigation of

---

[2]While deficiency payments themselves do not affect agricultural chemical use rates under frozen base yields, the Acreage Reduction Programs implemented in conjunction with the deficiency payment scheme still influence per acre use of pesticides and fertilizers.

[3]Production systems referred to as *low input, sustainable,* or *alternative* are typified by enterprise, spatial and temporal diversity, and a substitution of land, labor, management, and information for agricultural chemicals. The crop systems typically employ rotations of leguminous crops to facilitate soil nutrition and reduce pest populations.

potential aggregate environmental effects is achieved through periodic, paid land diversion programs, under which the government actually "rents" idled land for one year.

Although total annual soil erosion and agricultural chemical use are reduced by short-term acreage reduction schemes, attempts to control commodity supply by temporarily restricting the amount of land that can be used to produce program crops further reinforce the incentives to intensify production practices. They may encourage increased per acre use of agricultural chemicals on land remaining in production, eliciting, in effect, a substitution of agricultural chemicals for the constrained fixed factor (see Hertel, in this volume). Trends in the demand for agricultural chemicals relative to acreage in production lend some support to this possibility, since, historically, annual demand for fertilizers and pesticides has fallen by only about 60 percent of same-year decreases in planted acreage (Daberkow, 1987).

It is also important to note that the short-term, aggregate environmental benefits of annual acreage reductions and paid land diversions are realized only when the agricultural sector is suffering from excess supply and low market prices, conditions that trigger the use of set-asides. In other words, environmental benefits of these programs accrue only when the agricultural sector is performing poorly. This observation has implications for the success of long-term Acreage Reduction Programs, which are reviewed later in this chapter.

Other Farm Programs

Various other programs aimed at supporting the farm sector also have secondary environmental effects. For example, the supply of many fruit, vegetable, and specialty crops is indirectly controlled through the use of marketing orders, rather than directly controlled through acreage or production limitations. Marketing orders use flexible standards to describe the grades given by produce. Thus, the orders control the amount of produce that reaches premium markets, such as fresh markets, at any given time. The marketing order system provides incentives to produce the highest-quality output possible in order to receive a high grade and its associated high price. It is actually the grading system, not marketing orders themselves, that can exacerbate environmental problems. If grades are determined on the basis of cosmetic standards, and cosmetic quality can be ensured only through the use of high rates of pesticides, then the production of the commodity will involve greater individual rates and total aggregate use of pesticides.

Disaster payments, subsidized crop insurance, and subsidized farm credit are income support measures that also reduce the risks farmers face from bad weather or low commodity prices. Reducing the risk of failure allows farmers to invest in soil conservation structures and experiment with alternative pest management techniques—an environmental plus. However, price and production guarantees offered by income support programs also provide incentives for increased production. Of particular relevance to the issue of

TABLE 2. Independent Short-Run Effects of Agricultural Policy on Environmental Quality

	Net effect on—			
Agricultural policy instruments that—	Total soil erosion	Loss of wildlife habitat	Rates of agrichemical use	Total use of agrichemicals
Raise commodity prices	↑	↑	↑	↑
Tie farm income support to production levels	↑	↑	↑	↑
Reduce risk	↑	↑	↓	↑
Subsidize credit	↑↓	↑	↑	↑
Require short-term acreage retirement	↓	No effect	↑	↓
Establish cosmetic standards	No effect	No effect	↑	↑

Note: Arrows indicate direction of net effect (increase or decrease) and do not imply whether the effect is "good" or "bad."

environmental quality is that these programs encourage increased production in locations ill-suited for some program crops. In locations where the source of unusual risk is, for example, highly variable rainfall or a high pest threat, farm programs that underwrite risk lead also to greater use of agricultural chemicals. Although the effects of crop insurance on resource use have not been empirically tested, one can speculate that the yield protection offered by the program has also encouraged some farmers to expand production on marginal lands that may be more erosive. In any case, without the availability of risk-reducing safety nets, production would be distributed much differently, concentrating in regions with a natural competitive advantage and likely occurring at lower average aggregate production costs with lower rates of fertilization, pesticide use, and soil erosion.

Long-Term Net Effects of Farm Programs

The short-run, independent environmental effects of selected agricultural policy instruments are summarized in table 2. But as explained, farm programs are implemented in conjunction with one another and, having been in place for such a long time, are likely to have both long-term and synergistic effects.

One long-term consequence with environmental implications is the influence of farm programs on the direction of technological advance in agriculture. The provision of price supports and income protection through quotas on land has profound effects on the direction of agricultural research as well as on the magnitude of benefits derived from such research (Ruttan, 1982). Commodity-pricing policies, in particular, distort the incentives faced by private investors in agricultural research (Alston, Edwards, and Freebairn, 1988).

The fact that program commodity prices have for fifty years been maintained at high levels—levels determined in part by production costs—means that commodity prices have remained high relative to agricultural input prices. But at the same time wage rates have been rising, making labor a relatively more expensive input, and the use of land as a factor of production has been constrained by Acreage Reduction Programs. Thus, there has been consistent, policy-induced encouragement for research on, development of, and marketing of inputs for labor- and land-saving production of major commodities. Agricultural input industries have responded magnificently to these incentives, developing a parade of improved seeds, farm machinery, and agricultural chemicals to meet the demand for such inputs by practitioners of intensive agricultural production of program commodities.

Finally, farmers' expectations of and reliance on continuing government price supports affect long-run investment in the adoption of intensive production technology. For example, in their study of the effects of the feed grain and wheat programs on irrigation in Nebraska, Just, Lichtenberg, and Zilberman (forthcoming) found that high price supports accelerate substantially the rate of irrigation adoption as well as the rate of groundwater depletion.

To summarize, the net effects of the current, long-established set of traditional farm programs have been to encourage the development of, and sustain, an agricultural system that favors intensive production of relatively soil erosive and chemical intensive commodities, distributed across regions that include those ill-suited for and most vulnerable to environmental damage from the byproducts of that production system. These unintended environmental effects result from a set of agricultural programs that have, however, been successful in stabilizing food prices and supporting farm income levels. Let us not forget that the agricultural-environmental policy interaction is a two-way street. While agricultural policy has implications for environmental quality, environmental policy also affects the performance of the agricultural sector, as discussed below.

ENVIRONMENTAL POLICY AND THE AGRICULTURAL SECTOR

A range of tactics has been employed or proposed by the federal government, states, and localities to reduce the level of environmental damage resulting from agricultural activities. Alternative environmental policy instruments may be classed as relying primarily on (1) subsidies, (2) regulation, or (3) taxes, tariffs, and fees. Each approach has different implications for individual farmers and the agricultural sector.

Direct and Indirect Subsidies

Agricultural policies and programs recognize that agricultural production can have adverse environmental effects and have traditionally addressed that concern by providing support for the adoption by farmers of practices that reduce environmental threat. This support is provided through (1) direct subsidies to farmers and research on and extension of environmentally benign technologies and (2) indirect incentives that subsidize the development of human capital.

The Agricultural Conservation Program of the U.S. Department of Agriculture (USDA), under which the government shares with farmers the cost of installing on-farm soil conservation structures, has as long a history as do current commodity programs. Other USDA programs offering farmers subsidies for conservation activities include the Forestry Incentives, Water Bank, and Great Plains Conservation programs. The sum of federal expenditures on these subsidy programs totaled $218 million in 1988 (USDA, Economic Research Service, 1988). A result of these programs has been the transfer, over time, of a significant amount of conservation capital into the farm sector. These programs were originally designed with particular care to distribute this transferred capital equally throughout the agricultural sector. As a consequence, they came under attack during the early 1980s for having failed to target appropriately subsidies to those regions and production operations on which soil conservation would yield the greatest social benefits (Strohbehn, 1986).

Public sector research has performed an important function by filling the technological gaps left by private research and development, which must, by its very nature, focus on the development of products for which there is broad market demand and whose returns can be captured as profit. Public education, extension, and technical assistance follow up by demonstrating the benefits and assisting farmers in the adoption of socially beneficial technologies. In the past, USDA, other federal research institutions, and the land-grant universities have been highly successful in developing and extending to farmers technologies such as conservation tillage and integrated pest management (IPM) that, when practiced, reduce the probability of environmental damage from agricultural production. More recently, these same public institutions have initiated efforts to research, develop, and extend low input sustainable agriculture systems.

These efforts on the part of the agricultural establishment have the potential to complement farm income goals. Agricultural research and extension have certainly proved successful in developing new technologies and making information available about them, but their success in counteracting policy-distorted market signals that beg for production intensification and its associated environmental problems is less clear. The extent of conservation tillage more than doubled between 1980 and 1986 (USDA, Economic Research Service, 1988). And IPM has evolved during the last twenty years from a fringe-movement philosophy to a well-accepted, if not widely employed, management practice. Evidence suggests, however, that increases in the use of conservation tillage have more to do with rises in the cost of energy than with the public provision of encouragement for its adoption. IPM, while accepted as a legitimate approach to agricultural pest control, has not been so widely adopted that it has had any perceptible effect on the aggregate use of agricultural pesticides.

Although subsidies and public research and extension have resulted in environmentally beneficial change in certain aspects of production decision making, they have had little or no effect on the basic structure of agricultural production or on the market-directed trend in production technology. While discouraging to many, this is not too surprising. The mere *availability* of new technology, however innovative, is insufficient to predicate adoption. Economic incentives must also exist. Macroeconomic and market conditions will largely determine whether incentives provided through subsidies can actually effect a change in farming systems (Nielsen, Miranowski, and Morehart, 1989). And to the extent that commodity programs influence market conditions, they can have additional—positive or negative—impacts on the expected returns to adoption of new technology.

The U.S. Environmental Protection Agency (EPA), the nation's environmental regulator, has supported others' provision of "carrots" to the agricultural sector, even as it wields sticks against a few of that sector's tendencies. For example, EPA's implementation of the Water Quality Act of 1987 and its proposed comprehensive national strategy for reducing pesticide contamination of groundwater (U.S. Environmental Protection Agency, 1987) reflect a federal-state approach that emphasizes voluntary compliance with suggested guidelines. The authority remains with EPA and the states, however, to institute regulations or other means of modifying behavior if the voluntary action of subsidized, publicly informed farmers is insufficient to reduce environmental threat to reasonable levels. Many states have already taken the regulatory approach.

Regulation

Regulatory action can be taken to restrict the availability of environmentally hazardous agricultural inputs or prohibit the use of environmentally damaging production practices; it can be uniformly applied to all farmers or targeted to specific types of farming operations or to particularly sensitive regions of production. The most prevalent adverse effect of agroenvironmental regulation in general and targeted regulation in particular is a change in the distribution of income within the farm sector to, by definition, the disadvantage of some farmers. Other effects include an increase in the average cost of agricultural production for some or all farmers, with the possibility of rising consumer prices and decreased trade competitiveness. Still, the regulatory approach has been employed under circumstances when perceived environmental costs are high, and it is infeasible, as in the case of non-point-source pollution, to measure the contributions of individuals to an accumulated hazard. Because of the difficulty and high administrative costs of attempting to enforce environmental regulation of agricultural practices, occurring as they do over widely dispersed, multiple production units, it is generally more feasible to direct regulation toward the availability of observable agricultural inputs. Pesticide regulation fits this description. Its consequences serve as an illustration of the effects of environmental regulation on the agricultural sector.

After responsibility for pesticide regulation was given to EPA in 1970, all uses of nine pesticides were banned, and some uses were canceled or restrictions imposed on the use of twenty-six other products. These bans are enforceable because the sale of the products, either generally or for restricted uses, is also illegal. While many individual farmers have been stung by these actions, pesticide regulation has not had adverse aggregate economic consequences. The immediate effect of pesticide cancellation or restriction is to raise the cost of production to all farmers formerly using any banned material for which there is no same-cost, equally effective substitute. The rise in average production cost means that at any given commodity price, farmers can produce less of the affected commodity. As a result, supply shifts, commodity prices go up, and some or all costs of regulation, in essence, are passed on to consumers.

All farmers who produce the affected commodity, whether they used the banned pesticide or not, realize the induced rise in the commodity's price. Depending upon the extent of the price increase and the market for the affected commodity, extra revenue may or may not be gained from the price rise; if gained, it may or may not compensate users for the increased cost of production. In the case of commodities produced largely for export, commodity price rises induced by regulation can reduce aggregate farm revenue because the increased price reduces U.S. trade competitiveness. On the other hand, because domestic demand for agricultural commodities is relatively unresponsive to price, regulations that increase the cost of production can, paradoxically, increase aggregate farm revenue from production of commodities largely consumed in the United States. The upshot is that whether or not aggregate farm income rises, net revenue changes to a greater degree for some farmers than others and will decline (or decline most) for those farmers most heavily dependent on the banned material (Osteen and Kuchler, 1987). According to the "polluter pays" principle, this is as it should be. The farmers whose practices caused environmental damage have incorporated the costs of reducing the damage into their accounting, and consumers pay some of the tab for improved environmental quality as they purchase higher-priced goods. Furthermore, regulation provides those otherwise missing signals for innovative, private sector development of safer alternative technologies. But, there are some problems.

If the adversely affected farmers include "small farmers" or if, as a result of targeting or a coincidence of regional concentration, the regulations primarily affect "small farmers," then the outcome flies in the face of a public that wishes to protect small farmers as well as the environment. Furthermore, the extent to which regulation can engender a true improvement in environmental quality depends on the opportunities for input substitution. It is entirely possible that the process of substituting factors for the banned material will lead to different types of environmental damage of equivalent or greater magnitude (Archibald, 1988; Osteen and Kuchler, 1987). In both regards, public sector research has an important, complementary role. For example, IPM research has not only provided safe substitutes for banned pesticides; the availability of IPM technology also helps mitigate the distributional effects of pesticide regulation because it most helps the same group of farmers that is adversely affected by regulatory actions (Antle, 1988).

Of course, farmers and consumers are not the only groups potentially affected by environmental regulation. Regulations that prohibit or restrict the use of particular farm inputs will adversely affect the agribusinesses that manufacture, distribute, and sell the regulated materials. Over the long run following a regulatory action, firms specializing in the production and sale of substitute inputs or services will gain revenue. However, regulatory actions that result in decreased commodity production will also decrease demand for the services of input suppliers and businesses that process and sell end products.

Taxes, Tariffs, and Fees

The imposition of taxes or user fees on fertilizers and pesticides has often been proposed as a means of reducing the threat of water contamination (see, for example, Fleming, 1987). The state of Iowa has, in fact, imposed a 50-cents-per-ton sales tax on fertilizer and increased license and registration fees for state pesticide dealers, although these moves are aimed primarily at raising revenues to fund monitoring and research of water quality and are too low to affect behavior.

From a theoretical standpoint of economic efficiency, taxes, tariffs, and fees associated with polluting inputs, practices, or effluent are preferred to regulation as an environmental policy approach (Segerson, 1989). Their conceptual superiority arises from the fact that if a tolerable standard can be established for a pollutant and if the tax rate can be set at a level that reduces aggregate discharge to meet that standard, each farmer can independently decide, according to his or her production situation, whether and how much to modify practices in lieu of paying the tax. In theory, the resulting level of environmental contamination would be optimal from both the societal and the farm sector viewpoints.

In reality, however, it has proved to be difficult, if not impossible, to determine the optimizing tax rate or fee schedule for agricultural pollutants (Shortle and Dunn, 1986). Because agricultural pollution is of the non-point-source variety, effluent charges are infeasible. Because the biophysical processes and pathways by which many agricultural activities become manifest as off-farm pollution are highly variable and/or not well understood, neither standards nor appropriate tax and fee structures can easily be devised.

Aside from the practical problems, even if one forgoes optimizing and seeks instead to set an input tax rate that leads to some unspecified but "significant" reduction in fertilizer or pesticide use, there is evidence that such a tax would have to be very high indeed. The demand for agricultural chemicals is relatively unresponsive to changes in chemical prices. Recent studies, as reported in Capalbo and Vo (1988), find that a 10 percent increase in chemical price leads only to a 2.5–5.0 percent decrease in chemical use. Much like prohibitionary regulation, taxation increases production costs and imposes upward pressure on commodity prices (Taylor and Frohberg, 1977). It thus can have similar, adverse effects on short-run trade competitiveness and agribusiness industries. Taxation, too, has implications for the distribution of farm income. Broadly imposed taxes on agricultural inputs will most severely affect the group of farmers most heavily dependent on the taxed inputs, while farmers using low rates of the inputs could gain total revenue. The income distributional consequences of a differential tax are even greater. While there is no reason to suspect that small farmers would necessarily be more adversely affected by input taxation than would large farmers, there also is no reason to expect that they would be excluded from the most negatively affected set.

Environmental Policy Summary

The predominant net effects on the agricultural sector of various environmental policy approaches are summarized in table 3. This qualitative comparison of environmental policy instruments assumes that each alternative action improves environmental quality, since environmental protection or enhancement is the rationale for the policy action. Under that particular assumption, table 3 facilitates an understanding of some of the trade-offs between environmental quality and other measures of well-being. However, it is important to keep in mind that while an individual environmental policy action can be expected to improve a specific goal, it is possible that the action will exacerbate a different environmental problem. The potential trade-offs between surface water quality and groundwater quality form a relevant case in point since policies that independently induce reduction in agricultural chemical use may encourage more extensive land use with a subsequent increase in surface water sedimentation (Crowder, Ribaudo, and Young, 1988). A major distinction between the set of agricultural policies listed in table 2 and the environmental policy options shown in table 3 is that while the former are to a large degree centrally coordinated with one another in recognition of interrelated agricultural market phenomena, the latter do not generally address interrelated environmental effects, a problem related to the diffusion of responsibility for environmental policy among various congressional committees, government agencies, and levels of government. Despite this general qualification, a range of potential agricultural-environmental trade-offs may be identified.

A major consequence of environmental regulation is a shift in property rights. Both the regulatory and taxation approaches are disadvantageous to some subset of farmers. Moreover, any individual farmer who contemplates a tax or regulatory constraint on production may tend to view the prospect in terms of its effects on production cost, uncertain in a volatile market that he

TABLE 3. General, Net, Short-Run Economic Effects of Alternative Environmental Policies

Environmental policies that—	Average cost of agricultural production	Commodity prices	Aggregate farm income	Farm income distribution	Nonfarm agribusiness income	Taxpayers and government budget
Prohibit or restrict the use of agricultural inputs or production practices	↑	↑	↓↑[a]	△	↓	No effect[b]
Tax agricultural inputs	↑	↑	↓↑[a]	△	↓	↑
Provide subsidies to alter production practices	No effect	No effect	No effect	[c]	[d]	↓[e]
Fund research on and extension of environmentally beneficial technology	↓	↓	↓	[c]	[d]	↓[e]

Note: Arrows indicate direction of net effect (increase or decrease) and do not imply whether the effect is "good" or "bad." A triangle indicates change in income distribution.

[a]Effect on total revenue depends upon price elasticity of demand for affected commodities.

[b]Except administrative and enforcement costs.

[c]Depends upon whether subsidies are targeted and whether production practices or technologies are scale-neutral.

[d]Depends upon whether new or substituted technologies generate the same private sector revenues as those they replace.

[e]Or budget level maintained through redirection of funds from other activities.

or she will recoup those costs through increased prices. In times of surplus production, environmental policies that raise production costs will cut the size of the surplus without affecting government-supported price levels. And for commodities traded in the world market, regulation that successfully improves environmental quality may also dampen trade opportunities in the short run.[4]

Of the various approaches, subsidization has the highest public cost and the least certain effect on environmental quality. The failure of subsidization to bring about widespread change has caused state governments, in particular, to turn toward market-intervening measures and environmental regulation to induce or impose desired changes. The rapidly growing body of evidence and public concern about agricultural pollution suggests that the mix of environmental policy will tend to favor mandatory controls over subsidies (Libby and Boggess, 1989; Reichelderfer and Hinkle, 1989).

INCREASING THE COMPATIBILITY OF AGRICULTURAL SUPPORT AND ENVIRONMENTAL QUALITY

A comparison of table 2 (which shows the general effects on environmental quality of policy instruments applied to maintain farm income and control commodity prices) and table 3 (which shows the general effects on farm prices and income of policies directed toward environmental protection) points up complementarities and conflicts between the policy sets. The major complementary effect is that environmental regulation can, under some circumstances, enhance the commodity price support and aggregate income goals of agricultural policy.

The major conflict arises from the diametrically opposed signals that agricultural and environmental policies send to farmers and agribusiness. On the one hand, the current set of farm programs strongly encourages the development and use of land-saving agricultural technologies, including, importantly, fertilizers and pesticides. On the other hand, environmental regulation induces and direct environmental subsidies and some aspects of public research and extension attempt to encourage a shift toward new, more environmentally benign production technologies. The net effect is that each policy set cancels out some of the influence of the other. In the meantime, the costs of implementing opposing efforts increasingly drains the public coffers.

Studies on the efficiency, effectiveness, and economic consequences of pesticide regulation demonstrate the potential for counteracting effects of

[4]It is entirely conceivable that environmental policy actions which raise production costs and inhibit trade opportunities in the short run may enhance future trade competitiveness by maintaining the natural resource-based comparative advantage of U.S. agricultural production vis-à-vis trade competitor countries where continued resource degradation leads to reduced agricultural productivity.

independent public programs. Specifically, farm programs are shown to reduce the potential commodity price-enhancing benefits of pesticide regulation that would otherwise occur in the absence of already imposed price supports (Lichtenberg and Zilberman, 1986).

There are other, somewhat less obvious conflicts. The authority for environmental regulation increases farmers' uncertainty even as various farm income support measures attempt to reduce the riskiness of farm planning and management. The income distributional effects of agricultural and environmental policy may also be at odds. The majority of commodity program benefits accrue to a minority of the largest farmers (Gardner, Hoagland, and Kramer, 1984). The fear that environmental policies may shift this income advantage through their distributional effects may explain part of the general opposition to them.

As increasing concern with the federal budget deficit demands that these costly program inconsistencies be eliminated, any or all of three different approaches to increasing compatibility between agricultural and environmental policy may be considered: (1) integrating farm and environmental programs; (2) modifying farm programs; and (3) developing new agroenvironmental policy. A range of options exists within each of these overall approaches, all with different implications for the farm sector and environmental quality. In the following sections, the observed success of former efforts and hypothesized advantages and disadvantages of proposed approaches for enhancing policy consistency are reviewed.

Integrating Environmental and Agricultural Programs

Past recognition that historical soil conservation and commodity programs were inconsistent with one another led to the design of a particularly innovative set of conservation provisions for the Food Security Act of 1985 (FSA). The provisions arose from the observation that some commodity program participants were cultivating highly erodible lands and/or converting wetlands to cropland; in other words, they were creating environmental damage at the same time they were enjoying the benefits of generous farm program payments. The FSA conservation provisions overcome this program inconsistency by linking agricultural and conservation objectives in a single set of legislative innovations.[5] Because the programs created under these

[5]The conservation provisions of the Food Security Act of 1985 establish a Conservation Reserve Program (CRP) and require new forms of compliance by farm program beneficiaries under sodbuster, swampbuster, and conservation compliance programs. The CRP offers annual rental payments to land owners or operators retiring highly erodible and other environmentally critical lands from crop production for ten years and also provides up to half the cost of establishing a soil conserving cover on the retired land. The sodbuster program requires that farmers who convert highly erodible native lands to cropland must do so under an approved conservation plan or forfeit eligibility for any and all USDA program benefits. In similar fashion, the swampbuster program denies farm program eligibility to any farmer who converts

provisions have formed a model for proposals that would integrate agricultural and environmental objectives in a 1990 farm bill, their observed short-run and expected long-run consequences warrant close examination.

Long-Term Acreage Retirement. Let us look first at the Conservation Reserve Program (CRP). This long-term acreage retirement scheme, aimed at both reducing soil erosion and controlling commodity supply, has been remarkably successful in rapidly reducing soil loss from cropland. During the first three years of the CRP, 30 million highly erodible acres were enrolled at an average government cost of $49 per acre and total soil savings of 609 million tons per year (USDA, Agricultural Stabilization and Conservation Service, 1989). This early success has led to proposals to expand the CRP and create similar long-term acreage retirement schemes targeted to other environmentally sensitive croplands, such as those overlying vulnerable groundwater sources. But before the CRP is used as the model for the design of new long-term acreage retirement programs, it is important to recognize that there is a limit to how much land can be retired at acceptable public cost. Serious doubts have arisen, in fact, about whether the CRP can be successful in enrolling the full 40-million-acre minimum mandated by its authorizing legislation. Such doubts have two sound bases.

First, while the CRP was originally envisioned as a substitute for a portion of acreage that would otherwise be retired through annual Acreage Reduction Programs (ARPs), ARPs have instead become strong competitors for CRP enrollment. At effective CRP rental rates and high deficiency payment rates, it is more advantageous for many farmers to continue short-term acreage retirement through commodity program participation than to enroll land in the CRP. Either CRP rental rates (and the associated public cost of CRP) must go up or deficiency payment rates must go down to make CRP an attractive enough option to bring substantially more acres into the program. Now, consider what circumstances could lead to a decline in deficiency payment levels. There are two possibilities: (1) a further decline in target prices could be legislated, or (2) market prices could rise sufficiently to make deficiency payments small. If the latter occurs, the increased market returns to keeping land in production—not the ARPs against which CRP competes—will be the more attractive option. The more profitable it is to cultivate, the higher CRP rental rates have to be to induce acreage retirement.

This conundrum is further complicated by the second reason for doubts about the ultimate CRP enrollment level, namely, that the eligible land that has not been enrolled in the CRP is more valuable than the land already enrolled. The first parcels of land enrolled in the CRP were likely those with the lowest reservation price demanded by their owners or operators. The average CRP rental rate has since been creeping upward, reflecting the simple fact that in order to obtain larger quantities of a good the average

---

wetlands to crop production. Conservation compliance requires that farmers with highly erodible land in production begin to implement approved conservation plans on such lands by January 1, 1990, and complete implementation by 1995 or lose eligibility for farm program benefits.

price must rise. If we also consider that the land enrolled initially may have lower productivity than the remaining eligible acreage, attempting to enroll not only more but also better-quality land implies that even steeper rental rate increases will be necessary. So, if the CRP is approaching its maximum obtainable level of enrollment at present rental rates and if it and similar proposed extensions continue to be integrated into existing farm programs, successful implementation of new programs similar to CRP would be very expensive.

It is not clear whether the benefits would justify the high cost of new or expanded environmentally targeted, long-term acreage retirement schemes implemented in conjunction with current farm programs. On the one hand, such voluntary programs are an effective means for supplementing farm income both directly, through payment for the environmental services gained from the retirement of vulnerable lands, and indirectly, through the commodity price-supporting aspect of acreage retirement. Furthermore, the benefits to society of improved environmental quality are likely to be high in value. However, the full, potential societal benefits of long-term acreage retirement cannot be achieved because of two counteractive forces. First, because the CRP and its proposed clones have multiple objectives (commodity supply control and environmental quality enhancement), no one goal can be fully met without trading off performance in achieving the other goal. For example, using the CRP to maximize erosion reduction would require giving up some of its supply control objective. Similarly, maximizing the supply control effects of the CRP would require significant forfeiture of the program's potential to reduce soil erosion.

Implementation of the CRP thus far has reached neither the potential soil erosion reduction nor the potential supply control benefits that could have been achieved at equivalent public cost (Reichelderfer and Boggess, 1988). Like other complex, discretionary programs, the CRP's performance is more a function of the way it is implemented than of the way it is legislated. If the program employed a true bid system that reflected the value of land and targeted land to meet a specific or prioritized set of its multiple objectives, its cost-effectiveness and public benefits could be greatly improved (Phipps, 1987).

Although farmers' incomes are supported through the CRP, idling a large portion of acreage normally planted to crops can reduce the revenues of businesses selling inputs that complement land. This adverse effect on input industries is not exclusive to long-term acreage retirement schemes. It occurs also during years with extensive annual acreage reductions. For example, the Payment-in-Kind Program, which idled 78 million acres in 1983, significantly reduced tractor sales as well as fertilizer and pesticide sales (Kuchler and Vroomen, 1987). The CRP's multiyear planning and implementation period may actually help reduce the uncertainty of input industries with respect to inventory and production needs for anticipated acres planted.

Finally, long-term acreage retirement reinforces the long-term trend in development and use of land-saving agricultural inputs. It is not known to what extent increased environmental hazards resulting from continued intensification on land remaining in production would offset some of the environmental benefits of large-scale acreage retirement. The trade-offs between land-intensive and land-extensive production may require more study before too much more land is tied up in long-term retirement programs.

Compliance. Sodbuster, swampbuster, and conservation compliance are marvelous ideas for assuring that commodity program participants adhere to minimum standards for protecting society's interests in soil conservation. Their conceptual appeal and political success have thus made them models for proposals that include "wellbusting," a form of water quality compliance with respect to management practices on land near wellheads, and "chemical compliance," aimed at requiring farm program participants to use specified fertilization and pesticide use practices. The merit of these proposals rests with the policy consistency they would impose. Whether they could lead to successful enhancement of either environmental or agricultural policy goals is an entirely different question.

The consistency aspect of compliance schemes, while laudable, can prove also to be their downfall. For example, sodbuster, swampbuster, and conservation compliance were not just made consistent with other farm programs; they were also inextricably linked with them by virtue of the fact that the enormity of the penalty for noncompliance is a function of the attractiveness of the other programs' benefits. This relationship works fine when there are conditions of surplus and low commodity prices, as when the legislation was formulated. But now, with depleted stocks, rising commodity prices, and excellent prospects for high, near-term market-determined farm income, the tight linkage between programs looks much less desirable.

As commodity program payment levels decline, either for legislated or unpredicted reasons, the penalty for noncompliance also diminishes. By tying incentives for these programs to the existence of other farm programs, their success in protecting the environment becomes a direct function of unrelated programs' benefits.

Furthermore, compliance schemes ignore the contribution of farm program nonparticipants to environmental hazards, even though nonparticipants also benefit from the commodity price-enhancing effects of commodity programs. Finally, if too many conditions are put upon farm program participants, compliance schemes may begin to interfere with participation rates in agricultural programs, reducing the effectiveness of the same policies with which they have been integrated.

It is important to point out that all the disadvantages of these otherwise potentially successful program options arise from the fact that they represent accommodations to the existing system of commodity programs. As such, they address the symptoms rather than the causes of incompatibility between environmental policy and agricultural policy. They are effective only when government support is necessary to prop up the farm sector. If, as a result of drought, a drop in the value of the dollar, increased foreign demand for U.S. agricultural products, or other reasons, commodity market prices surge to higher levels, production will expand despite these provisions. Under such conditions, the integrated programs lose their appeal. It is when the farm sector is thriving, a time when public concern for the environmental effects of agriculture tends to peak, that the integrated agroenvironmental programs are most likely to fail.

Modifying Farm Programs

Informal debate abounds on the feasibility and relative merits of various changes in farm programs to enhance their compatibility with environmental policy. Alternative modifications range from changing or eliminating individual components to overhauling the entire system of farm programs. The general implications of a range of proposed actions are reviewed below.

Continued Declines in Target Prices. The FSA of 1985 legislated annual reductions in target prices as a way to conserve public funds and make U.S. commodity prices more competitive in world markets. An environmental side benefit of this prescription has been a reduction in the output price effect on production levels and intensity. The 1990 farm bill could either abort or continue the gradual decline in target prices. Continued declines would sustain reductions in the incentive for intensive production of program commodities; halting the target price declines would renew incentives for intensive production of program commodities even when market conditions do not warrant it.

Similarly, the 1990 farm bill could either repeal or allow continuation of the freeze in base yields used to calculate deficiency payment levels under any target price scheme. Unfreezing base yields would be expected to elicit immediate increases in per acre use of irrigation water, fertilizers, and pesticides, as commodity program participants attempt to maximize deficiency payments.

Reformation of the Base Acreage System. The base acreage accounting system has been singled out for its effects on the economic feasibility of on-farm diversification of land use. However, because base acreage is a tool rather than an end in itself, it is difficult to consider changes in that system independent of the acreage reduction and deficiency payment scheme it serves. To do so requires that a substitute system be devised.

One suggestion would maintain the current base acreage system but improve flexibility by allowing farmers to plant nonprogram crops on the proportion of base retired in any year through ARPs. Examination of the recent 0/92 program[6] suggests this does not always work. As long as defi-

[6]The 0/92 provision of the current wheat and feed grains programs allows producers to devote all or part of their wheat or feed grain base acreage to conserving uses and still receive deficiency payments based on 92 percent of the crop's base acreage. Very few wheat or feed grain producers have taken advantage of this option.

ciency payment rates are high, retiring the acreage and collecting associated payments is worth more than planting alternative crops. As deficiency payment levels drop, nonprogram crop production on retired acres may seem relatively more attractive to program crop producers. Of course, at the same time the new subsidized competition threatens long-time producers of nonprogram crops. A related proposal that partially mitigates the competitive pull of program crops but also reduces the severity of competitive threat to nonprogram crop producers would limit the proportions of base acreage on which deficiency payments would be received and on which nonprogram crops could be produced. Setting these proportions at levels that eliminate the penalty against farmers who wish to diversify their plantings but that do not at the same time reduce participation in commodity programs would be difficult.

Another proposed approach to the base acreage problem is to revert to the "normal crop acreage" concept of base that was employed before 1981. Under a normal crop acreage, base is not specific to a given program crop but applies to all program crops and could be expanded to allow a limited set of nonprogram crops to be planted in any combination the farmer desires after an annual acreage reduction has been determined. This scheme maximizes flexibility but does nothing to guarantee that anything but a single, most profitable crop would be planted.

An interesting twist to the normal crop acreage proposal is to vary its deficiency payments in direct correspondence to the goal of greater flexibility in cropping patterns. This goal could be achieved by granting annual additions to or charging annual excises against the standard deficiency payment, using a formula that rewards farmers who diversify their operations and penalizes those who retain monocultural or bicultural production practices. As a means of reforming the base acreage system to achieve a single additional objective, this approach has some conceptual appeal. However, the practical aspects of its implementation would need to be worked out before its effects on the performance of farm income support programs can be estimated.

A final proposal for reformation of the base acreage system would fix all current, recorded base acreages at their current levels and proceed to use the production (bushel, pound, or hundredweight) equivalent from those acres as the accounting tool for making supply control decisions. This option precludes Acreage Reduction Programs per se, because annual reductions would be made on production that resulted from former base, rather than units of land. If acreage bases were converted to fixed "bushel bases," land could be used less intensively, and farmers could take more advantage of environmentally benign technologies.

Acreage Reduction by Environmental Criteria. An alternative approach would be to design annual acreage reductions for short-term supply control purposes so that the land set aside would strictly be land on which production poses a particular environmental hazard. Such targeting of ARPs would eliminate a large portion of the current counteractive effect between Acreage Reduction Programs and environmental regulation. However, the income distributional implications of such a scheme are enormous. An effect would be to reward owners and operators of environmentally sensitive cropland by restricting payment for acreage reduction to that group. This approach could even create a new market for the sale of environmentally sensitive lands. Furthermore, targeted annual acreage reduction could not be carried out within the current system of commodity price and farm income support programs.

Provision of Farm Income Support for Environmental Services. One alternative to attempts to accommodate current farm programs in designing agroenvironmental policy would be to scrap the current set of programs and redesign them in such a way as to retain their farm income support benefits while directly achieving environmental goals. In other words, design a set of farm programs that accommodate society's environmental quality objectives, instead of the other way around. This goal could be achieved by supporting farm income through payments for the environmental services proffered by farmers rather than making income support payments for adjustments to commodity production.

How might such a program work? One can imagine a range of services for which the urban public might be willing to pay and that farmers might be willing to accept that are more related to production practices than to levels of production. For instance, income support could be granted in the form of subsidies to farmers who demonstrate the practice of low input sustainable agriculture, who reduce fertilizer and pesticide use to some preset standard, or who employ any of a range of defined "best management practices." The key to success of such measures would be that this program would have to be the only form of direct income support available. Income support payments associated with acreage and production levels would have to be eliminated in order to prevent the counteractive effect they would have on the subsidy incentives. Commodity prices under such a system would be determined by market conditions, rather than being manipulated for income support purposes. In order to maintain income support levels during a transition from one system to another, initial income payment levels for environmental services would have to be set at rates that give support equivalent to that under the former system. As commodity markets equilibrate, payment schedules could be revised to vary more directly with the value of environmental services provided by farmers responding more freely to true market signals.

The system described above for linking farm income support to resource and environmental stewardship could be classed as a form of what has been called decoupling, depending as it does on breaking the link between income support and production levels. The potential environmental quality advantages of this particular form of decoupling are large and long term. Besides offering the opportunity to purchase environmental services, decoupling would eliminate the historical bias toward land-saving technological change and foster an environmentally beneficial shift in the location of agricultural production activities. No doubt this type of program would have to be very carefully devised to avoid retention of excess resources in the agricultural sector and prevent surplus production with accompanying low commodity prices.

Nevertheless, economic studies suggest change in chemical input and irrigation water requirements with a shift in farm program strategy. It was particularly clear in the 1970s that a shift to a free market situation could result in larger environmental gains than long-term land retirement or production quota alternatives (Miranowski, 1975). But that was before full adoption of agricultural chemicals—particularly herbicides—had occurred among all program commodities and across all production regions. As we approach the 1990s, the short-term effects of commodity program elimination on environmental quality are less clear. Analysis by Miranowski, Hrubovcak, and Sutton (forthcoming) suggests that commodity policy reform would certainly reduce the adverse environmental effects of production but would offer limited opportunities for further improvements in environmental quality. Of course, over the long run the outcomes of policy reform all are predicated upon farmers' expectations and analysts' assumptions regarding export demand. The implication is, however, that pursuit of a free market solution to agricultural support may accomplish the same environmental objectives that input taxation would, with less intervention and positive rather than negative effects on farming profitability.

Substituting environmental services for production levels as a basis for farm income support, of course, has a range of extended economic ramifications. U.S. and world agricultural prices would fall, and while the volume of U.S. exports would increase, export value would decline, input industries could experience a loss in revenues, and commodity prices could become more volatile (Phipps, Rossmiller, and Meyers, in this volume). In addition, such an extreme overhaul of current farm programs is likely to meet with a fair degree of political resistance, implying as it does a radical shift in both income and wealth distribution within the agricultural and agribusiness sectors.

Preferential Crop Insurance. Among the less dramatic proposals for marginal reform in farm programs is the idea that federally subsidized crop insurance be provided specifically for the purpose of reducing the risk faced by farmers who attempt a transition from traditional to low input sustainable agriculture systems. The motivation for the proposal is that low input sustainable systems are perceived as more risky. This approach would seek to subsidize the technology's adoption by covering that additional risk through public funding. The practical obstacles to such a scheme are almost overwhelming. For example, its administrators would have to devise objective procedures for determining whether claims for yield loss could legitimately be attributed to the implementation of low input sustainable systems as opposed to normal variation in weather conditions. Such procedures would have to overcome the inherent disincentive for appropriate crop management provided to those who subscribe to the insurance. For instance, if one's yield is guaranteed against losses to pests that occur during a presumed phase-down of pesticide use rates, what is to prevent the insured from failing to use even minimal pesticide applications? And how could program administrators judge the difference between the outcome of poor management, generally, and good management under a transition to low input sustainable systems?

Aside from the practical problems of operating it, a preferential crop insurance scheme could be extremely expensive and would, at the same time, continue to reinforce the current tendency toward ill-suited location of crop production activities. Its expense arises from a simple fact of logic: If risk is substantially increased by a change in production system, then the market value of the insurance premium needed to cover that risk would be very high. Further, since farmers are unlikely to purchase high-cost insurance, the federal subsidy required to make the insurance reasonably priced to farmers would also be very high in value. Whether that public cost is justified depends upon the extent to which the risk-reducing incentives to produce on marginal land lead to environmental hazards that partially offset the environmental gains made through increased adoption of low input sustainable systems. Like many other proposals that, on the surface, have great conceptual appeal, the likelihood that preferential crop insurance would ultimately be successful in its aims is entirely a function of its implementation.

Developing Agroenvironmental Policy

There has been a much shorter period of experimentation with alternative environmental policy instruments than with various farm program instruments. Thus, the list of proposals for modifying environmental programs to increase compatibility with agricultural objectives is shorter and more tenuous than the preceding consideration of farm program alternatives. One popular approach would be to redirect agricultural research funding. An additional possibility could involve a change in the target of and incentives provided by input taxation proposals.

Redirecting Agricultural Research. This proposal would direct a greater proportion of the existing level of public sector agricultural research toward development of alternative, low-input, or sustainable farming technologies. The rationale offered is that "government research funding should not encourage continued overproduction" and should correct for the disincentives provided (in part by commodity programs) to the private sector for "alternative" technological development (Fleming, 1987). Research redirection is also seen as a means of providing alternative technologies that allow a shift in production systems.

This argument, however, fails to recognize that there is already a large stockpile of alternative agricultural technologies that have been developed and now sit on the shelf, awaiting political or economic conditions that support their adoption. Throwing more money at the issue, independent of any other policy change, will not overwhelm the basic economic disincentives for adoption of alternative technology. Over the long run it might make more sense for the public sector to identify and enhance the factors that lead to technologies being pulled off the shelf and used than to add more products to the shelf. Should the political and economic disincentives provided by intervention in the agricultural market be reduced, the private sector should respond by supplementing public research to meet new demand for alternative technologies.

Advances in biotechnology hold particular promise for the profitable development of products that maintain productivity while reducing the environmental risk or remediating the environmental consequences of agricultural production. By the same token, some avenues of biotechnological research, such as the development of crop varieties resistant to broad-spectrum herbicides, may increase the potential for conventional agroenvironmental problems.

New Approaches to Input Taxation. The effects on agricultural input industries of potential changes in agricultural or environmental policy are often overlooked, since these industries are not generally the object of public concern. This oversight is a mistake because environmental policy could be reoriented simultaneously to enhance input industry performance and to improve environmental quality.

Direct taxation of agricultural chemical sales or use results in decreases in demand for those chemicals, with subsequent loss of input industry profit. Although such tax schemes would induce some private sector efforts to discover safer alternatives, incentives for private sector research redirection could be intensified through a manufacturer tax and rebate scheme. For example, taxing the manufacturers of environmentally damaging inputs could, with appropriately specified tax rates, have exactly the same effects on farm demand for chemicals as does taxation of the farmer-user or chemical distributor. But direct taxation of the manufacturer increases the chance that a portion of the tax can be recouped through total revenue gains arising from higher wholesale prices. If, in addition, the tax revenues were then rebated to manufacturers who can demonstrate innovative development and marketing of alternative technologies, dual incentives would exist for research redirection.

As agroenvironmental policy matures as a field of study, it is likely that new and innovative approaches could be devised for conscious, long-term influence on the course of agricultural technology change.

SUMMARY AND CONCLUSIONS

The failure of agricultural commodity markets to incorporate the environmental costs of agricultural production has led, among other things, to a range of interventions to internalize those social costs. Intervention in the form of environmental regulation has been less effective and more costly, both to farmers and to society, than it otherwise would be if government intervention in agricultural markets did not simultaneously support commodity prices and link farm income support to production levels. The resultant conflict in policy objectives can be addressed either through efforts to integrate farm and environmental programs or through independent reform of agricultural policy, environmental policy, or both.

No single, best alternative strategy is obvious. Subsidization of environmentally benign production practices—both direct and indirect—is a high-cost option that is likely to have limited success unless the counteracting incentives induced by agricultural policy for maintaining current production patterns are eliminated. Integration of environmental goals within the construct of current farm programs substantially improves policy consistency but is effective in meeting environmental objectives only during periods when commodity market prices are low—not during periods of agricultural expansion when the threat of increased environmental hazards is greatest. Basic reform of agricultural policy toward a free market situation offers the least costly and most effective opportunity for meeting both agricultural and environmental objectives. However, the consequences in terms of income redistribution of the latter option greatly restrict its near-term political feasibility.

If agricultural legislation is not reformed to address directly the environmental consequences of current farm programs, then the pace of environmental legislative actions affecting agriculture is likely to accelerate. Farmers and farm policy decision makers now face a choice. They must either take independent action to incorporate environmental concerns into their decisions or prepare to face externally imposed constraints. Modifying farm programs to allow for joint achievement of environmental and agricultural policy goals may be less traumatic for the agricultural sector and less costly to the public sector than the alternative adjustment to additional, restrictive environmental regulation.

ACKNOWLEDGMENTS

The author gratefully acknowledges the contributions of Tim Phipps, whose collaborative work is reflected in the paper's contents and whose critical review of an earlier draft helped improve the material. Acknowledgment and thanks go also to Kristen Allen, for perceptive review and preliminary edit of an earlier draft; Monika Potemra, for her skill in formatting and typing the paper; and Ross C. Korves, Dean Kleckner, Allen Rosenfeld, and other members of the Food and Agriculture Committee of the National Planning Association who provided insightful comment and reaction that proved valuable in revising the earlier draft. Despite all this helpful input, the areas emphasized, views expressed, and conclusions drawn within the paper are the sole responsibility of the author.

REFERENCES

Alston, J. M., G. W. Edwards, and J. W. Freebairn. 1988. "Market Distortions and Benefits from Research," *American Journal of Agricultural Economics* vol. 70, pp. 281–288.

Antle, John M. 1988. *Pesticide Policy, Production Risk, and Producer Welfare: An Econometric Approach to Applied Welfare Economics* (Washington, D.C., Resources for the Future).

Archibald, S.O. 1988. "Incorporating Externalities into Agricultural Productivity Analysis," pp. 366–394 in S. Capalbo and J. Antle, eds., *Agricultural Productivity Measurement and Explanation* (Washington, D.C., Resources for the Future).

Batie, Sandra S. 1988. "Agriculture as the Problem: New Agendas and New Opportunities," *Southern Journal of Agricultural Economics* vol. 20, pp. 1–12.

Capalbo, S. M., and T. T. Vo. 1988. "A Review of the Evidence on Agricultural Productivity and Aggregate Technology," pp. 96–137 in S. Capalbo and J. Antle, eds., *Agricultural Productivity Measurement and Explanation* (Washington, D.C., Resources for the Future).

Crosson, Pierre. 1989. "The Long-Term Adequacy of Agricultural Land in the United States: Economic and Environmental Issues," draft manuscript, Resources for the Future, Washington, D.C.

Crowder, Bradley M., Marc O. Ribaudo, and C. Edwin Young. 1988. "Agriculture and Water Quality." U.S. Department of Agriculture, Economic Research Service, Agricultural Information Bulletin no. 548 (Washington, D.C., August).

Crutchfield, Stephen R. 1987. "Controlling Farm Pollution of Coastal Waters" *Agricultural Outlook* AO-136 (Washington, D.C., U.S. Department of Agriculture, Economic Research Service), November, pp. 24–25.

Daberkow, Stan G. 1987. *Agricultural Input Industry Indicators in 1974–85.* U.S. Department of Agriculture, Economic Research Service, Agricultural Information Bulletin no. 534 (Washington, D.C., November).

Fleming, Malcolm H. 1987. "Agricultural Chemicals in Ground Water: Preventing Contamination by Removing Barriers Against Low-Input Farm Management," *American Journal of Alternative Agriculture* vol. 2, no. 3, pp. 124–130.

Gardner, Patrick H., William Hoagland, and Randall Kramer. 1984. "The Distribution of Benefits from the 1982 Federal Commodity Programs," U.S. Senate, Committee on the Budget, Senate Print no. 98-238 (Washington, D.C., Government Printing Office), November.

Just, Richard E., Erik Lichtenberg, and David Zilberman. Forthcoming. "Effects of the Feed Grain and Wheat Programs on Irrigation and Groundwater Depletion in Nebraska," in Richard E. Just and Nancy Bochstael, eds., *Commodity and Research Policy in Agricultural Systems* (Heidelberg, West Germany, Springer-Verlag).

Kuchler, Fred, and Harry Vroomen. 1987. "Impacts of the PIK Program on the Farm Machinery Market," *Journal of Agricultural Economics Research* vol. 39, no. 3 (summer), pp. 2–11.

Libby, Lawrence W., and William Boggess. 1989. "Agriculture and Water Quality: Where Are We and Why?" Proceedings of a symposium presented by the Agricultural, Resource, and Environmental Policy Research Consortium, Washington, D.C., April.

Lichtenberg, Erik, and David Zilberman. 1986. "Problems of Pesticide Regulation: Health and Environment Versus Food and Fiber," in T. T. Phipps, P. R. Crosson, and K. A. Price, eds., *Agriculture and the Environment* (Washington, D.C., Resources for the Future).

Miranowski, J. A. 1975. "The Demand for Agricultural Crop Chemicals Under Alternative Farm Program and Pollution Control Solutions" (Doctoral dissertation, Harvard University, Cambridge, Mass.).

Miranowski, John A., James Hrubovcak, and John Sutton. Forthcoming. "The Effects of Commodity Programs on Resource Use," in Richard E. Just and Nancy Bochstael, eds., *Commodity and Research Policy in Agricultural Systems* (Heidelberg, West Germany, Springer-Verlag).

Nielsen, Elizabeth G., John A. Miranowski, and Mitchell J. Morehart. 1989. *Investments in Soil Conservation and Land Improvements: Factors Explaining Farmers' Decisions,* Agricultural Economic Report no. 601 (Washington, D.C., U.S. Department of Agriculture, Economic Research Service), January.

Osteen, C., and F. Kuchler. 1987. "Pesticide Regulatory Decisions: Production Efficiency, Equity, and Interdependence," *Agribusiness* vol. 3, pp. 307–322.

Osteen, C., and Phillip Szmedra. 1989. *Agricultural Pesticide Use Trends and Policy Issues,* Agricultural Economic Report no. 622 (Washington, D.C., U.S. Department of Agriculture, Economic Research Service).

Phipps, Tim. 1987. "The Conservation Reserve: A First Year Report," *Resources* no. 86, pp. 14–16.

Public Opinion. 1987. "Opinion Roundup" (Washington, D.C., American Enterprise Institute), March–April and May–June.

Reichelderfer, Katherine, and William G. Boggess. 1988. "Government Decision Making and Program Performance: The Case of the Conservation Reserve Program," *American Journal of Agricultural Economics* vol. 70, no. 1 (February), pp. 1–11.

Reichelderfer, K. H., and M. K. Hinkle. 1989. "The Evolution of Pesticide Policy: Paving the Way for Environmental Interests and Agriculture," in C. Kramer, ed., *The Political Economy of U.S. Agriculture: Challenges for the 1990s* (Washington, D.C., Resources for the Future).

Ribaudo, Marc O. 1986. *Reducing Soil Erosion: Offsite Benefits,* Agricultural Economic Report no. 561 (Washington, D.C., U.S. Department of Agriculture, Economic Research Service), September.

Ruttan, V. W. 1982. *Agricultural Research Policy* (Minneapolis, University of Minnesota Press).

Segerson, Kathleen. 1989. "Incentive Policies for Control of Agricultural Water Pollution." Proceedings of a symposium presented by the Agricultural, Resource, and Environmental Policy Research Consortium, Washington, D.C., April.

Shortle, J. S., and J. W. Dunn. 1986. "The Relative Efficiency of Agricultural Source Water Pollution Control Policies," *American Journal of Agricultural Economics* vol. 68, pp. 668–677.

Strohbehn, Roger, ed. 1986. *An Economic Analysis of USDA Erosion Control Programs: A New Perspective.* Agricultural Economic Report no. AER–560 (Washington, D.C., U.S. Department of Agriculture, Economic Research Service), August.

Taylor, C. R., and K. K. Frohberg. 1977. "The Welfare Effects of Erosion Controls, Banning Pesticides, and Limited Fertilizer Application in the Corn Belt," *American Journal of Agricultural Economics* vol. 59, pp. 25–36.

Tiner, R. W., Jr. 1984. *Wetlands of the United States: Current Status and Recent Trends.* U.S. Department of the Interior, Fish and Wildlife Service (Washington, D.C., Government Printing Office).

U.S. Department of Agriculture, Agricultural Stabilization and Conservation Service. 1989. "Conservation Reserve Program" (mimeographed distribution summarizing CRP activity through the 8th sign-up), June 14.

U.S. Department of Agriculture, Economic Research Service. 1988. *Cropland, Water and Conservation Situation Outlook Report.* Agricultural Resources, AR-12, U.S. Department of Agriculture, Economic Research Service (Washington, D.C., September).

U.S. Environmental Protection Agency. 1987. "Agricultural Chemicals in Groundwater: Proposed Pesticides Strategy" (Washington, D.C., Office of Pesticides and Toxic Substances), December.

Young, Robert A., and Gerald L. Horner. 1986. "Irrigated Agriculture and Mineralized Water," in T. T. Phipps, P. R. Crosson, and K. A. Price, eds., *Agriculture and the Environment* (Washington, D.C., Resources for the Future).

AGRICULTURAL TRADE LIBERALIZATION
The Multilateral Stake in Policy Reform

DAVID BLANDFORD, HARRY DE GORTER, PRAVEEN DIXIT, and STEPHEN MAGIERA

Agriculture is a major item on the agenda of the current Uruguay Round of multilateral trade negotiations under the General Agreement on Tariffs and Trade (GATT). The prominence of agriculture reflects the critical state of world agricultural trade and the unsustainability of the current agricultural policies of many industrial countries due to their effect on world prices and budgetary costs. Countries are concerned about the high costs of agricultural support and the spread of subsidies and countersubsidies. There has been a growing realization that many existing policies are both inefficient and self-defeating. The GATT negotiations provide an opportunity to achieve a multilateral reduction in protective support and in the trade distortions it creates. Nevertheless, there remain significant problems of how to modify existing policies in order to reduce trade distortions and how to meet domestic objectives within the constraints imposed by a GATT agreement.

The purpose of this chapter is to demonstrate that the United States and other industrial countries have a major stake in reaching a multilateral agreement on agricultural policy reform in the GATT. Such an agreement will result in significant savings to taxpayers and consumers and an increase in national economic welfare. It will make domestic policy reform less painful for agriculture but also impose restrictions on the types of policies that can be used to achieve domestic objectives. This chapter will identify the implications of an agreement to reduce trade-distorting policies for the major countries and will discuss the characteristics of minimally trade-distorting policies. Because of their importance in industrial countries, farm income support and market stabilization policies are the primary focus of this analysis.

THE FORM OF A GATT AGREEMENT FOR AGRICULTURE

No consensus has yet been reached within the GATT on how far to reduce trade-distorting policies in the current round or on how phased reductions in agricultural protection would be achieved. A number of proposals have been made and are being discussed. The debate includes the question of whether or not reductions should be based on an aggregate measure of support, whether national plans detailing policy adjustments should be used, and whether the tariffication[1] of some existing trade policies should be adopted as a means of freezing and subsequently reducing the level of import barriers.

It is assumed here that a GATT agreement will require the phased reduction of trade-distorting policies but that modification rather than elimination of current policies will be allowed if this less drastic change reduces trade distortions. We do not evaluate short-run measures through which the trade-distorting effects of existing policies could be ameliorated temporarily. We do not analyze the phase-in of longer-run reforms, either through a negotiated reduction in an aggregate measure of support or through the reduction of tariffs following the tariffication of existing policies. Such details will need to be worked out in a comprehensive GATT agreement and will influence the scope and speed of domestic policy reform in industrial countries. However, since there exist a large number of possible approaches to reform and an equally large number of adjustment formulas that could be applied, a detailed consideration of these issues is not possible until the likely content of the agreement becomes clearer. Instead, we focus on the end point of reform—the complete elimination of trade-distorting policies—to show why the United States and other industrial countries have a major stake in an agreement to reduce trade distortions and to identify what this would imply for domestic agricultural policies. In particular, we discuss the conversion of price support programs to a system of direct payments for a limited quantity of production as a means of providing income support that could be acceptable as part of the multilateral negotiations.

THE STAKES IN THE MULTILATERAL REFORM OF AGRICULTURAL POLICIES

Industrial countries pay heavily to support agriculture through direct transfers from taxpayers to producers and indirect transfers from consumers to producers. One measure of such transfers is the producer subsidy equivalent (PSE) and its counterpart, the consumer subsidy equivalent (CSE). These two indicators have been discussed and estimated widely in recent years. The PSE measures the gross producer subsidy created by government programs. It incorporates direct and indirect transfers to agriculture from (1) taxpayers, through input, output, and marketing subsidies and (2) consumers, through policies that increase domestic consumer prices. The CSE measures the explicit and implicit consumer taxes created by such policies. PSEs and

[1]Tariffication is the conversion of existing measures, e.g., import quotas, to an equivalent tariff that can be frozen and subsequently reduced.

CSEs are expressed as a percentage of the total value of production and consumption, respectively.

It is important to note that the PSE is a gross measure of agricultural support. It does not represent the net effect of support on producer incomes or on international trade distortions (de Gorter and McClatchy, 1984; de Gorter, 1988). The net effects depend on the impact of policies on production costs, output, and producer revenues and these, in turn, depend on the incentive effects and the transfer efficiency of the policy measures. The issue of transfer efficiency is fundamental, since the PSE overstates the benefits that farmers receive from current policies and hence any compensation that may have to be paid to producers if these policies are eliminated.

PSEs provide a means for ranking nominal protection across countries and commodities. Recent information on aggregate PSEs by country is given in figure 1. The estimates show that Japan supports its agricultural producers the most, followed by the European Community (EC), Canada, and the United States. Among the major industrial countries, Australia and New Zealand have the lowest levels of support. In most countries, the PSEs increased during the early 1980s. The PSE for the European Community rose from 30 percent in 1982–1983 to 50 percent in 1986. In the United States the corresponding increase was from 22 to 36 percent. Only New Zealand, which has implemented major unilateral reforms in agricultural and other types of policies since 1984, has reduced the level of support for agriculture in the 1980s.

The costs of producer support have to be borne through either higher food prices or higher taxes. The distribution of costs varies considerably among countries. In the European Community and Japan, policies that tax consumers account for well over two-thirds of the support to agriculture. This distribution is shown in figure 2, where the PSEs in figure 1 are broken down into their consumer and taxpayer shares. In the United States it is taxpayers who pick up most of the bill for supporting agriculture.

The elimination of existing agricultural policies in the industrial market economies would lead to significant net gains in economic welfare in industrialized countries. Estimates of these gains for 1986, shown in figure 3, suggest that net welfare in the United States would increase by roughly $9 billion. For the major industrial countries as a whole, the gain would be roughly $33 billion. Trade liberalization would increase world agricultural prices substantially. A number of estimates of the world price changes resulting from complete elimination of current support policies are given in table 1. The estimates differ among studies because of the different base years chosen. The level of protection varies with the level of world prices, and the percentage effects of eliminating protection also vary with the base. World prices—especially for dairy products and beef—would rise. Government support is high for these commodities, and industrial country trade is a major part of world trade. The effect of trade liberalization for grains was predicted to be modest until the United States and the European Community increased the use of subsidies. The estimates for 1986 reflect the effects of

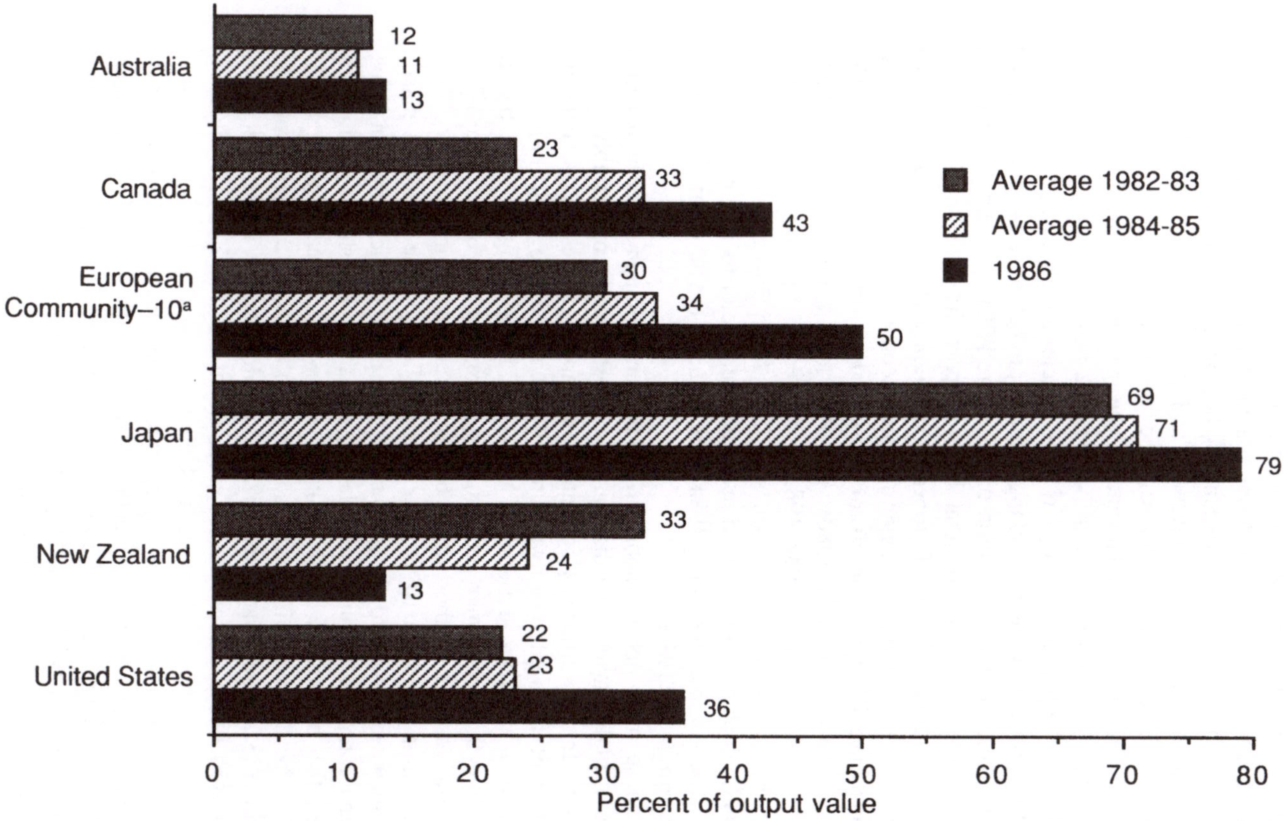

FIGURE 1. Agricultural producer subsidy equivalents in major industrial countries. *Source:* Based on data from USDA (1988).

[a]Excludes national programs.

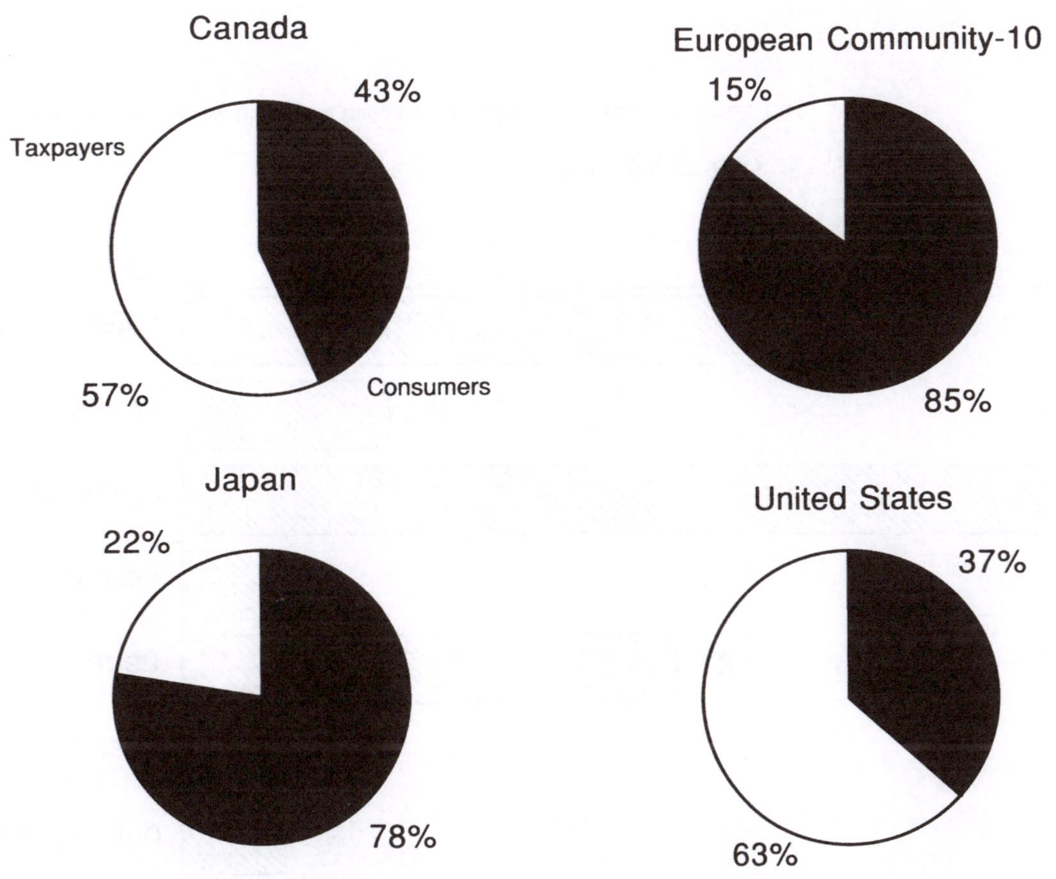

FIGURE 2. Consumer and taxpayer contributions to agricultural support. *Note:* Data are for 1982–1986, except for the EC, for which data are for 1982–1985. *Source:* Based on data from USDA (1988).

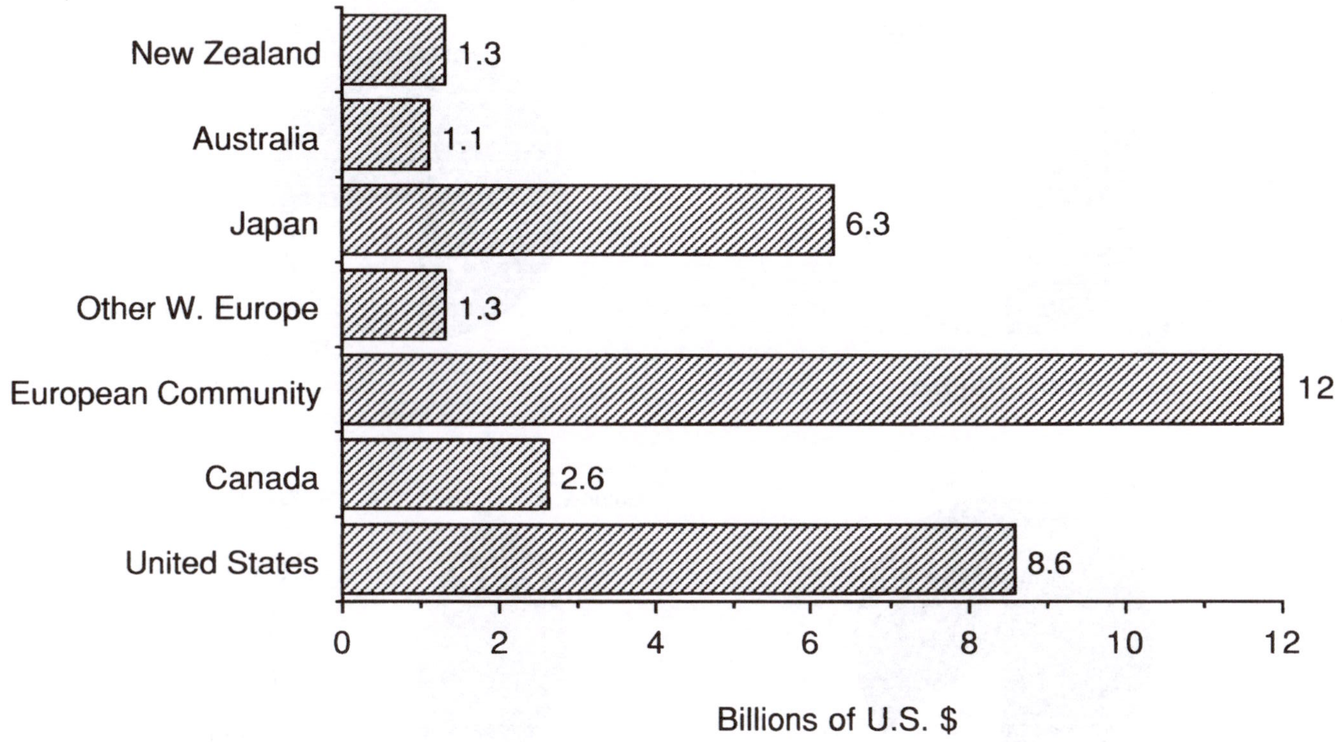

FIGURE 3. Change in economic welfare from trade liberalization in industrial countries, 1986.

TABLE 1. Predicted Effects of Industrial Country Trade Liberalization on World Agricultural Prices
(percent change in world price)

Study	Period	Wheat	Coarse grains	Rice	Beef[a]	Dairy products	Sugar
OECD (1)	1979–1981	−1	−3	1	15	44	10
Tyers and Anderson (2)	1980–1982	10	3	11	27	61	11
Roningen and Dixit (3)	1984	7	11	18	22	30	48
Tyers and Anderson (4)	1985	2	1	5	16	27	5
Roningen and Dixit (5)	1986	30	23	24	18	50	39
Parikh et al. (6)	2000	18	11	21	17	31	n.a.

Note: n.a. means not available.

Sources: Data from (1) Organization for Economic Cooperation and Development (1987); (2) Tyers and Anderson (1986); (3) Roningen and Dixit (1987); (4) Tyers and Anderson (1986); (5) Roningen and Dixit (1989); (6) Parikh et al. (1988).

[a]All ruminant meat in sources (2), (3), (4), and (5); bovine and ovine meat in source (6).

the increased subsidization of grain production and exports in these countries in recent years.

Figure 4 breaks down the contribution of various countries to the predicted increase in world prices of selected commodities in 1986. The European Community and the United States would be by far the most important contributors to the change in world prices through trade liberalization. These countries are the large actors in world agricultural trade, and changes in their production, consumption, and trade have a great effect upon world prices. It is estimated that the elimination of trade-distorting supports in the European Community alone would raise world prices by an average of 8 percent. This figure is about 40 percent of the estimated increase that would be realized if all industrial countries eliminated agricultural support. EC policies have the largest price impacts in the dairy, sugar, ruminant meat, and wheat markets.

U.S. agricultural policies and those of other countries affect the total costs of support in the industrial countries. Many policies depress consumption and increase production. Imports fall, exports rise, and world prices are depressed. Lower world prices then increase the costs of maintaining a given level of domestic support. Figure 5 shows that in the United States and the European Community a significant proportion of the apparent support provided to producers merely offsets the price-depressing effects of the policies of other countries. *Offsetting support* is defined as the level of compensation to farmers in a country that offsets the decline in producer welfare due to the lower world prices resulting from agricultural policies in other countries. Significant consumer and taxpayer savings could be made if policies were

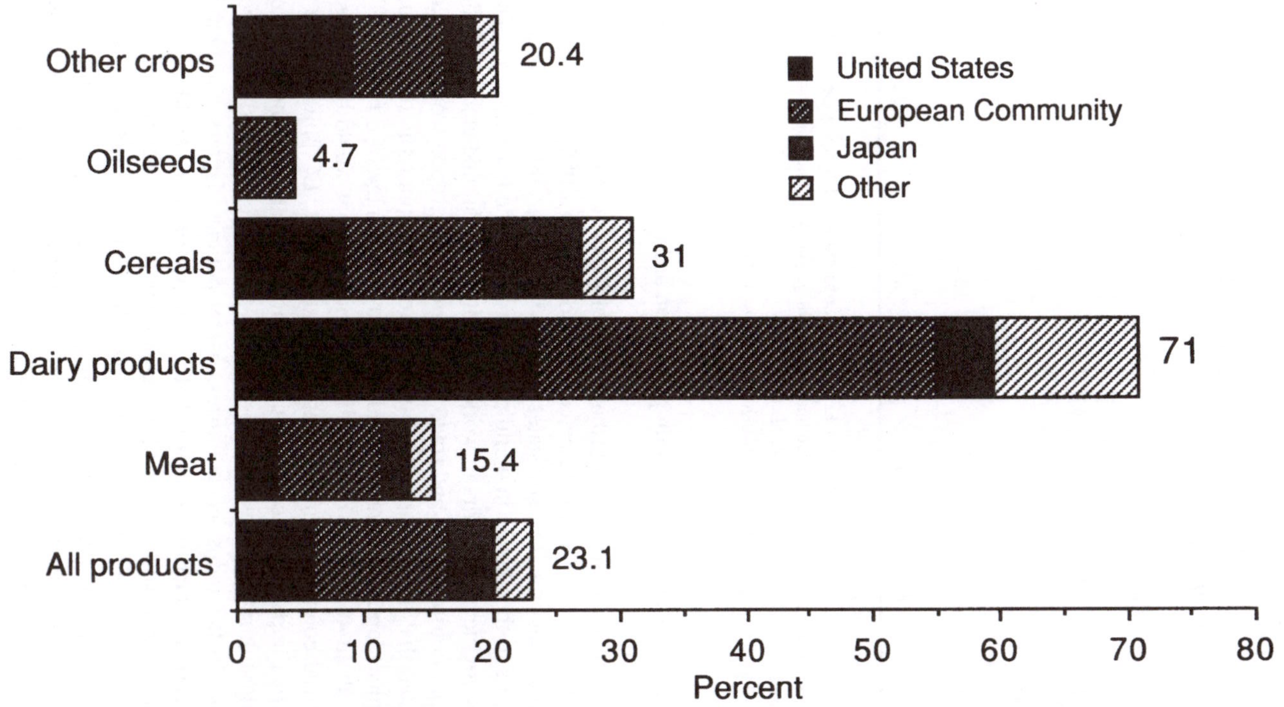

FIGURE 4. Contribution of industrial countries to changes in world agricultural prices under free trade, 1986.

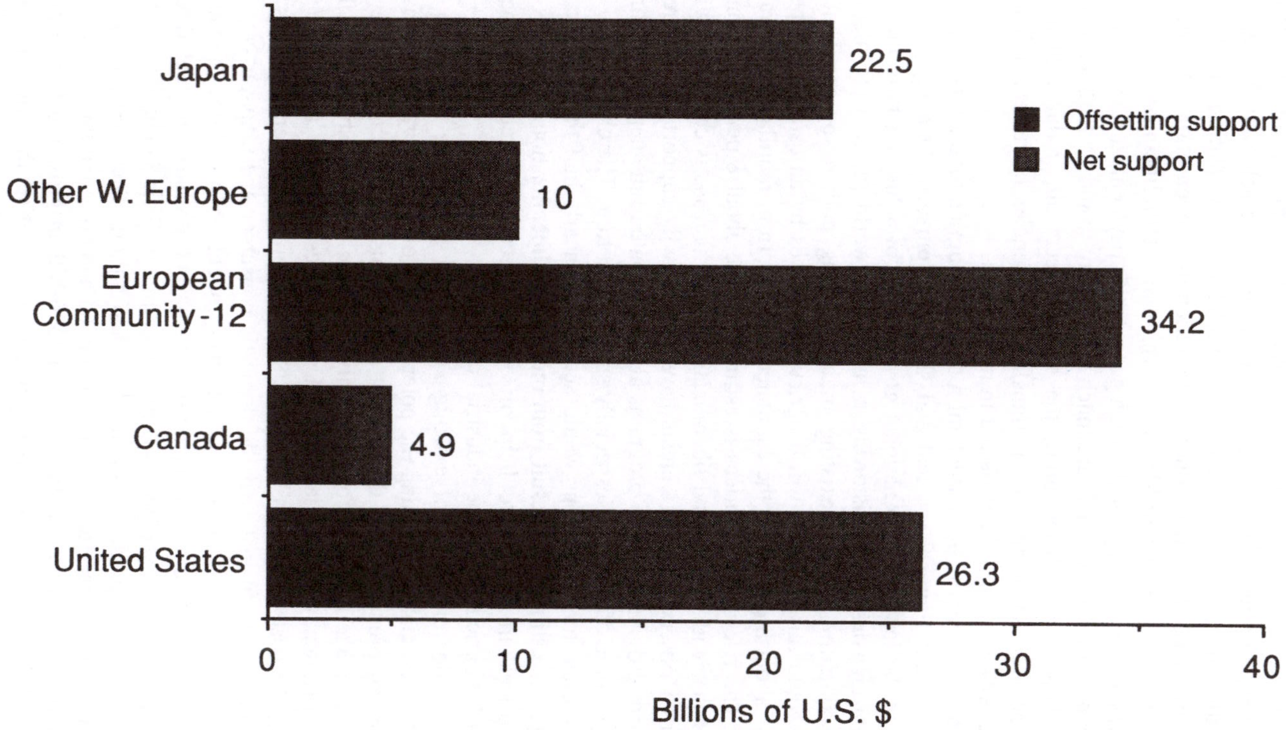

FIGURE 5. Gross and net support from agricultural policies in 1986. *Source:* Gross support values are from USDA (1988).

liberalized unilaterally, but the adverse effects on domestic producers would be severe because world prices would not rise as much as they would under multilateral free trade. Figure 6 shows that the compensation required to keep producer incomes at 1986 levels would be greater if countries liberalized unilaterally than if multilateral liberalization were achieved. For example, if the United States had undertaken unilateral liberalization in 1986, it would have needed to pay producers $26 billion in compensation. This figure would have been reduced to $16 billion with multilateral liberalization. Since 1986 was a year of exceptionally high government payments to U.S. farmers, these compensation estimates are much higher than would be required under current conditions. Furthermore, the actual effects on producer incomes would be even lower because of transfer inefficiencies in current policies. Transfer efficiency will be explored in more detail.

The mutual benefits of multilateral reduction in agricultural support shown in table 1 and figure 3 are real and should be pursued. One key benefit would be the lower cost to national governments of achieving farm income goals through less trade-distorting measures, if world prices were higher. Unilateral elimination of existing trade-distorting policies by the United States would raise average world agricultural prices by an estimated 5 percent, or about a quarter of the estimated increase from multilateral liberalization. The rise in cereal prices is estimated to be about 8 percent, or one-third of that with multilateral liberalization. This increase is consistent with our earlier observation that cereals have been heavily supported by current U.S. farm programs. These programs also contribute to the depressed world prices of some other crops, especially sugar and cotton. The change in U.S. policies alone for these crops would account for nearly half the increase in world prices estimated to result from multilateral liberalization. Increases in world prices are needed to help offset the effects of lower government support for agriculture in the industrial countries.

The transfer efficiency of existing agricultural policies is an important issue. It is crucial in assessing the potential costs of multilateral trade liberalization for producers and the amount of compensation that might have to be paid to make liberalization politically palatable. Partial estimates of the transfer efficiency of policies in the United States for 1986 are shown in table 2. The costs and benefits are calculated assuming unilateral changes in U.S. policy only. In that year the domestic consumer and taxpayer cost of agricultural price and income support was roughly $36 billion, of which about 72 percent ($26 billion) came from taxpayers, primarily through direct government payments and price support expenditures. This $36 billion yielded a gross transfer to producers of just over $26 billion. The other 28 percent was "cost" due to inefficiencies in production and consumption or transfers generated by the support policies. Of the $26-billion gross transfer to producers, a further $11 billion was required simply to offset the price-depressing effects of agricultural support policies in other industrial countries. The ratio of benefits to costs when this loss is subtracted falls from over 70 percent to just over 40 percent. Of this presumed net benefit (net of the

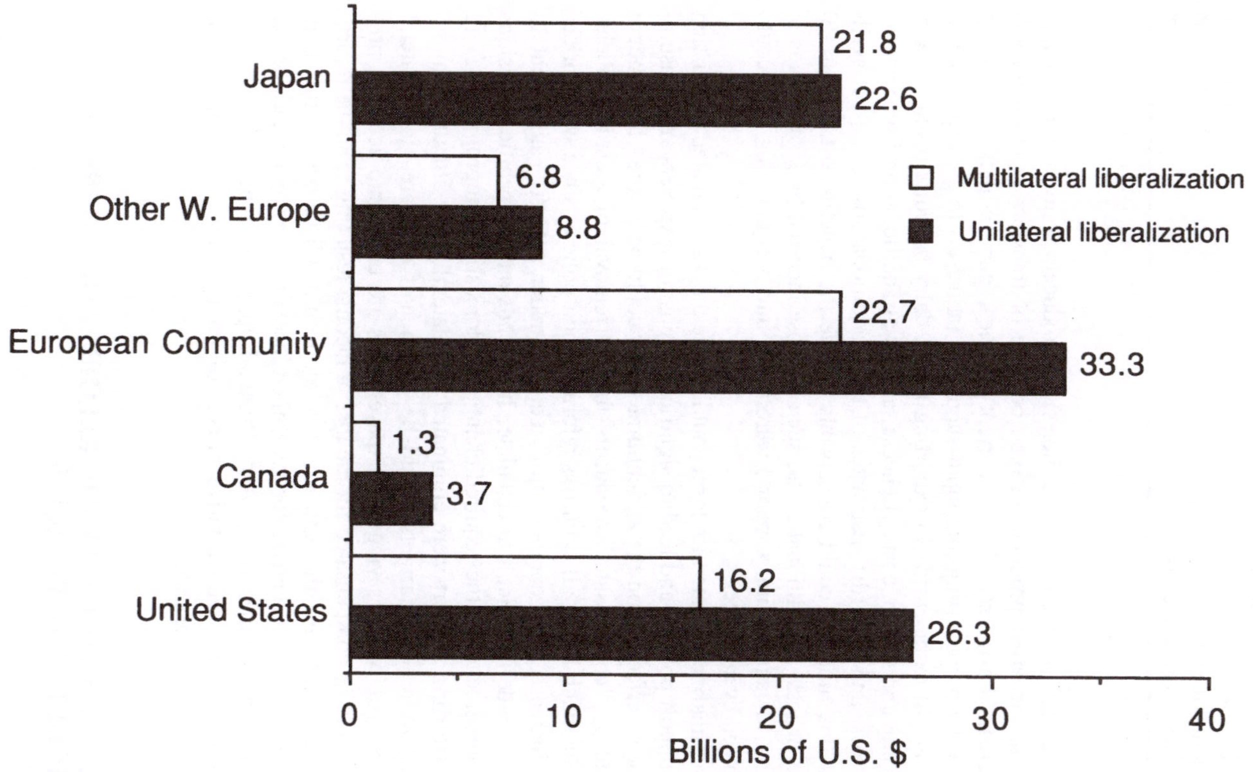

FIGURE 6. Gross producer compensation requirements with trade liberalization in 1986.

TABLE 2. Estimated Costs and Benefits of U.S. Agricultural Policies, 1986

Cost or benefit	Amount
Taxpayer and consumer costs	\$36.3 billion
Producer benefits (maximum)	\$26.3 billion
Less policy offsets	\$11.3 billion
Net benefits (maximum)	\$15.0 billion
Ratio of net benefits to costs	41 percent

impact of all other countries' policies) to producers, only a part actually results in increased producer income because of transfers to upstream and downstream businesses. Under current policies, part of the transfers are captured by output-using and input-supplying industries (e.g., export subsidies result in extra profits to exporting firms, dairy supports result in extra processing and storage costs of cheese, as well as additional profits to cheese processors, and so forth). Estimates of these additional losses are not available at this time but could conceivably reduce the benefit-cost ratio to 30 percent or less. If additional costs, such as those incurred by farm groups in lobbying for agricultural support, were taken into account, this ratio could be reduced even further.

To summarize, unilateral reform could reduce the waste created by existing U.S. support policies and lead to significant gains for taxpayers and consumers, but its effects upon the agricultural sector could be severe. The amount required to "compensate" producers for the removal of existing policies would be much less if all countries reformed their policies since world prices would increase. The compensation required under a GATT agreement to liberalize trade is difficult to calculate. It depends on the specific provisions of the agreement and the required changes in domestic farm programs. If all trade-distorting support were eliminated multilaterally, the amount required to compensate U.S. farmers would be cut in half compared to unilateral policy reform. When savings in taxpayer costs are taken into account, producers could be compensated for the loss in income resulting from the reform of farm programs and the nation as a whole would still be better off. There are therefore great incentives for the United States to enter into a multilateral agreement to liberalize trade. A multilateral approach is needed to reduce the severity of the redistributive effects of policy reform and make such reform politically feasible.

IMPLICATIONS OF TRADE REFORM FOR CURRENT AGRICULTURAL POLICIES

The major types of income support measures currently employed in industrial countries are identified in table 3. The degree to which income support measures distort trade is determined by how much the policies affect produc-

TABLE 3. A Ranking of Agricultural Support Policies by Trade Distortions

1. Production subsidies (including price supports with deficiency payments, stabilization payments, and government procurement) and taxes (including coresponsibility levies, superlevies, and so forth).
2. Input subsidies (for credit, fertilizer, irrigation, crop insurance, interest rates, and so forth) and taxes (e.g., land).
3. Production-marketing quotas (including two-price schemes, maximum guaranteed quantities, and so forth) and input controls (e.g., acreage set-asides, and so forth).
4. Consumption taxes, quotas (rationing), or subsidies (including domestic price ceilings with consumer deficiency payments and domestic food aid).
5. Marketing taxes (e.g., value-added tax) and subsidies (e.g., transportation subsidies).
6. Imperfect market structure (producer-consumer monopolies, parastatal marketing boards, and so forth).
7. Import-export taxes and subsidies (ad valorem and specific) and under/overvalued exchange rates.
8. Import-export quotas (including voluntary export restraints, state trading, and orderly marketing arrangements).
9. Contingent export-import subsidies and taxes (including export enhancement programs, domestic price supports with variable import levies–export restitutions, and countervailing and antidumping duties).

tion and consumption. For a given level of support, border measures are more trade distorting than most domestic policy measures because they affect both production and consumption directly. Domestic support measures, such as production or input subsidies, are less trade distorting. They affect production but do not have a major impact on consumption. Such subsidies can be targeted directly to farmers, rather than providing indirect support by changing output prices. Direct income subsidies are both a more efficient means of supporting farm incomes and are less trade distorting.

Among border measures, tariffs and subsidies are less distorting than quantitative restrictions because they allow changes in international prices to be reflected domestically. Production and consumption will vary in response to changes in world market conditions. Quantitative restrictions, variable levies, and price-fixing schemes do not allow such response and are likely to be more trade distorting. Other types of quantitative restrictions, such as output or marketing quotas, are more distorting than production subsidies because they affect domestic prices and consumption. They can be less trade distorting than border measures, but whether they are depends on the relative size of their effects on prices and trade.

A second consideration in judging the distortions resulting from existing policies is their effect on the stability of international trade and prices. This issue is complex, particularly when countries use a number of measures, each of which influences production, consumption, storage, and trade. In general, however, policy measures that limit the response of domestic consumption, stocks, and production to changes in international market prices will tend to destabilize international prices. Thus, a variable levy or import quota is more destabilizing internationally than a tariff. Furthermore, some policies essentially export domestic instability. Export subsidies to dispose of domestic stocks, such as those used by the European Community and the United States, are in this category.

The implication of the trade-distortion ranking and the policies' effects on stability is that variable levies and quantitative controls on trade through explicit quotas or state trading are prime candidates for reform *if* the main goal of reform is to reduce trade distortions. So also are the export subsidies used by the European Community and the United States. The protection provided by these measures could be replaced by an equivalent tariff (in the import case) or production subsidy (in the export case), and these measures could then be progressively reduced in order to liberalize trade under a GATT agreement.

IMPLICATIONS OF TRADE REFORM FOR FUTURE AGRICULTURAL POLICIES

If multilateral negotiations are to be used to reduce the trade distortions resulting from domestic policies, a critical question is whether there are policy instruments that allow national governments to continue to pursue domestic objectives, such as support for farm income or stabilization of domestic markets, and at the same time reduce or eliminate trade distortions. For income support, economists have long advocated decoupled income support (direct payments unrelated to the current level of production). In practice, it is difficult to identify such distortion-free transfers except for unanticipated lump sum transfers. Although such decoupled options have desirable characteristics in theory and should be encouraged, several major shortcomings can be identified. First, fully decoupled income supports are likely to be unattractive to policymakers and farmers since they represent a radical departure from traditional commodity-based price support policies. Farmers have a psychological preference for obtaining a fair price, preferably from the market, for what they do rather than for who they are. A welfare stigma is associated with income transfers that are unrelated to production. It is unlikely that fully decoupled policies will ever be sufficiently acceptable politically to replace existing support instruments completely. Second, decoupling is a general principle, not a program. Its implementation would require modification of existing policies on a case-by-case basis. There are no clear-cut criteria under which decoupled policies can be implemented by governments or monitored through an international body such as the GATT as part of a trade agreement. It is conceivable, for example, that decoupled payments to producers could be financed by consumer taxes on food. Because of its effects on consumption, however, such a program would not have the desired effect of minimizing trade distortions.

Because farm income support is such a critical issue in most industrial countries, considerable attention will inevitably be directed to the characteristics of support policies that will achieve support objectives and be compatible with the GATT. However, there are other policy objectives that may have to be met. One of the most important is domestic market stability. Many countries profess to place great importance on the stabilization of agricultural prices. Some are concerned about national "food security," which can have several interpretations but may be equated with maintaining access to adequate supplies of food at reasonable prices. How can these objectives be met in an environment of freer trade?

Trade liberalization itself would tend to stabilize international prices. Figure 7 shows that the variability of prices would tend to decrease under free trade, although not by much except for wheat, dairy products, and perhaps rice. As indicated earlier, many countries export their domestic instability through the use of export subsidies or state trading. By insulating domestic markets from world price changes, they pass on to others the burden of adjusting to fluctuations in world supply. These destabilizing influences would be reduced with reforms of domestic policies made necessary by a GATT agreement. The agreement could also place specific limitations on the use of destabilizing policy measures—for example, export embargoes or import prohibitions. But even with these reforms, domestic markets might not be sufficiently stable to satisfy domestic objectives. Additional national or multinational measures could deal with volatility. Countries could use public stocks to stabilize markets. The acquisition and release of these stocks would be based on fixed rules and would be linked directly to changes in international prices. Nations could enter into bilateral or multilateral purchase or sales agreements to provide some degree of supply security. As in the case of public stocks, purchases and sales under such agreements would be made at world prices. If farm income instability is a major concern, insurance-type schemes could be used. In order to minimize potential trade distortions, however, any government subsidy element in such a scheme would have to be capped.

Several additional food and agricultural policy objectives could be met in a manner consistent with liberal trade. Environmental or natural resource conservation aims are becoming increasingly important in industrial countries. Tax and subsidy policies aimed at conservation or environmental protection are compatible with liberalized trade if these policies merely equalize the private and social costs of the use of natural resources. Similarly, the provision of public goods such as food safety would not be trade distorting, provided that domestic and foreign sources of supply were accorded equal treatment in terms of health and safety restrictions and inspection requirements. Other public goods, such as extension services and support of research and development, which typically involve public expenditures but result in a gain for society as a whole, could continue to be a part of the domestic agricultural policy environment under the GATT. The line between a trade-influencing and a trade-distorting policy is not always easy to draw, but it must be drawn in the GATT negotiations.

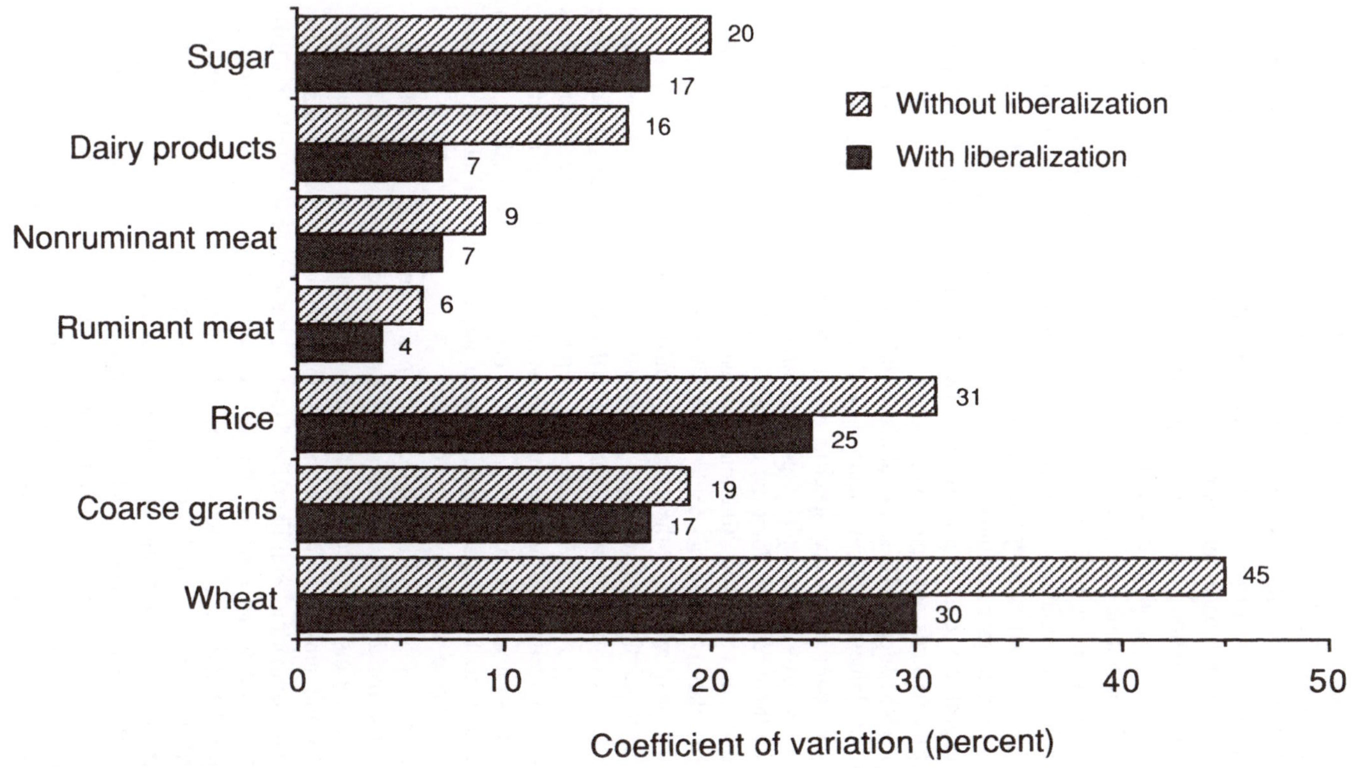

FIGURE 7. Estimated effect of trade liberalization in industrial countries on world price variability. *Source:* Based on data from Tyers and Anderson (1986).

REQUIRED ADJUSTMENTS IN AGRICULTURAL POLICY

A General, Internationally Workable Approach to Adjustment

One approach that has similarities to the U.S. grains and cotton and the EC sugar programs but minimizes trade distortions is Production Entitlement Guarantees (PEGs) (Blandford, de Gorter, and Harvey, 1988, 1989; Blandford et al., 1989). (For a general description of a PEG program, see Phipps, Rossmiller, and Meyers, in this volume). If PEGs are to be a part of the reform of agricultural policies under the GATT, the quantity or portion of production eligible for support must be determined as part of the negotiations. As an illustration, however, figure 8 shows the estimated adjustment of current world prices toward free trade levels for selected commodities, with PEGs established at 80 percent (PEG80) of 1986 production. The actual level of producer support *per unit* is kept at that actually estimated for 1986 through the PSE, with all other market intervention eliminated so that consumers pay market prices.

Figure 8 shows that all commodities would experience a substantial movement toward free trade prices if PEGs were introduced at this level. With the exception of rice and sugar, a PEG80 scheme would move prices to more than 98 percent of their free trade level. Even for rice and sugar, whose prices are heavily distorted by domestic support programs, PEG80 prices are estimated to be roughly 93 percent of free trade prices. Hence, if historical levels of support are maintained but the form of support is switched to a direct payments system based on a fixed quantity of production, most of the benefits resulting from freer trade can be realized.

A system of direct payments such as PEGs could provide a largely decoupled and explicit program for providing income support with minimal trade distortions. It could be implemented independently of the type of agreement reached in the GATT on how to phase out trade-distorting policies (e.g., through tariffication or the use of phased reductions in existing policies based upon an aggregate measure of support). It would provide policymakers with necessary flexibility, since this type of scheme could be either a transitory mechanism for compensation and adjustment assistance under a negotiated agreement on the reduction of trade-distorting policies or a more permanent vehicle for income support within such an agreement. PEGs provide a means for governments to reduce trade distortions while maintaining farm income supports. They also provide a mechanism for compensating producers for income losses from the adoption of environmentally sound production practices, or to meet other environmental objectives. PEGs allow countries to realize mutual gains from freer trade through increased world prices and reduced consumer prices. They involve a known and, therefore, potentially limitable, level of budgetary expenditures. They are consistent with the concept of traditional commodity programs that provide support based on production. PEGs could provide substantial national flexibility in the targeting of support in terms of commodities, farms, farmers, or regions. Finally,

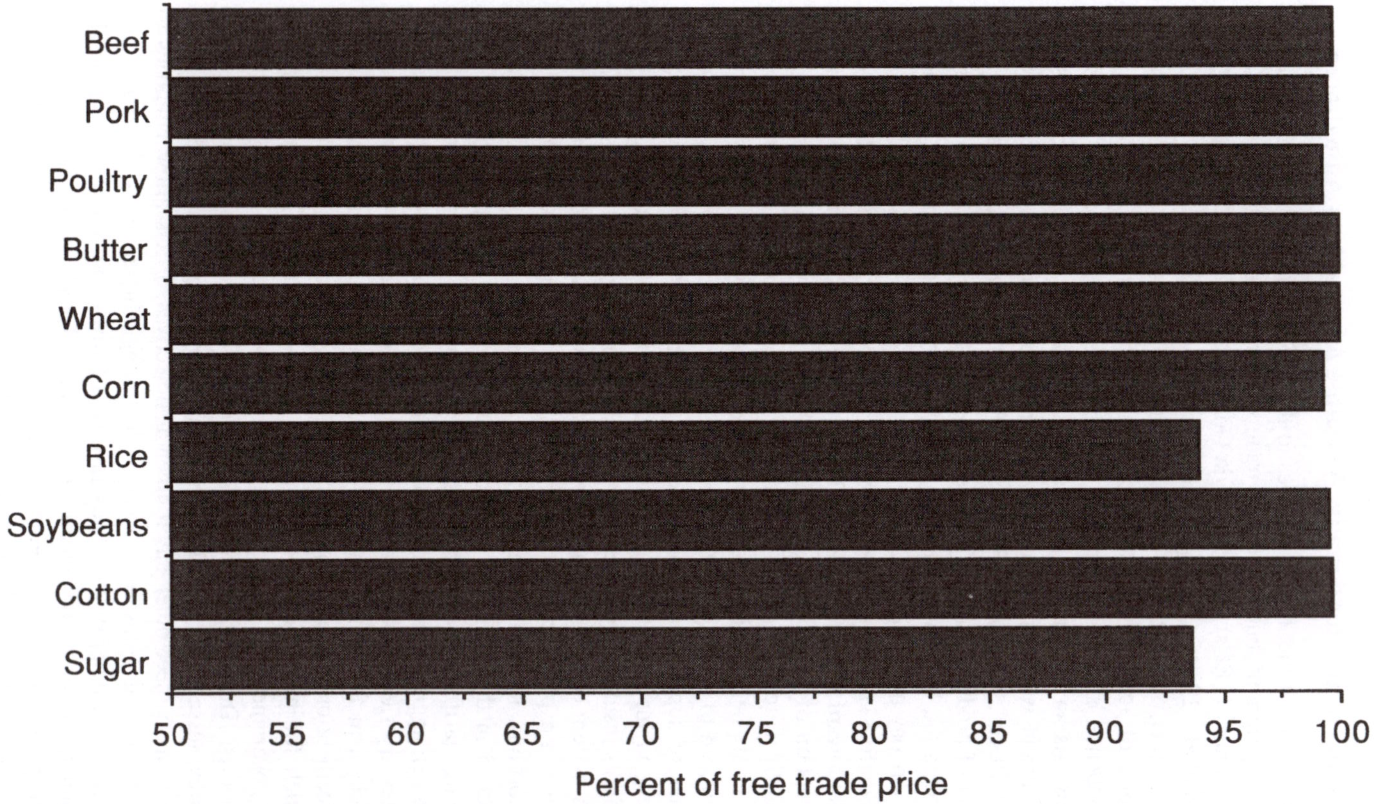

FIGURE 8. Adjustment of international prices toward free trade levels under a PEG80 in 1986.

they are a more efficient vehicle for achieving compensation, adjustment assistance, and support than most existing programs.

Specific adjustments in the policies of industrial countries that would have to be made to implement a PEG program would depend on the exact terms of a GATT agreement on agriculture. Tariffication, for example, would require a radically different approach than the phased elimination of trade-distorting support based on an aggregate measure. The general direction of domestic reform, however, can be identified under the assumption that trade-distorting forms of support are phased out and replaced by decoupled payments or a PEG-type scheme, at least during a period of transition.

The United States

For U.S. commodity programs reform would require the phaseout of protection from import quotas for dairy products and sugar, support prices (dairy support prices, loan rates, target prices, and so forth), import tariffs and export subsidies (e.g., export enhancement programs), and acreage and production controls.

For grains, rice, and cotton, U.S. target price/deficiency payment programs have evolved substantially in recent years toward a largely decoupled model. Under the Food Security Act of 1985, each farm has an established base acreage and program yield upon which a farmer's payments are based. To make current programs compatible with a GATT agreement to eliminate trade distortions, all acreage reduction and limitation provisions and subsidies through Commodity Credit Corporation (CCC) loan programs would need to be eliminated. The government could still hold stocks for security reasons, but all purchases and sales would have to be made at market prices. Farm-level base acreage and program yields would be reduced in line with the final terms of the GATT agreement (e.g., under a PEG80, to 80 percent of a historical base) and unconditionally frozen. Related policies such as export subsidies and CCC surplus disposal through Payment-in-Kind certificates would be eliminated. Payment limitations could be maintained, tightened, or eliminated as U.S. political conditions demanded.

For other supported commodities, notably sugar, dairy products, tobacco, and peanuts, existing measures that distort consumer prices, such as import quotas, would be phased out. For example, for dairy products, the government could issue a production "base" to each farmer determined by some fraction of the farmer's historical production of raw, whole milk. Payments could be set at the current level of price support. The fluid price differential, import quotas and tariffs, and CCC support purchases would all be eliminated.

The Boschwitz-Boren Bill (the proposed Family Farm Protection and Full Production Act of 1985), which was debated by Congress in 1985, closely approximates a largely decoupled scheme for crops. It proposes freezing payment bases at or near current program levels and making future payments at a declining rate, independent of output or input decisions. (See Phipps,

Rossmiller, and Meyers, in this volume, for a summary of the proposed bill.) The Boschwitz-Boren approach has so far not found sufficient support for enactment either in Congress or among farm and commodity organizations, but its time may come as part of multilateral agricultural policy reform. Although a PEG program would not necessarily require a phase down of the level of support, it would require the extension of the Boschwitz-Boren approach to other commodities currently receiving support in the United States.

The European Community

For European farm policy, reform would require the phasing out of variable import levies, export restitutions, production quotas, and producer coresponsibility levies, with an equalization of consumer and world market prices. The now accepted principle of limiting output receiving support payments in the cereals sector via "maximum guaranteed quantities" could be used to implement a PEG scheme. Current production quotas in the milk and sugar regimes could evolve into payment quotas with no restrictions on producer output decisions. For meat, the EC could issue PEGs to farmers based on an agreed portion of historical production in order to maintain current income levels based on direct payments.

Canada

In Canada reform would require the elimination of import quotas and transportation subsidies. Stabilization payments would have to be limited by replacing marketing quotas with PEGs (in the case of crops in western Canada) and the establishment of PEGs for all other sectors currently receiving stabilization payments. As in Europe, production quotas would have to be transformed into payment quotas.

Japan

In Japan reform would imply the elimination of import quotas, tariff quotas, tariffs, and the management and manipulation of imports and domestic stocks by state trading agencies. Stocks could be held for domestic stabilization and food security purposes but would have to be sold at prices linked to world market prices. Existing deficiency and land diversion payments would have to be phased into decoupled direct payments or PEGs. Input subsidies would be eliminated.

CONCLUDING REMARKS

The liberalization of international agricultural trade through the GATT process would lead to significant gains in the economic welfare of industrial countries. In the United States substantial consumer and taxpayer savings would outweigh producer losses from freer trade. The phased elimination of current trade-distorting agricultural income support measures in all countries would increase world market prices and reduce adjustment costs. Nevertheless, mechanisms such as a PEG scheme would have to be found to provide at least transitory compensation or adjustment assistance to those parties adversely affected by trade liberalization.

Although the exact form a GATT agreement may take is not yet known, any significant reduction in trade-distorting support measures will impose constraints on domestic policy choices. Domestic objectives such as income support, market stabilization, environmental goals, food safety, and the provision of public goods can be met within the constraints imposed by a GATT agreement but will require changes in the policy mechanisms employed. The goal of farm income support can be realized through the use of direct payments unrelated to production or based on a fixed production quantity. Market stabilization could be achieved through public stockholding and international supply and purchase guarantees. However, in both cases, transactions would be linked closely to world market prices.

If these measures are used, domestic policy objectives will be achievable and at lower cost than at present. The efficiency of many existing policy measures is low because of leakages due to economic waste and the accrual of benefits to upstream and downstream agribusiness firms rather than to farmers. In addition, a substantial part of current farm subsidies merely offsets the effects of other countries' policies on prices. The task of achieving policy goals could be met more easily and effectively if each goal were addressed directly. The efficiency of income transfers to farmers could improve substantially if alternative schemes—such as producer entitlement guarantees or completely decoupled payments—were adopted. Such programs are consistent with internal policy needs and are also minimally trade distorting. They are a natural evolution of existing policy instruments rather than a radical departure from these instruments. The discipline imposed by a GATT agreement on agriculture presents challenges to domestic policymakers but in the end may simplify their task of achieving domestic policy goals efficiently.

REFERENCES

Blandford, D., H. de Gorter, and D. Harvey. 1988. "Production Entitlement Guarantees (PEGs): A Minimally Distorting Method of Farm Income Support." Paper presented at the International Agricultural Trade Research Consortium (IATRC) symposium, Bringing Agriculture into the GATT, Annapolis, Md., August 18–19.

———. 1989. "Farm Income Support with Minimal Trade Distortions," *Food Policy* vol. 14, no. 3 (August), pp. 268–273.

Blandford, D., B. Gardner, H. de Gorter, and D. Harvey. 1989. "Production Entitlement Guarantees (PEGs): Farm Income Support with Minimal Trade Distortions," *Choices* (first quarter).

de Gorter, H. 1988. "Analyzing Trade Distortions Due to Agricultural Policies." Paper presented at the International Agricultural Trade Research Consortium symposium, Bringing Agriculture into the GATT, Annapolis, Md., August 18–19.

de Gorter, H., and D. McClatchy. 1984. " 'Rates of Distortion' as an Alternative to 'Rates of Protection' in Analyzing the Trade Effects of Agricultural Policies." Appendix I to a paper presented at the International Agricultural Trade Research Consortium symposium, Airlie House, Va., June.

Organization for Economic Cooperation and Development. 1987. *National Policies and Agricultural Trade* (Paris).

Parikh, K. S., G. Fischer, K. Frohberg, and O. Gulbrandsen. 1988. *Toward Free Trade in Agriculture.* Food and Agricultural Program (Laxenburg, Austria, International Institute of Applied Systems Analysis).

Roningen, V. O., and P. Dixit. 1987. "Impact of Removal of Support to Agriculture in Developed Countries." Unpublished paper (Washington, D.C., U.S. Department of Agriculture, Economic Research Service).

———. 1989. "Economic Implications of Agricultural Policy Reform in Industrial Market Economies." U.S. Department of Agriculture, Economic Research Service, Staff Report no. AGES 89-36 (Washington, D.C., August).

Tyers, R., and K. Anderson. 1986. "Distortions in World Food Markets: A Quantitative Assessment" (Canberra, National Center for Development Studies, Australian National University, January).

U.S. Department of Agriculture, Economic Research Service. 1988. "Estimates of Producer and Consumer Subsidy Equivalents: Government Intervention in Agriculture, 1982-86." Staff Report no. AGES 880127 (Washington, D.C., April).

THE COMPETITIVENESS OF U.S. AGRICULTURE IN THE 1990s

ALAN BARKEMA, MARK DRABENSTOTT, and LUTHER TWEETEN

Competitiveness, long taken for granted, has now become a byword in U.S. agriculture. A number of events in the 1980s brought this about. For one, U.S. farm exports slumped through much of the decade. Also, the exchange value of the dollar underwent a dramatic cycle, sharply affecting the prices of U.S. farm products in world grain markets. And agricultural technology advanced rapidly in other producing nations, narrowing the historical edge of the United States in agricultural productivity. In short, the U.S. agricultural export supremacy that was assumed in the 1970s was questioned in the 1980s.

Competitiveness will remain a critical issue for U.S. agriculture in the coming decade. U.S. farmers were fully integrated into international markets in the 1980s. With that process complete, U.S. agriculture must now either compete effectively or shelve substantial amounts of its extensive resources.

Three principal factors appear likely to determine the competitiveness of U.S. agriculture in the 1990s: (1) the size and productivity of the U.S. farm resource base; (2) the infrastructure supporting farm production, including farm inputs, transportation, and processing; and (3) macroeconomic, farm, and trade policy decisions.

This paper evaluates the ability of U.S. agriculture to compete in the 1990s on these three factors. It concludes that U.S. agriculture will retain its share in world food markets under a number of possible scenarios. The competitive edge that has been widely assumed in the past, however, will be much narrower in the decade ahead. Much of the U.S. advantage will flow from agricultural research and infrastructure. Investments in both areas will likely be needed to maintain competitiveness.

COMPETITIVENESS AND TRENDS IN U.S. AGRICULTURAL TRADE

Competitiveness is difficult for economists to gauge precisely. It is closely related to comparative advantage, the cornerstone of economic trade theory. At issue is which products should be produced domestically and which should be imported. A nation is said to have a comparative advantage in commodities that it can produce and export at the highest return per unit of resource fixed (nontraded) to the country (such as labor and land) in a well-functioning world economy and in the absence of market distortions (see Tweeten, 1989). According to the theory of comparative advantage, each country should specialize in the production and export of those goods produced with resources in relatively abundant domestic supply. Likewise, each country should import from its trading partners those goods produced with resources in relatively scarce domestic supply. Scarce domestic resources are then freed for other, more profitable uses.

The concept of comparative advantage describes the trading patterns that would occur as countries assess the relative costs of producing or trading in a world free of price distortions imposed by government policy. The real world, however, is replete with trade, agricultural, and tax policies that distort markets, effectively overriding the determination of comparative advantage. Thus, a nation is said to have competitive advantage in those commodities in which it makes the greatest return per unit of resources fixed to the country, given the actual economic environment including policy-induced distortions. Distortions to expand trade may make trade profitable to export firms but may reduce national income. Thus policies which enhance the competitiveness of one industry may not be in the interest of the nation as a whole. In brief, comparative advantage applies to a world of efficient, well-functioning, undistorted markets, and competitiveness applies to the world as it actually is.[1]

The competitiveness of U.S. agriculture was never questioned during the 1970s, when U.S. farmers boosted production and expanded market share in the face of strong growth in world food demand. Throughout the decade, world grain production and consumption grew at a brisk rate of just under 3 percent per year (table 1). But the pattern of production and consumption around the globe was not uniform. In importing countries a surge in consumption—fueled by strong income growth and increasing populations—outpaced grain production. Growth in consumption among exporting countries was nearly flat, however, and left large reserves of grain available for export. The result of the mismatched location of growth in production and growth in consumption was a swift expansion in world grain trade at an average annual rate of nearly 6 percent (table 1).

As a result of its huge and elastic resource base, U.S. agriculture was uniquely positioned to meet the increased world food demand and reap the benefits of the explosion in world trade. During the 1970s the United States aggressively expanded its plantings of feed grains, wheat, soybeans, and cotton by nearly 70 million acres, or about a third (figure 1). The expansion in acreage, combined with rising yields, doubled U.S. production of the

[1]For a more complete and mathematically rigorous discussion of the theory of comparative advantage, see Leamer, Edward E., 1984, *Sources of International Comparative Advantage, Theory and Evidence* (Cambridge, Mass., The MIT Press).

TABLE 1. Annual Rates of Growth in World Grain Production, Consumption, and Trade, 1970–1979 and 1980–1985

(percent)

	1970–1979				1980–1985[a]			
	Major importers[b]	Major exporters[c]	United States	World	Major importers[b]	Major exporters[c]	United States	World
Harvested area	0.6	1.5	1.9	0.9	−1.4	0.6	−1.0	−0.5
Yield	1.9	1.7	2.1	2.0	4.1	1.8	3.0	3.1
Production	2.5	3.2	4.0	2.9	2.7	2.4	2.0	2.7
Consumption	3.2	0.7	0.2	2.6	1.2	0.1	3.0	1.8
World trade	—	—	—	5.9	—	—	—	1.2

Note: Dashes mean data not applicable.

Sources: Importer and exporter growth rates are derived from unpublished data provided by the U.S. Department of Agriculture, Foreign Agricultural Service, Washington, D.C. World growth rates are derived from *World Agricultural Supply and Demand Estimates,* U.S. Department of Agriculture, Economic Research Service and Foreign Agricultural Service, Washington, D.C., various years.

[a]The data period ends in 1985 to exclude the effects of the U.S. Food Security Act of 1985. The act sharply curtailed U.S. grain production (which fell about 20 percent from 1985 to 1987), thereby slowing growth in grain production worldwide.

[b]Major importers include Brazil, China, eastern Europe, Egypt, Japan, Korea, Mexico, Saudi Arabia, Taiwan, and the USSR. Members of the European Community are included as net importers in the 1970s.

[c]Major exporters (excluding the United States) include Argentina, Australia, Canada, South Africa, and Thailand. Members of the European Community are included as net exporters in the 1980s.

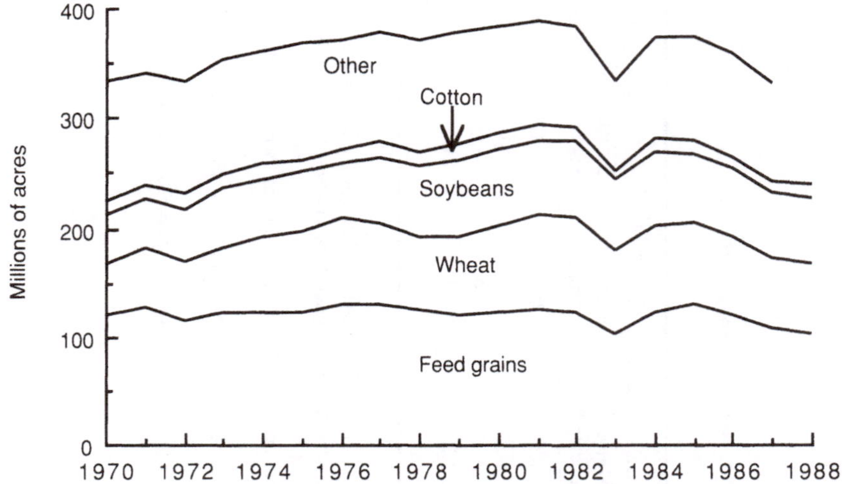

FIGURE 1. Cumulative U.S. planted acreage, 1970–1988. *Sources:* USDA, Economic Research Service and Foreign Agricultural Service (various years) and USDA, National Agricultural Statistics Service (various years).

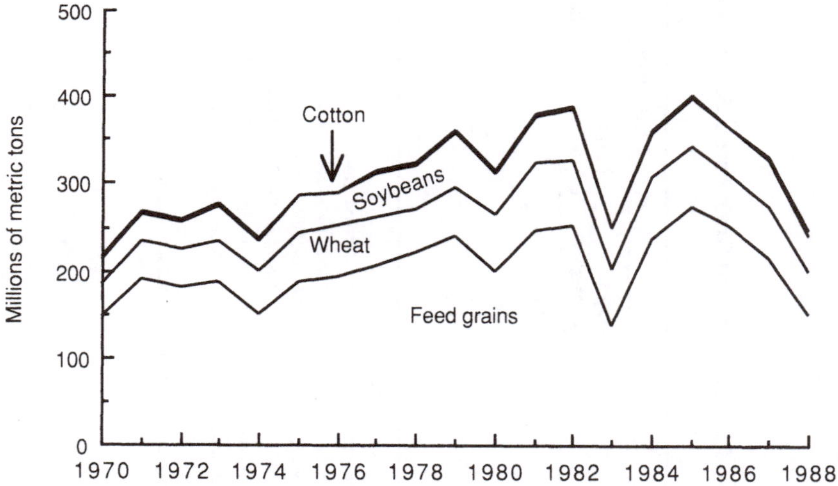

FIGURE 2. Cumulative U.S. crop production, 1970–1988. *Note:* Cotton production follows soybean production so closely that the distinction is not apparent on a graph of this scale. *Sources:* USDA, Economic Research Service and Foreign Agricultural Service (various years) and USDA, National Agricultural Statistics Service (various years).

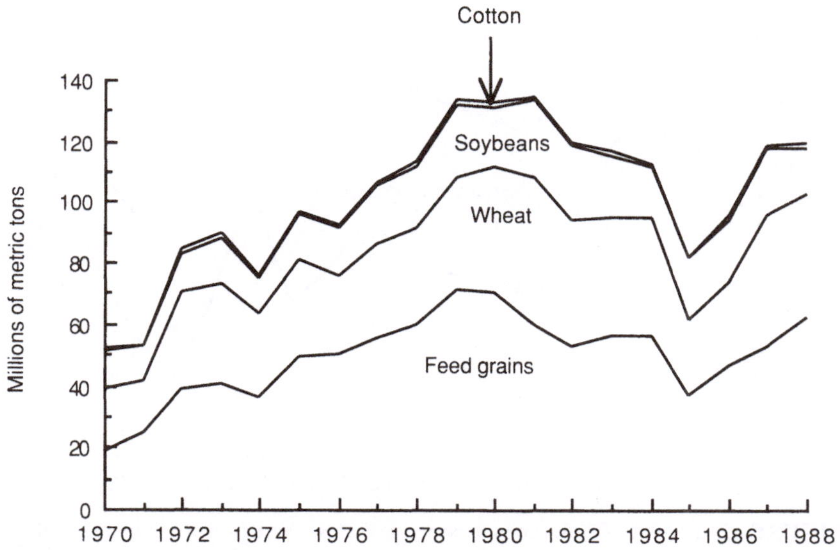

FIGURE 3. Cumulative U.S. exports, 1970–1988. *Note:* Cotton production follows soybean production so closely that the distinction is not apparent on a graph of this scale. *Sources:* USDA, Economic Research Service and Foreign Agricultural Service (various years) and USDA, National Agricultural Statistics Service (various years).

major crops in little more than a decade (figure 2). Throughout that period, a steadily increasing share of the rapidly expanding production was shipped abroad. The proportion of U.S. production exported grew from about a fourth to well over two-fifths as exports of the major crops soared from about 50 million metric tons in 1970 to nearly 135 million metric tons in 1981 (figures 3 and 4). U.S. agriculture's share of the world market rose by well over a third, from 44 percent at the beginning of the decade to 61 percent in 1979.

The explosion in world grain trade during the 1970s fizzled in the early 1980s.[2] Plummeting foreign demand triggered by sluggish foreign income growth, a soaring U.S. dollar, and rapidly expanding grain production abroad took a huge toll on U.S. farm exports. Real economic growth fell in 1982 for the Organization for Economic Cooperation and Development countries and slowed to a crawl for the developing world, where debt stress and poor economic performance replaced the robust growth of much of the previous decade. The lackluster performance of foreign economies resulted in diminished income for purchases of U.S. farm products. High rigid commodity loan rates under 1981 farm legislation also reduced the ability of the United States to compete in international agricultural markets.

[2]This discussion covers the period from 1980 to 1985. Beginning in 1986, growth in world grain production slowed because of the sharp curtailment of U.S. grain production with the implementation of the Food Security Act of 1985. Under the supply control provisions of the act, U.S. grain production fell about 20 percent from 1985 to 1987.

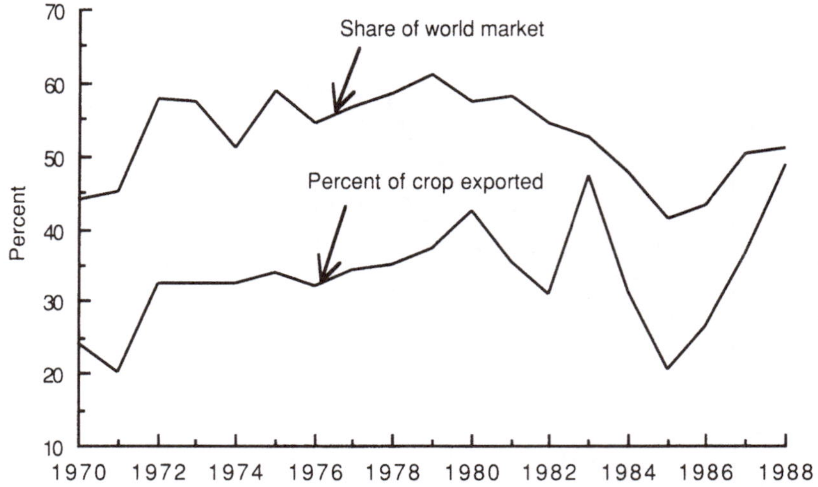

FIGURE 4. Portion of U.S. crops exported and U.S. share of world market, 1970–1988, noncumulative. *Sources:* USDA, Economic Research Service and Foreign Agricultural Service (various years) and USDA, National Agricultural Statistics Service (various years).

Also in the 1980s, grain yields, supported by rapid advances in production technology, rose even faster than in the 1970s in importing countries and the United States and other exporting countries (table 1). Expanded production enabled some countries that were net importers in the 1970s—especially the members of the European Community—to become net exporters in the 1980s. Overall, world grain production continued to expand at almost the same rate that it had in the 1970s, despite a slight worldwide decline in acreage. But growth in world grain consumption plummeted. This was especially pronounced among importing countries, where food demand was constrained by sluggish income growth and rising debt burdens. As a result of increased production worldwide and slack demand among the importing countries, world grain trade grew at a much slower pace than it had in the preceding decade.

The impact on U.S. agriculture of the abrupt contraction in the world market was enormous. U.S. exports of the major crops dwindled from nearly 135 million metric tons in 1981 to only 82 million metric tons in 1985 as U.S. producers' share of the diminished world market fell to less than 45 percent, the smallest in more than fifteen years (figures 3 and 4). The industry's adjustment to the new world market was to cut back production sharply by pulling 55 million acres out of production by 1988.

Thus, a shrinking world market in the early 1980s brought U.S. agriculture face to face with its foreign competitors. The experience left the industry with a significant portion of its resource base lying idle and with doubts about its ability to compete in a tougher world market. Although the costly 1985 farm bill (the Food Security Act of 1985) engineered a modest turnaround in U.S. farm exports during the past two years, discount pricing and outright export subsidies are not adequate substitutes for long-term competitiveness.

COMPETITIVENESS: FARM-LEVEL FACTORS

At the level of the farm, competitiveness will be influenced by U.S. agriculture's resource base and how productive it is and can become. An initial assessment of farm-level competitiveness can be made by comparing costs of production among countries. Future competitiveness trends can then be gauged from the outlook for productivity and yield advances.

A Comparison of Production Costs

The United States traditionally has been very competitive in such agricultural commodities as grains and oilseeds. It has the largest mass of first-class agricultural land in the world and enjoys a favorable climate. Well-educated and highly skilled family-farm operators, supported by a strong transportation, research, storage, and informational infrastructure, further strengthen U.S. agricultural competitiveness.

Costs of production are frequently compared across countries in an attempt to gauge competitiveness. While such comparisons provide one static glimpse of how well the United States compares to other major producers, competitiveness is subject to a manifold set of factors that can change quickly over time. This set includes exchange rates, differing costs across regions, and other factors that are discussed below. Thus, cost comparisons must be viewed cautiously.

A comparison among countries of the costs of production and transportation to major food importing ports, such as Rotterdam and Japan, provides insights into the competitiveness of U.S. commodities in world markets. Producers earn more net income by producing than not producing if at least variable costs of production are covered. Hence, short-term landed costs (returns above variable costs plus transportation), as shown in table 2, provide clues to competitiveness when all costs cannot be varied. Below this price, producers have reason to exit markets.

In the short run, Argentina and Thailand can produce and ship corn to Rotterdam and/or Japan more cheaply than the United States can. Argentina, Australia, Canada, France, and the United Kingdom can produce and ship wheat to Rotterdam and/or Japan more cheaply than the United States can. No country can produce and ship soybeans more cheaply.

In the long run, commodities must compete more broadly for use of resources. For example, the United States may be able to produce soybeans at a greater profit over variable costs than Brazil in the short run. But in the long run, U.S. soybean production must compete with corn. If U.S. corn production is more profitable than U.S. soybean production, land prices will

TABLE 2. Landed Costs (Returns Above Variable Costs Plus Transportation) of Corn, Wheat, and Soybeans in Rotterdam and Japan for Various Exporting Countries

(U.S. dollars per bushel, 1986 price and exchange rates)

	Corn		Wheat		Soybeans	
Port/exporting country	Short-term costs	Long-term costs	Short-term costs	Long-term costs	Short-term costs	Long-term costs
Rotterdam						
Argentina	2.40	3.41	2.45	3.55	3.66	5.54
Brazil	3.23	5.14	7.67	9.83	4.95	7.03
South Africa	—	4.14	3.53	5.09	—	—
Thailand	2.49	3.38	—	—	—	—
Australia	—	—	2.81	4.90	—	—
Canada	—	—	2.70	5.55	—	—
France	3.21	5.98	2.26	5.44	—	—
United Kingdom	—	—	2.74	4.70	—	—
United States	2.45	3.99	3.03	5.51	3.42	7.63
Japan						
Argentina	2.75	3.76	2.83	3.93	4.04	5.92
Brazil	3.68	5.59	8.15	10.31	5.43	7.51
South Africa	—	4.43	3.84	5.40	—	—
Thailand	2.29	3.18	—	—	—	—
Australia	—	—	2.64	4.73	—	—
Canada	—	—	2.93	5.78	—	—
France	4.40	6.69	3.02	6.20	—	—
United Kingdom	—	—	3.47	5.43	—	—
United States	2.79	4.33	3.40	5.88	3.78	7.99

Note: Dashes mean data not available or not applicable.

Source: Reprinted with permission from Rask, Norman, 1987, "Production and Marketing Costs for Corn, Wheat, and Soybeans in Major Export Countries," unpublished paper presented at the extension meeting of the Department of Agricultural Economics and Rural Sociology, Ohio State University, Columbus, January.

rise. When the higher land price is included in the long-run total cost of producing soybeans, U.S. soybean production costs may be higher than Brazil's, despite Brazil's higher variable costs. Thus, total costs of production including land and transport provide useful insights into long-term competitiveness.

According to the landed-cost estimates in table 2, Argentina and Thailand have a long-run advantage over the United States in corn production and transport. Argentina, South Africa, Australia, Canada, France, and the United Kingdom are competitive in wheat production and transport to Rotterdam and/or Japan. And Argentina and Brazil are competitive in soybean production and transport.

Landed costs such as those shown in table 2 have numerous shortcomings as measures of competitiveness. One is that they do not take into account changes in exchange rates, which can cloud cost comparisons. The U.S. dollar has fallen sharply against the Japanese yen and German mark since 1986 but has not declined as much against currencies of other countries important to agricultural trade. Hence, the estimates in table 2 would not necessarily be very different with updated exchange rates. The major deficiency of landed-cost estimates is that they fail to account for the vast variation in costs of production among regions within each country and from year to year. What region or producer should represent production and transportation costs? Some low-cost regions have only modest production capacity and do not represent a nation's overall capabilities.

A more complete assessment of relative production costs across different regions or countries would include a comparison of production costs throughout a range of production levels. Unfortunately, data to support such an analysis are limited. A somewhat dated but nevertheless unique study by the U.S. Department of Agriculture (1985) of U.S. costs in 1974 and 1981 showed that the variable cost of approximately 60 percent of total U.S. corn production in both years was less than the average variable cost for the crop as a whole. The portion of the total U.S. wheat crop produced at less-than-average variable cost increased from about 40 percent in 1974 to about 55 percent in 1981. These limited data suggest that a substantial portion of the U.S. corn and wheat crops may be produced at costs lower than those in competitor countries with moderately lower average production costs.

Also important to a more complete production cost analysis is the fact that the production and export volumes of U.S. wheat, corn, and soybeans are several times larger than the production and export volumes of other major exporters, especially the low-cost producers Argentina, Brazil, and Thailand. Many nations compete effectively with the United States at low prices by utilizing fertile agricultural land requiring few inputs and a low labor cost. But they are incapable of expanding output substantially without costs rising well above those in the United States, because additional output requires fertilizers, chemicals, irrigation, and cropping of less fertile soils—all costly where transportation and other infrastructure are undeveloped.

The United States produces and exports far greater volumes than the current least-cost producer of each major crop, and production costs of a significant portion of U.S. output may well be below costs in competitor countries. Argentina, for example, can produce corn, wheat, soybeans, and beef more cheaply than the United States but falls far short of being able to meet world export demand for farm commodities. The United States annually produces approximately twenty times as much corn as Argentina, six times as much wheat, and five to seven times as many soybeans (table 3). Clearly, U.S. agriculture would maintain a significant presence in world markets even with a moderate reduction in production volume, because the reduction would lower average production costs by limiting high-cost domestic production.

TABLE 3. Production and Export Volumes of Principal Exporting Countries, 1986–1988
(millions of metric tons)

	Production			Net exports		
Crop/country	1986	1987	1988	1986	1987	1988
Corn						
Argentina	9.25	9.00	5.30	4.03	4.38	1.60
China	70.86	79.82	75.00	3.78	4.30	4.00
South Africa	7.15	7.08	11.00	1.44	0.75	4.40
Thailand	4.31	2.74	4.50	2.92	0.80	1.90
United States	209.56	179.65	125.00	38.20	43.98	53.34
Wheat						
Argentina	8.93	8.80	7.60	4.44	3.71	3.30
Australia	16.12	12.44	14.10	15.64	9.86	10.00
Canada	31.38	25.95	15.66	20.78	23.50	11.70
EC-12	72.04	71.43	75.02	28.04	27.77	32.49
United States	56.93	57.36	49.29	27.32	43.33	40.82
Soybeans						
Argentina	7.00	9.90	8.30	1.35	2.09	2.00
Brazil	17.30	18.05	21.00	3.29	2.71	4.50
United States	52.80	52.33	41.88	20.60	21.83	14.97

Note: Volumes for 1987 are estimates; volumes for 1988 are projections.
Source: U.S. Department of Agriculture, various years, *World Agricultural Supply and Demand Estimates,* Economic Research Service and Foreign Agricultural Service, Washington, D.C.

U.S. Labor and Multifactor Productivity Trends

Ample evidence supports the view that the United States currently is competitive in the production of major agricultural commodities. But future trends in comparative advantage depend on a number of factors, including the productivity of agriculture versus other industries in the United States, and productivity gains of American agriculture versus agriculture in other countries.

Figure 5 compares U.S. labor productivity in the farm versus the nonfarm sector from 1950 to 1986. Farm labor productivity has risen steeply relative to nonfarm labor productivity. Figure 6 compares multifactor productivity between the farm and nonfarm sectors in the United States from 1950 to 1986. Multifactor productivity, which shows the ratio of all output of a sector per unit of all production inputs, has also risen in the U.S. farm sector compared to the nonfarm sector. Rising labor and multifactor productivity in the farm versus the nonfarm sector indicates continuing gains in relative efficiency for farming within the domestic economy, but does not necessarily indicate comparative advantage internationally if agricultural productivity is rising more rapidly in other countries than in the United States.

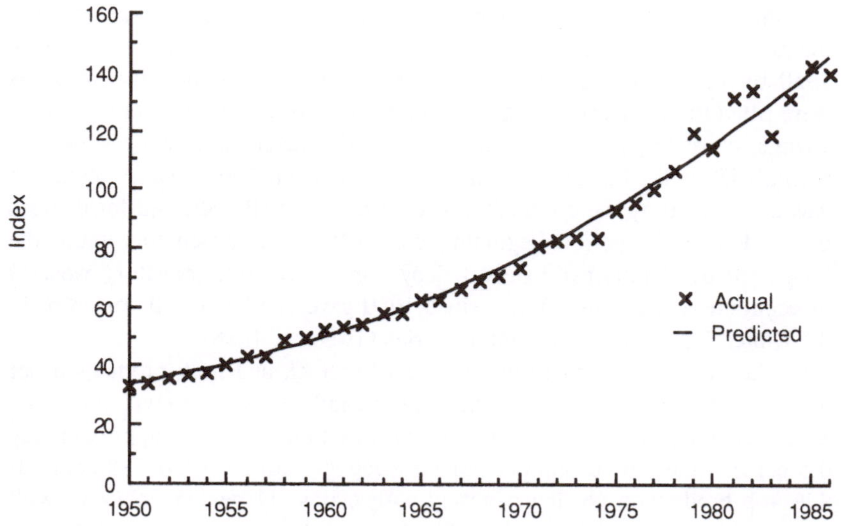

FIGURE 5. U.S. labor productivity in the farm sector as a percentage of the nonfarm sector, 1950–1986, with 1977 = 100. *Source:* Based on data from the Council of Economic Advisors and U.S. Department of Labor (various years).

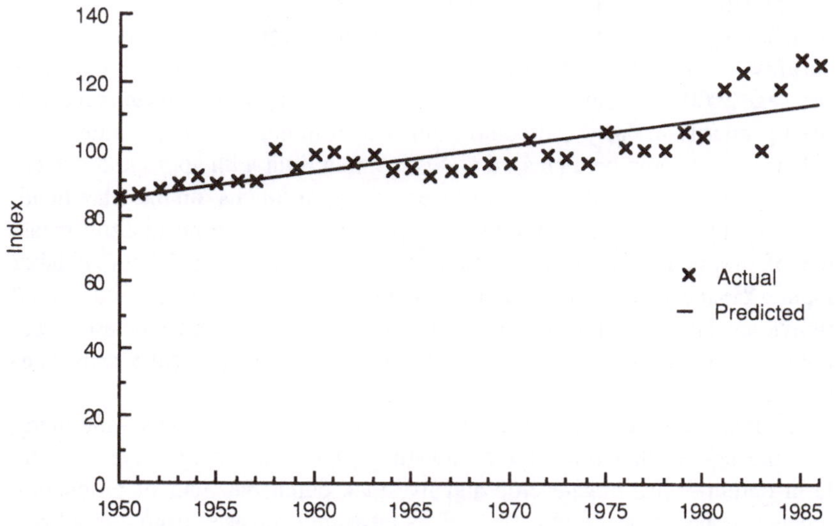

FIGURE 6. U.S. multifactor productivity in the farm sector as a percentage of the nonfarm sector, 1950–1986, with 1977 = 100. *Source:* Based on data from the Council of Economic Advisors and U.S. Department of Labor (various years).

Although weather and other random shocks have obscured yield trends, the rate of gain in the productivity of American agriculture has slowed since 1950 by nearly all measures, as shown in table 4. Productivity measures were fitted to time trends by four equations that allowed for constant absolute change, constant percentage change, or variable percentage changes over the period 1950 to the most recent year of available data. The "best fit" equation was determined by the coefficient of determination (R^2 adjusted for degrees of freedom). The period beginning with 1950 was chosen to capture the long-term trend minimally distorted by unrepresentative transitory weather or economic conditions. Trends since 1960 gave similar results for crops in the United States and in foreign countries (Tweeten, 1988).

Of the productivity measures, only all livestock and milk production per hour of labor input did not display significantly slowing growth (table 4). Livestock and milk production per hour showed rapid gains of approximately 6.5 percent and 7.9 percent per year, respectively, for the 1950–1986 period. All crop production per hour showed early gains (11 percent in 1950) with extensive mechanization, but by 1986 the trend had slowed to a 6.4 percent annual rate (table 4), about the same as for both livestock and poultry production per hour.

If these trends continue, by the year 2000 the rate of increase in all livestock production per breeding unit and all crop production per acre will slow to half the trend rate of 1950. That tendency for annual gains to halve is apparent for a number of measures in table 4.[3]

Productivity gains in pigs per litter and meat animal production per hour of labor input were never very large over the 1950–1986 period and fell steadily. If this trend continues, these indicators will show rates of gain by the year 2000 only half those of 1950. Productivity gains for animals have been primarily in dairy and poultry rather than in hogs and beef cattle.

Farm labor could be increased almost without limit with appropriate incentives. Cropland and crop yield enhancement opportunities, on the other hand, are more limiting. Hence, productivity per acre is a more meaningful measure of production capacity constraint than is production per hour of labor input. Yield trend annual increases in production per acre slowed for all individual crops as shown in table 4. By 1986, yield trend increases were only 1.7 percent for wheat, 2.3 percent for corn, and 0.9 percent for soybeans and cotton.

The most comprehensive estimate of overall productivity gains—and therefore the best explanation of a rightward shift in the supply curve for total farm output—is aggregate crop and livestock output per unit of production inputs, the last measure in table 4. This measure decreased from a trend rate of 2.6 percent in 1950 to 1.5 percent in 1986. If the trend continues, aggregate farm output supply will increase only 1.3 percent in the year 2000.

[3]A specific algebraic form can force a pattern of percentage changes. However, the "best fit" determined the choice of algebraic function.

TABLE 4. U.S. Productivity Trends for Livestock and Livestock Products and for Crops

Item	Equation[a]	R^2	Percent increase per year 1950	1970	1986	2000 est.[b]
All livestock production per hour of labor input[c]	ln-linear	.998	6.51	6.51	6.51	6.51
	ln-ln 1950	.998	6.57	6.51	6.45	6.41
All livestock production per breeding unit[c]	ln-ln 50	.968	2.06	1.47	1.20	1.03
Milk production per dairy cow[c]	ln-linear	.995	4.81	2.45	1.76	1.41
Pigs per litter[c]	ln-ln 50	.688	0.40	0.29	0.24	0.20
Meat animal production per hour of labor input[c]	ln-linear	.980	0.10	0.07	0.06	0.05
Milk production per hour of labor input[c]	ln-linear	.988	7.86	7.86	7.86	7.86
Poultry production per hour of labor input[c]	ln-ln 50	.998	11.00	7.85	6.39	5.50
All crop production per hour of labor input[c]	ln-ln 50	.998	11.00	7.85	6.39	5.50
All crop production per acre[c]	ln-ln 50	.953	2.60	1.86	1.51	1.30
Wheat production per acre[d] (with ln harvested acres)	ln-ln 50	.916	2.97	2.12	1.73	1.49
Wheat production per acre[d] (without ln harvested acres)	ln-ln 50	.898	2.82	2.02	1.64	1.42
Corn production per acre[d]	ln-ln 50	.885	3.98	2.84	2.31	1.99
Soybean production per acre[d]	ln-ln 50	.748	1.57	1.12	0.91	0.79
Cotton production per acre[d]	ln-ln 50	.655	1.58	1.13	0.92	0.79
Total output of crops and livestock per all production inputs[c]	ln-ln 50	.962	2.59	1.85	1.51	1.29

Source: Tweeten, Luther, 1989, "Technological Progress, Productivity, and the Production Capacity of American Agriculture," paper presented to the conference on Surplus Capacity and Adjustment Problems of American Agriculture, St. Louis, Mo., January.

[a]Selected from following equations giving highest coefficient of determination (R^2 adjusted for degrees of freedom):

linear	$Yi = a + bT$
ln-linear	$\ln Yi = a + bT$
ln-ln 50	$\ln Yi = a + b\ln T$ (T = 50, 51, . . . 86)
ln-ln 1950	$\ln Yi = a + b\ln T$ (T = 1950, 1951, . . . 1986)

where Yi is productivity measure and T is time trend.

[b]Only 1950–1986 predictions are within range of data.

[c]Estimated from annual data for 1950 through 1986.

[d]Estimated from annual data through 1988.

Of course, U.S. agriculture's aggregate supply curve could also shift to the right if input prices decline. But real prices are likely to increase for labor, fertilizers, energy, and other inputs. Restraints on nitrogen fertilizer and pesticide usage for environmental protection could further slow advances in productivity.

Productivity gains in agriculture appear to be characterized by technological revolutions (Tweeten, 1987). These are difficult to predict because they are the result of complex forces including luck, institutions, and the long lags associated with investment in education and science. The beginnings of the next revolution, originating from biotechnology (recombinant DNA, tissue culture, growth stimulants, and so on) and computers (artificial intelligence, information systems, and the like), are observable now, and the promise of nuclear fusion power and superconductors is dimly perceived. At issue is not whether, but when, these will affect agriculture.

A panel of scientific experts assembled by the U.S. Congress, Office of Technology Assessment (1986) predicted gains in milk production per dairy cow averaging 4 percent per year from 1982 to 2000, assuming the adoption of bovine somatotropin.[4] That prediction now seems overly optimistic on the basis of the trend apparent in table 4. Vaccines, growth hormones, ice-minus bacteria to protect against frost, and corn and soybeans resistant to pesticides are genetically engineered products now emerging as proven technologies. But these and other products of the emerging scientific age are unlikely to have a large impact on farm productivity before the year 2000. Much adaptation and environmental impact research awaits. Thus, the slowdown in productivity shown in table 4 is likely to continue into the early 1990s. By the year 2000, however, rates may begin to rise as emerging technology comes on line.

Yield Trends in the United States and Other Countries

If the rate of advance in agricultural productivity was slowing more in other nations than in the United States, this would indicate that the competitiveness of U.S. agriculture was not declining internationally. Expressing U.S. crop yields as a percentage of foreign crop yields is one way to measure whether such a shift is occurring. It is well to note that yields are only a proxy for productivity. They may be influenced by sound investments in education and research or by market-distorting policies such as price supports and input subsidies. Figures 7 through 11 show yields of major crops in the United States as a percentage of foreign yields.

The U.S. wheat trend yield was 150 percent of foreign trend yield in 1961, the first year that reliable foreign data became available (figure 7). The

[4]For a more thorough assessment of the United States' position among other countries in developing new agricultural technologies and a summary of the expected impact of these technologies on U.S. agriculture, see U.S. Congress, Office of Technology Assessment, 1986, *Technology, Public Policy, and the Changing Structure of American Agriculture*, OTA-F-285 (Washington, D.C., Government Printing Office), March.

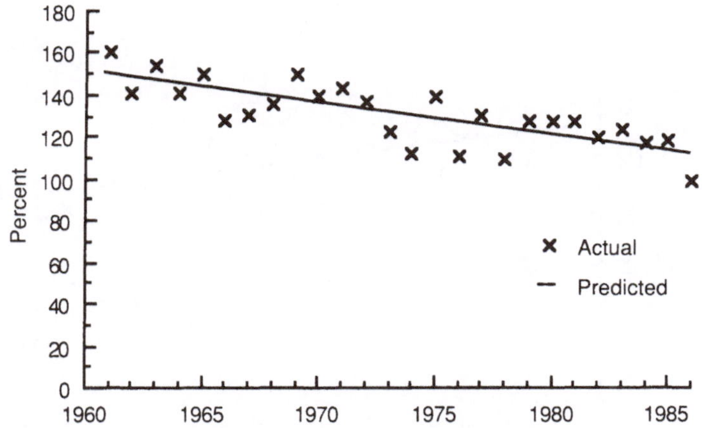

FIGURE 7. U.S. wheat trend yield as a percentage of foreign trend yield, 1961–1986. *Source:* U.S. Department of Agriculture, Economic Research Service (various years).

Green Revolution and other manifestations of the adoption of technology in foreign countries caused foreign yields to increase more rapidly than yields in the United States. Continuation of the trend shown in figure 7 will bring foreign yields up to U.S. levels in the 1990s. Although yield trend is just one of many dimensions of dynamic comparative advantage, it does suggest a declining competitive position for U.S. wheat.

Coarse grain yields for the United States as a percentage of those in other countries display an uneven but generally rising pattern from 1961 to 1986 (figure 8). U.S. trend yield, 2.6 times that in other countries in 1961,

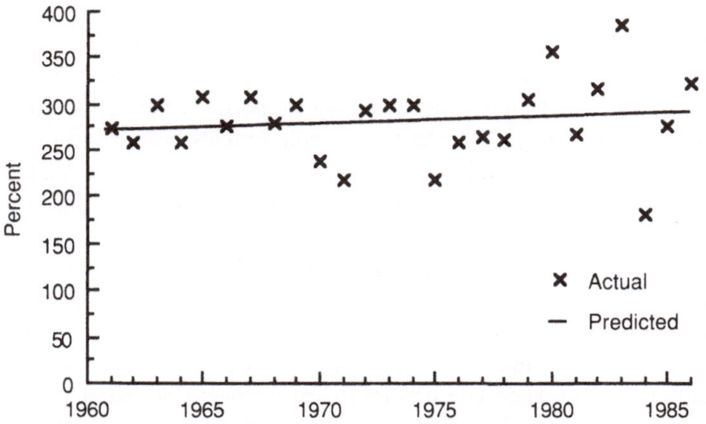

FIGURE 8. U.S. coarse grain trend yield as a percentage of foreign trend yield, 1961–1986. *Source:* U.S. Department of Agriculture, Economic Research Service (various years).

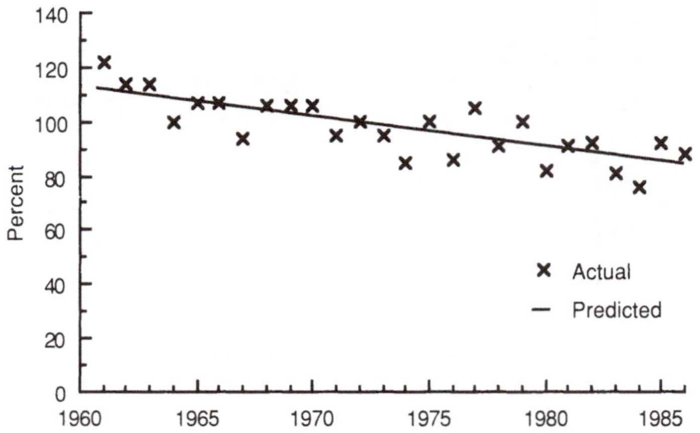

FIGURE 9. U.S. soybean trend yield as a percentage of foreign trend yield, 1961–1986. *Source:* U.S. Department of Agriculture, Economic Research Service (various years).

increased to a ratio of nearly 3.0 by 1986. This indicates historically strong and slightly increasing U.S. competitiveness.

In 1961 the United States displayed a trend yield for soybeans 15 percent above foreign yields (figure 9). Foreign countries reached parity with U.S. yields by the early 1970s and surpassed the United States in the 1980s. By 1986, trend yield in the United States was approximately 90 percent that of foreign countries. The declining U.S. share of the world soybean market corroborates the trend apparent in figure 9. Nevertheless, other factors, such as deficiency payments to U.S. farmers to produce grains rather than soybeans, probably contributed to the trend.

Cotton trend yields for the United States versus foreign countries display an even sharper decline than soybean trends, although the United States continues to enjoy a 40 percent cotton yield advantage (figure 10). The linear equation used in all yield ratios does not properly fit the cotton yield trend. The decline in U.S. yields relative to foreign yields bottomed in the 1970s. Thereafter, the yield relationship has been rather flat and the United States continues to enjoy an approximately 40 percent yield advantage over other countries.

Tobacco yields have been declining in the United States versus foreign countries (figure 11). U.S. trend yield averaged 240 percent of that of other countries in 1961 but only 180 percent by 1986. Because tobacco production is relatively labor-intensive, rising labor costs in the United States reduce comparative advantage. The problem is exacerbated by mandatory controls maintaining tobacco prices well above free-market levels and restraining plot size, thereby limiting mechanization efficiency. As with soybeans, policy changes could improve the competitive position of U.S. tobacco in world markets.

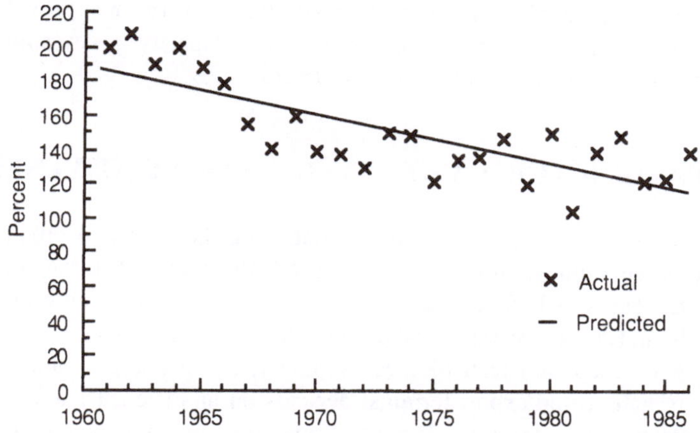

FIGURE 10. U.S. cotton trend yield as a percentage of foreign trend yield, 1961–1986. *Source:* U.S. Department of Agriculture, Economic Research Service (various years).

In summary, the growth rate in agricultural productivity in the United States appears to have begun to slow down. The decline in domestic productivity gains and the acceleration in foreign productivity growth suggest that U.S. agriculture's competitive advantage in some commodities has slipped in recent years. Technological innovations just coming into view or as yet unseen may boost competitiveness in the future. The benefits U.S. agriculture may derive from future technological innovation, however, may also be enjoyed by other countries that adopt the same technology. Therefore, to

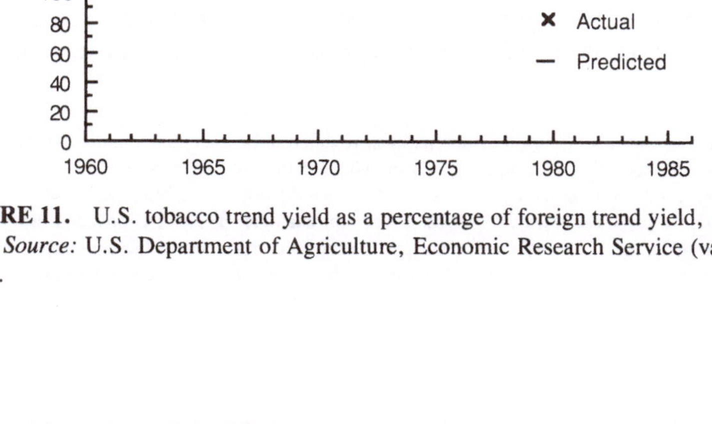

FIGURE 11. U.S. tobacco trend yield as a percentage of foreign trend yield, 1961–1986. *Source:* U.S. Department of Agriculture, Economic Research Service (various years).

reap the future benefits of new technology, the industry must be willing to maintain an effective research and development capacity, a key component of U.S. agriculture's underlying infrastructure.

INFRASTRUCTURE: A KEY TO U.S. COMPETITIVENESS

The previous section suggested that maintaining the ability to produce and market large volumes of commodities efficiently is critical to the continued competitiveness of U.S. agriculture. Maintenance of production efficiencies depends, in turn, on research and development to produce a steady stream of innovative production technologies. Similarly, moving large volumes from the farm gate to the export terminal depends on an efficient marketing and transportation infrastructure. These key infrastructure components—research and development and marketing and transportation—have formed a solid but sometimes forgotten foundation for U.S. agriculture in the past. Recently, this base has shown signs of weakening. Such deterioration could jeopardize the agricultural industry's future in world markets.

The Agricultural Research Infrastructure

Technological innovation holds much promise to improve productivity growth, reduce real production costs, and maintain future competitiveness for U.S. agriculture. For example, the United States is currently believed to lead other countries in the development and application of biotechnology, the newest and one of the most promising areas of agricultural research (Office of Technology Assessment, 1986). But a recent slowdown in funding for agricultural research and development activities could adversely affect the industry's future productivity. In the absence of domestic gains in productivity, continued productivity growth abroad could eventually cause further erosion in the competitiveness of U.S. agriculture.

The principal sources of the U.S. lead in agricultural technology are well-established public and private sector research and development capabilities. The public sector research establishment includes the U.S. Department of Agriculture (USDA) and the state land-grant universities and agricultural experiment stations. The Cooperative Extension Service, a joint effort of the USDA and the state agricultural experiment stations, forms a third leg of support beneath the public agricultural research effort and is charged with disseminating the results of public agricultural research.[5] Private sector re-

[5]The focus of this discussion is the national research infrastructure committed specifically to agricultural research. With agriculture's growing reliance on developments in biotechnology, however, production agriculture is also likely to benefit from basic biological research funded or conducted by other research establishments—for example, medical schools and the National Institutes of Health. Nevertheless, the more traditional components of the agricultural research infrastructure will continue to bear responsibility for developing basic research discoveries from

TABLE 5. Decade Average Growth Rates in Real Agricultural Research and Extension Spending
(percent)

	Research			
Decade	Public	Private	Total	Extension
1950s	3.4	—	—	2.8
1960s	4.2	3.5	3.8	1.8
1970s	2.3	3.8	2.9	3.7
1980–1984	0.9	2.7	1.9	1.8

Note: Dashes mean not available.
Source: Data from Huffman, Wallace E., and Robert E. Evanson, 1989, "The Development of U.S. Agricultural Research and Education: An Economic Perspective," draft manuscript, Department of Economics, Iowa State University, Ames, Iowa. Used with permission of the authors.

search, on the other hand, is conducted by private businesses that hope to profit from the commercialization of their research.

Two closely related trends have recently emerged in spending on agricultural research in the United States. First, growth in total agricultural research expenditures has slowed in recent years, and second, a larger share of agricultural research now occurs in the private sector. Adjusted for inflation, growth in public research expenditures in the 1980s has been less than a fourth of the rate in the 1960s (table 5). Similarly, growth in public extension funding fell by half in the 1980s, after strong growth in the 1970s. Although growth in private research expenditures also slowed in the 1980s, the slowdown was not as sharp as in the public sector. As a result, the private sector enlarged its share of annual real agricultural research expenditures to an average of 60 percent in the 1980s from an average of 55 percent in the 1970s (figure 12).

The private sector's relatively consistent record of agricultural research support has helped stabilize agricultural research spending during the recent period of weak public research support. The impact of the farm recession of the early 1980s on state budgets contributed to the erosion of public support for research. This was partially offset, however, by the growing interest of private firms in new commercial opportunities in biotechnology. Private sector spending on biotechnology research is believed to be two to three times the level of public spending (Phillips, 1985). Nevertheless, private sector research cannot fully replace public sector research because several research activities are unique to the public sector. These include the advance-

nonagricultural laboratories and applying them to production problems on the farm. For a more complete discussion of the history and functions of the U.S. agricultural research infrastructure, see Huffman, Wallace E., and Robert E. Evanson, 1989, "The Development of U.S. Agricultural Research and Education: An Economic Perspective," draft manuscript, Department of Economics, Iowa State University, Ames, Iowa.

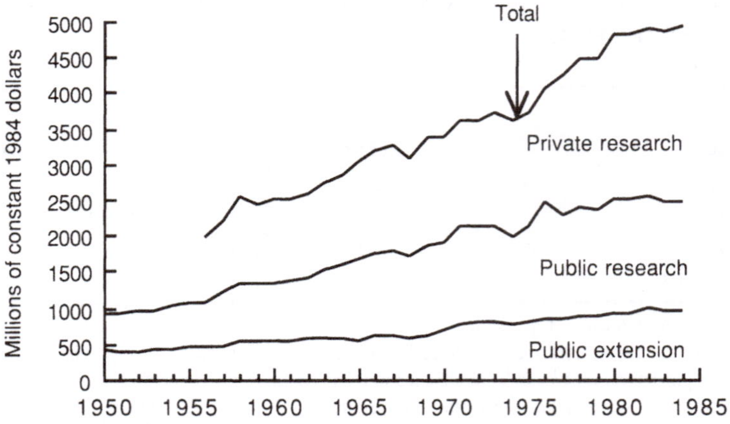

FIGURE 12. Cumulative real U.S. spending on agricultural research and extension, 1950–1984. *Source:* Huffman and Evanson (1989). Used with permission.

ment of basic knowledge that may bear no immediate commercial reward, the development of generic applications of new technology that may not be readily patentable, and the initial training of new scientists. For example, commercial interest in basic plant science research is especially lacking. Although most early advances in biotechnology have occurred in the animal sciences—for instance, the development of growth-stimulating hormones in dairy, beef, and swine production—biotechnological advances in the plant sciences may eventually prove to be at least as significant. But the basic research that must underlie any future biotechnological advances in the plant sciences may not be sustainable if profits remain elusive. Similarly, the public sector is more likely to generate input- and cost-saving technologies that do not necessarily generate the added sales that would finance private sector research. In short, public and private sector research activities are not perfect substitutes; rather, they are often complementary. The public sector plays a key role in basic research, while the private sector specializes in applied research.

With the growing interest of private, multinational firms in agricultural research and development activities, the transfer of new applied technologies around the globe is likely to accelerate in the future. In addition, foreign producers are becoming increasingly skilled at adapting technological innovations developed in the United States and elsewhere. As the U.S. lead in the rate of technology adoption narrows, U.S. farmers will be hard put to stay ahead of their foreign competitors on the technological treadmill. U.S. agriculture's future technological advantage over its competitors will likely be limited to the development of the basic science that underlies new commercial technologies. But in the absence of a renewed public commitment to agricultural research, that remaining advantage also may erode. A larger commitment to public sector research will likely be needed if U.S. agriculture is to improve its record of productivity growth and maintain its competitive position in world markets.

Agricultural Marketing Infrastructure

In addition to maintaining an efficient productive capacity, an exporting country must be able to deliver its produce into export channels in a timely fashion and in the volumes that the market demands. Thus, the development and maintenance of the marketing infrastructure that delivers supplies from the farm gate to the export terminal is a vital determinant of competitiveness.

Capital investments made during the 1970s and into the early 1980s built a U.S. commodity transportation system with sufficient capacity for the huge volumes of grain flowing into export channels in the early 1980s. The first component of the U.S. transportation infrastructure is the storage facilities that warehouse and maintain the quality of farm commodities as they await delivery to domestic and foreign consumers. Off-farm or commercial grain storage capacity in the United States has increased nearly two-thirds since the mid-1970s, to more than 250 million metric tons (U.S. Department of Agriculture, 1989). An additional 350 million metric tons of storage available on U.S. farms brings the total U.S. grain storage capacity to about 600 million metric tons, although on-farm storage capacity has declined slightly since 1986. With total annual production of wheat, feed grains, and soybeans of less than 400 million metric tons, the United States is currently able to hold the equivalent of more than one-and-a-half years of production in inventory awaiting the future needs of domestic and foreign consumers.

Inventories of U.S. commodities flow from holding facilities onto world markets through the second component of the marketing and transportation infrastructure—a network of rural highways, railroads, and inland waterways that is without parallel in the world. The first leg of this network, a system of rural, farm-to-market roads, has generally been well maintained throughout the last decade and a half. Total U.S. rural road and street mileage has remained nearly unchanged at about 3.2 million miles (U.S. Federal Highway Administration, various years). But the quality of the highway system has improved dramatically during that period, with the proportion of highways that are hard-surfaced growing a third, to nearly 2.2 million miles. The improvement in rural highways has at least partly offset a sizable decline in the size of the second leg of the transportation network, the nation's railroad system. Total railroad mileage in the United States in 1987 was about 147,000 miles, down well over a fourth since 1970 (Association of American Railroads, 1988). The third leg of the U.S. transportation network is the inland waterways system. The tonnage of wheat, corn, soybeans, and cotton transported along the nearly 26,000 miles of navigable inland waterways nearly tripled from about 21 million metric tons in 1970 to 63.1 million metric tons in 1982. Since then, the tonnage of these crops moving through inland waterways has declined about a fourth, to 47.1 million metric tons, as U.S.

farm exports slipped from their peak level (U.S. Department of Transportation, Division of Inland Waterways, unpublished data).

Grain industry analysts estimate that today the U.S. transportation network could accommodate the annual export of more than 200 million metric tons, or more than half the nation's peak annual grain and soybean production and nearly double the export volumes that placed severe strains on the system during the 1970s. Nearly half of the system's estimated peak capacity was idle during the sharp downturn in U.S. farm exports in the mid-1980s. As a result, new investment in grain export facilities and equipment declined, and the system's capacity has diminished in recent years.

The United States' investments in transportation infrastructure make a significant contribution to U.S. agriculture's competitiveness on world markets. Recent studies show that the costs of delivering U.S. commodities from farm gate to export terminal are strongly competitive with internal marketing costs in other countries. U.S. marketing costs are now thought to be roughly equal to those in Canada and the United Kingdom, modestly below marketing costs in Argentina and Australia, and well below marketing costs in Brazil, where transportation infrastructure is especially lacking (Ortman, Stulp, and Rask, 1986).

A slight advantage or disadvantage in the cost of moving U.S. crops to market is relatively insignificant when excess transportation capacity is available and export volumes are low. But the large excess U.S. transportation capacity indicates that significant expansion in domestic export volume could be readily absorbed with little increase in relatively low marketing costs. Few competing nations, if any, can match that capability. At larger export volumes, the contribution of a well-developed transportation infrastructure to U.S. agriculture's competitiveness in world markets would likely increase, since capacity constraints push marketing costs higher and curtail export expansion in other countries.

POLICY FACTORS AND COMPETITIVENESS

Even the best efforts to understand the competitiveness of U.S. agriculture become blurred when the effects of macroeconomic, farm, and trade policies are superimposed. The effects of policy on competitiveness have lately come under special scrutiny, because the Uruguay Round of trade negotiations has forced producing nations to examine the distortions that agricultural subsidies create. How do these policies affect competitiveness and what policy changes are in store that would alter U.S. agriculture's competitive position?

Macroeconomic Policy

Macroeconomic policy influences the competitiveness of U.S. agriculture in two ways. It directly affects the exchange value of the dollar, with obvious impacts on the prices of U.S. farm products in world markets. It indirectly

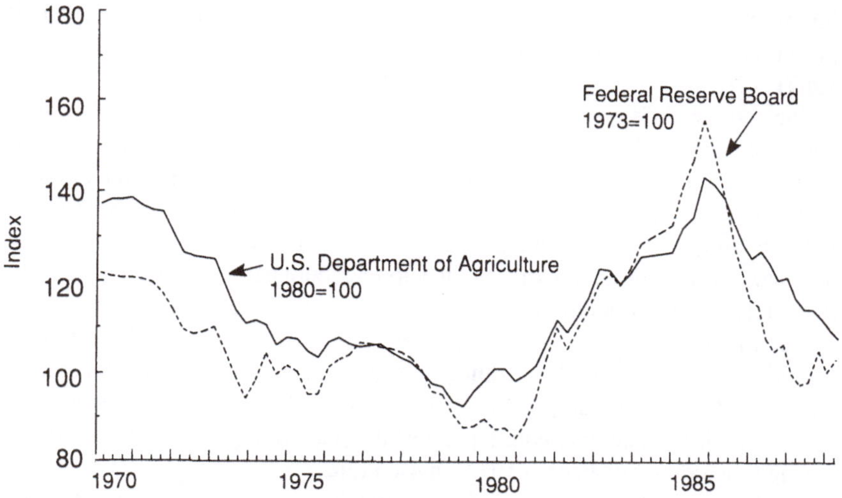

FIGURE 13. Weighted-average exchange rate of the U.S. dollar, 1970–1988, quarterly basis. *Sources:* Federal Reserve Board of Governors (various years) and USDA, Economic Research Service, Agriculture and Trade Indicators Branch.

influences interest rates, inflation, and rates of economic growth in the United States and abroad. Economic growth in developing countries will have a critical effect on the demand for U.S. farm exports and the degree to which U.S. agriculture can fully utilize its farm resources that depend on high levels of trade.

The exchange value of the dollar can have sharp effects on U.S. farm sales abroad, but for the most part, these appear to be cyclical and transitory. For example, from 1980 to early 1985, the nominal weighted exchange value of the dollar soared more than 75 percent as measured by the Federal Reserve index of ten major industrial currencies, and about 45 percent as measured by the USDA's index of more than fifty currencies important to agricultural trade (figure 13). But since 1985 the dollar has fallen sharply and is now nearly the same value as in 1981.

The effects of this cycle might be expected to average out: weak exports with a strong dollar and strong exports with a weak dollar. But the trade effects of a marked change in the dollar's value tend to linger. The dollar began to decline in 1985, but U.S. farm export volume did not recover until 1987, and farm export value did not really recover until 1988. The high dollar contributed to a reshuffling of world grain trade. Once U.S. competitors gained access to U.S. markets, U.S. producers faced a stiff challenge in regaining those markets, even with a weaker dollar.

The prevalence of fixed exchange rate regimes among many U.S. farm trading partners compounds the effects of a high-valued dollar. Many developing countries peg the value of their currencies to the dollar. Thus, when the dollar began to decline against major currencies, it declined much less against developing country currencies. In short, U.S. farm exports remained comparatively expensive in the developing world. This effect is especially important for commodities like wheat and live cattle, which are exported primarily to developing countries (Henneberry, Drabenstott, and Henneberry, 1987).

The indirect effect of macroeconomic policy on U.S. agriculture's competitiveness is difficult to measure. The impact on U.S. interest rates feeds a chain of economic reactions tied to the developing world. U.S. interest rates affect both the repayment terms on many third world loans and the rate of U.S. economic growth. Combined, these two effects have a major impact on the path of economic growth in developing countries, the prime growth market for U.S. agricultural exports. To the extent that U.S. agriculture's competitiveness depends on strong growth in world trade—which in turn enables the industry to take advantage of high trade volume—U.S. macroeconomic policy will be an important determinant of agricultural competitiveness. Therefore, macroeconomic policies that lead to high U.S. interest rates damage the competitiveness of U.S. agriculture.

A reduction in the U.S. budget deficit, and the likely decline in real interest rates that would follow, would benefit U.S. trade in general and U.S. farm exports in particular. When the U.S. federal budget is in deficit and the economy is at full employment, domestic saving is insufficient for public and private borrowing. The excess domestic demand for capital pushes U.S. real interest rates higher, attracting foreign funds to finance the domestic deficit. To benefit from high U.S. real interest rates, foreign investors purchase dollars in international financial markets for subsequent investment in dollar-denominated assets in the United States. The result is upward pressure on the exchange value of the dollar. The rising dollar, in turn, lowers the relative price of foreign goods and worsens the U.S. trade deficit.

Thus, the trade deficit mirrors the U.S. budget deficit. A sustained reduction in one is not likely to occur until the other is addressed. Reducing the federal budget deficit would reduce demand for funds in capital markets, leading to lower real interest rates and a lower-valued dollar. The lower-valued dollar would, in turn, contribute to an expansion in U.S. exports. The major portion of the export expansion would occur in agriculture and other sectors that are competitive in world markets. Alternatively, with little reduction in the U.S. budget deficit, the recent improvement in the U.S. trade balance is likely to stall.

Farm and Trade Policies

No policy is more intertwined with U.S. agriculture's competitiveness than farm policy. Farm programs generally serve to protect a nation's farmers. This veil of protection makes it especially difficult to assess a nation's underlying competitive position. Because many countries, including the United States, use trade policies to achieve farm objectives, it is useful to consider farm and trade policies together.

Farm policies vary in implementation around the world, but their effects are largely the same: higher domestic farm prices and bigger farm incomes.

TABLE 6. Levels of Farm Subsidy for Various Commodities, by Country

	Producer subsidy equivalents		
Country	50 percent or more	25–49 percent	0–24 percent
United States	Sugar	Sorghum, corn, wheat, dairy, cotton, rice	Soybeans, pork, poultry, beef, barley
Australia		Fluid milk	Wheat, barley, beef, cotton, sheepmeat, wool, rice, cane sugar, pork, poultry, manufactured milk
Canada	Dairy	Sugar	Corn, oats, soybeans, beef, flaxseed, pork, rye, wheat, rapeseed, poultry, barley
European Community	Beef	Durum wheat, dairy, sheepmeat, rapeseed, soybeans, rice, poultry, sugar	Corn, common wheat, pork, barley
Japan	Beef, rice, soybeans, pork, sugar, wheat, barley	Poultry	
South Korea	Rice, wheat, corn, barley, soybeans, beef, fluid milk	Pork	Poultry
Taiwan	Wheat, sorghum	Rice, beef, poultry, dairy, tobacco	Pork, corn, soybeans, sugar
India		Soy oil, peanut oil, rape oil	Rapeseed, peanuts
Nigeria		Wheat	Corn
Mexico	Corn	Soybeans, wheat, sorghum	Cotton
Brazil	Wheat	Cotton, rice	Poultry, dairy

Source: U.S. Department of Agriculture, 1987, *Government Intervention in Agriculture,* Foreign Agricultural Economics Report No. 229, Washington, D.C., April.

One measure of the degree to which farm policy "protects" agriculture from world market pressures is the percentage of farm income that crop producers receive from the government. Table 6 summarizes the degree of support producers receive in various countries for major commodities.

The information in table 6 supports three conclusions about U.S. farm policy and competitiveness. First, the United States falls in the middle in the degree of support it gives to the farm sector. It subsidizes less than Europe and Japan, but more than Australia and Canada.

Second, subsidy levels vary in the United States across commodities, leading to different farm policy effects on competitiveness. Sugar receives relatively more support than any other commodity, but feed grains, wheat, and cotton receive between a quarter and a half of their gross income from the U.S. government. The United States, therefore, is subsidizing agriculture to a considerable extent, and the corresponding effect on competitiveness must also be large, although it differs among commodities. Export expansion, however, is restrained by production controls. Some estimates indicate that more U.S. feed grains would be exported with no price supports and no supply controls than with current programs (World Bank, 1986).

Third, farm programs are well entrenched in both exporting and importing countries, greatly distorting world food markets. Japan, for example, insulates its producers and consumers from world markets through an extensive system of quotas, price supports, and trade barriers. These programs restrict Japanese food imports, leave more products on world markets at depressed prices, and generally limit the efficiency of the world food market. The observed world market, therefore, cannot provide full insight into how competitive U.S. agriculture might be in the absence of such farm policy distortions.

Farm policy will take one of three paths in the 1990s. It will remain as it is, become more protectionist, or become less interventionist and allow market forces to operate more fully. A strongly protectionist U.S. farm policy characterized by high support prices and restricted food imports would create a serious surplus problem for producers while raising retail food prices for consumers. High domestic farm prices under the protectionist policy would make U.S. farm products less competitive on world markets and ultimately necessitate the mothballing of a significant portion of U.S. farm resources. High domestic farm prices would also attract foreign production to U.S. markets. The United States already imports more than $21 billion of food annually. Thus, U.S. agriculture faces keen competition from foreign suppliers in some domestic markets now, and food import restrictions would become a necessary part of a protectionist domestic farm policy. In sum, any shift to a more protectionist farm policy would result in a significant contraction in domestic production and world food trade.

The Uruguay Round of the General Agreement on Tariffs and Trade, however, holds promise for greater market orientation. It is attempting to craft new agricultural trade rules, with many participants supporting a reduction in agricultural subsidies that distort farm trade. How would multilateral trade liberalization affect the competitiveness of U.S. agriculture? Trade liberalization models do not agree on an answer. One report suggests that the value of U.S. farm exports would rise modestly—about $4 billion (table 7) (Roningen, Sullivan, and Wainio, 1987). The increase would result from both greater export volume and higher world market prices.

An assessment of U.S. farm subsidies and farm resources supports the conclusion that the United States stands to gain some market share under

TABLE 7. Predicted Changes in the Value of Net Exports Worldwide Resulting from Multilateral Trade Liberalization
(millions of U.S. dollars)

Exported commodity	United States	Canada	European Community	Australia/ New Zealand	Japan	Centrally planned economies	Remaining countries worldwide	Total
Meat and eggs	7,929	238	−11,357	1,463	−3,006	617	3,832	−284
Dairy products	−1,323	−129	−6	1,387	−865	152	1,326	462
Food crops	−285	−55	122	160	−3,188	166	3,435	355
Feed crops	−565	343	−284	186	115	15	448	258
Oilseeds and products	182	−50	−115	18	55	4	−93	1
Other crops	−1,899	−102	−1,123	10	−396	67	2,638	−805
All commodities	4,039	249	−12,763	558	−7,285	1,021	11,586	−14

Source: Roningen, Vernon, John Sullivan, and John Wainio, 1987, "The Impact of the Removal of Support to Agriculture in Developed Countries," unpublished paper presented at the annual meeting of the American Agricultural Economics Association, East Lansing, Mich., August. Used with permission of the authors.

trade liberalization. Farm subsidies in the United States are smaller than in Europe and Japan but greater than in Australia and other Cairns Group countries.[6] The United States' productive soils, well-developed technology, efficient infrastructure, and reserve production capacity from diverted acres all suggest that the country will remain a strong long-run competitor in world markets.[7]

If the Uruguay Round does culminate in an agreement that reduces agricultural subsidies worldwide, the competitiveness of U.S. agriculture will become even more dependent on strong export demand and continuing advances in domestic productivity. An important assumption of the Roningen study just cited is that 1984 world trade levels would continue to prevail. Thus, the study's findings that the competitiveness of U.S. agricultural commodities improves only moderately is consistent with a weak world grain market. As shown in the preceding section, the United States derives its principal competitive edge by being able to deliver large volumes at low unit cost. When operating at only half of its export capacity, the United States cannot be fully competitive, even if farm and trade policies become more market oriented.

COMPETING IN THE 1990s

Three factors—farm-level productivity, agricultural infrastructure, and public policy—will interact to determine U.S. agriculture's competitive position in world markets in the next decade. The interactions among these factors can be set against a number of different backdrops of rates of growth in world food trade. Table 8 sets out four scenarios, showing two different rates of growth in world trade—sluggish and rapid—and two different farm policy directions—status quo and more market oriented. It then shows the expected outcome of each scenario in terms of public cost, U.S. agriculture's share of world markets, the U.S. cost of production relative to the rest of the world, and the competitiveness of U.S. agriculture.

---

[6]The number of grain-producing nations allied as the Cairns Group has fluctuated from eleven to sixteen. The following thirteen nations signed a December 1988 statement marking the midpoint of the Uruguay Round: Argentina, Australia, Brazil, Canada, Chile, Colombia, Hungary, Indonesia, Malaysia, New Zealand, the Philippines, Thailand, and Uruguay.

[7]Several aspects of the competitive posture of U.S. agriculture are examined in Ortman, Gerald F., Valter J. Stulp, and Norman Rask, 1986, "Comparative Costs of Agricultural Commodities Among Major Exporting Countries," working paper ESO 1325 (Columbus, Ohio State University, Department of Agricultural Economics and Rural Sociology); Barkema, Alan, and Mark Drabenstott, 1988, "Can U.S. and Great Plains Agriculture Compete in the World Market?" *Economic Review,* February, pp. 3-17; and Council for Agricultural Science and Technology, 1988, *Long-Term Viability of U.S. Agriculture,* Report No. 114, June, Ames, Iowa.

TABLE 8. Future Directions in the Competitiveness of U.S. Agriculture Under Different Trade Growth and Farm Policy Scenarios

Scenarios		Expected outcomes			
Rate of world trade growth	Farm policy trend	Public cost	Trend in U.S. agriculture's share of world markets	U.S. variable cost of production relative to rest of world	Competitiveness of U.S. agriculture
Sluggish	Status quo	High	Stable (if export subsidies continue)	High	U.S. uncompetitive (or can compete only with large subsidies)
Sluggish	More market oriented	Medium	Stable to weak	Moderately high	U.S. not very competitive
Rapid	Status quo	Medium to low	Stronger	Moderate to low	U.S. competitive
Rapid	More market oriented	Low	Stronger	Low to very low	U.S. very competitive

Sluggish Trade Growth

Macroeconomic policies that lead to high real interest rates slow economic growth both at home and in the developing world. Slow foreign growth leads to sluggish trade growth, in turn limiting one of U.S. agriculture's key competitive factors—the ability to deliver large volumes of grain at low unit cost. Sluggish trade growth deprives the United States of the advantages derived from its huge resource base and well-developed marketing and transportation infrastructure, advantages that are best realized at elevated levels of trade.

With sluggish world trade, the United States could choose one of two policy paths. A status quo policy would entail relatively high public costs as the United States continued to support domestic prices and subsidize exports. Export subsidies would allow U.S. agriculture to remain competitive to the extent that the industry maintains world market share. But domestic variable production costs would be high relative to the rest of the world, because high-marginal-cost acres would be kept in production by subsidized prices. In other words, the industry would "purchase" its competitiveness with expensive policy measures.

Adoption of a more market-oriented farm policy under sluggish trade growth would result in the loss of U.S. market share. As high-marginal-cost acres were retired, however, average production costs would decline. But U.S. variable production costs would still be high relative to those of other producers because foreign costs are low at low levels of output. In effect, weak world demand would prevent the United States from realizing the full competitive advantage of its huge resource base and efficient infrastructure.

Rapid Trade Growth

A U.S. macroeconomic policy of greater fiscal and trade balance would help lower real interest rates, stimulating economic growth here and abroad. A macroeconomic setting characterized by stronger growth in world food trade would boost U.S. agriculture toward its full competitive potential. With strong world demand, the United States could swing its huge resource and infrastructure advantages into play, delivering large volumes at competitive costs.

With more rapid trade growth and a status quo farm policy, the United States would become relatively more competitive, because foreign producers would have to move into the sharply rising portions of their marginal cost curves to expand production. U.S. agriculture would be most competitive with vigorous trade growth coupled to a market-oriented farm policy. Domestic resources would be used at high levels, while some marginal acres would be retired under the pressure of lower support prices. Costs of market-oriented public programs would be low. The relative cost of U.S. farm production would also be very low, because foreign producers would encounter the constraints imposed by a limited resource base and underdeveloped infrastructure.

In sum, U.S. agriculture would become more competitive as the growth in demand for U.S. farm exports increased. Improvement in export demand can be influenced by the future direction of macroeconomic policy in the United States and abroad. If export demand remains weak, the United States could "buy" competitiveness, or market share, by retaining its high-cost export subsidies and protectionist trade policies. Continuing current programs while demand is weak, however, will leave U.S. costs of production high relative to the rest of the world, raising questions about the ability of the United States to compete in the long run. Thus, U.S. agriculture must keep macroeconomic developments and farm and trade policies in perspective as it considers its future in world markets. Competitiveness may be purchased with high public outlays if world economic growth is slow. Only when demand growth is vigorous will market forces confer significant competitive advantage to the United States.

Even in a world of growing export demand for farm commodities, a phasing out of U.S. commodity programs would result in dislocation for some farms and regions currently sheltered from competition. Among domestic enterprises, dislocations would be most severe in tobacco, dairy (for manufactured product use), and sugar beet and sugarcane production. Across the nation, dislocations would be notable in the Great Plains, Southeast, and Great Lakes regions where agriculture is only marginally competitive.

If world market growth is strong, however, the economic gains from market adjustments would more than offset economic losses from a reduction in food market distortions. As a result, economic assistance could be provided to marginal producers unable to compete in the new market orientation. Such help could include personal, financial, and job counseling; job training; relocation assistance; and provision for early retirement. Research indicates that those who adjust to a changed economic environment feel they are better off in the longer run, but help is needed to ease the sometimes difficult transition to alternative economic activity.

ACKNOWLEDGMENTS

The authors thank Julie Stanley, a research associate at the Federal Reserve Bank of Kansas City, and Cindy Dishon, a research associate at Ohio State University, for their assistance in preparing this paper.

REFERENCES

Association of American Railroads. 1988. *Railroad Facts* (Washington, D.C.).

Council of Economic Advisors and U.S. Department of Labor. Various years. *Economic Report of the President* (Washington, D.C., Government Printing Office).

Federal Reserve Board of Governors. Various years. *Federal Reserve Bulletin* (Washington, D.C.).

Henneberry, David, Mark Drabenstott, and Shida Henneberry. 1987. "A Weaker Dollar and U.S. Farm Exports: Coming Rebound or Empty Promise?" *Economic Review* (May).

Huffman, Wallace E., and Robert E. Evanson. 1989. "The Development of U.S. Agricultural Research and Education: An Economic Perspective." Draft manuscript, Department of Economics, Iowa State University, Ames, Iowa.

Ortman, Gerald F., Valter J. Stulp, and Norman Rask. 1986. "Comparative Costs of Agricultural Commodities Among Exporting Countries." Working Paper ESO 1325 (Columbus, Ohio State University, Department of Agricultural Economics and Rural Sociology).

Phillips, Michael J. 1985. "Enhancing Competitiveness: Research and Technology in Agriculture," pp. 25–47 in *Competing in the World Marketplace: The Challenge for American Agriculture.* Proceedings of a symposium sponsored by the Federal Reserve Bank of Kansas City, October 31.

Roningen, Vernon, John Sullivan, and John Wainio. 1987. "The Impact of the Removal of Support to Agriculture in Developed Countries." Unpublished paper presented at the annual meeting of the American Agricultural Economics Association, East Lansing, Mich., August.

Tweeten, Luther. 1987. "Agricultural Technology—The Potential Socioeconomic Impact." *Bovine Practitioner* vol. 22, pp. 4-14, September.

———. 1988. "Implications for Agronomists Under Present Voluntary Supply Management Policies." Paper presented at the annual meeting of the American Society of Agronomy, Anaheim, Calif., November.

———. 1989. *Farm Policy Analysis* (Boulder, Colo., Westview Press), chap. 10.

U.S. Congress, Office of Technology Assessment. 1986. *Technology, Public Policy, and the Changing Structure of American Agriculture* OTA-F-285 (Washington, D.C., Government Printing Office), March.

U.S. Department of Agriculture. 1985. *Agricultural Food Policy Review: Commodity Program Perspectives.* Agricultural Economic Report No. 530 (Washington, D.C.), July, pp. 142–147.

U.S. Department of Agriculture, Economic Research Service. Various years. *Agricultural Outlook* (Washington, D.C.).

———. 1984. *Background for 1985 Farm Legislation.* (Commodity reports.) Agricultural Information Bulletin Nos. 467, 468, 471, 472, 473, 475, 476, 477 (Washington, D.C.).

U.S. Department of Agriculture, Economic Research Service and Foreign Agricultural Service. Various years. *World Agricultural Supply and Demand Estimates* (Washington, D.C.).

U.S. Department of Agriculture, National Agricultural Statistics Service. Various years. *Agricultural Statistics* (Washington, D.C.).

———. 1989. *Grain Stocks* (Washington, D.C., January).

U.S. Department of Labor, Bureau of Labor Statistics. Various years. *Handbook of Labor Statistics* (Washington, D.C.).

U.S. Federal Highway Administration. Various years. *Highway Statistics.*

World Bank. 1986. *World Development Report: 1986* (Washington, D.C.), p. 129.

U.S. TRADE POLICY AND THE GATT
Implications for Agriculture

DAVID BLANDFORD

International trade policy has always been a contentious issue in the United States. It was the imposition of a tariff that precipitated the unscheduled unloading of tea into Boston Harbor from ships of the British East India Company in 1773—an event popularly identified as marking the beginning of the American Revolution. British taxes and restrictions placed on exports from the colonies were one of the grievances cited in the Declaration of Independence, and their importance was reflected by the prohibition of export taxes in the Constitution of the new republic.

In more recent times, the controversy over trade policy in the United States has shifted toward the less revolutionary, but no less important, issues of protection versus free trade, the extent to which domestic policies should be constrained by international obligations, and the balance of power between the legislative and executive branches in the conduct of policy. These issues permeate current U.S. trade legislation, affect the relationship between trade and agricultural policies, and influence the role of the United States in the current multilateral trade negotiations under the General Agreement on Tariffs and Trade (GATT). This chapter discusses U.S. trade policy and its relationship to the negotiations, and the possible implications of a GATT agreement for U.S. agricultural policy.

THE GENERAL AGREEMENT ON TARIFFS AND TRADE

The GATT is an international treaty that was signed in Geneva, Switzerland, on October 30, 1947, by twenty-three countries (known as the contracting parties). It currently has ninety-seven signatories and is both a legal document and an organization. The document is composed of a set of original articles and subsequent additions, a series of protocols, and the tariff schedules of the member countries. It spells out the principles and rules to be used by the contracting parties in conducting their trade policies. The organization has a secretariat located in Geneva that helps to administer the agreement by sponsoring multilateral trade negotiations, monitoring trade flows and trade policies, and helping to resolve trade disputes between members.

The goal of the GATT is to raise living standards through "reciprocal and mutually advantageous arrangements directed to the substantial reduction of tariffs and other barriers to trade and to the elimination of discriminatory treatment in international commerce" (General Agreement on Tariffs and Trade, 1969, p. 1). To achieve these objectives, the GATT embodies four basic principles (U.S. Congress, Congressional Budget Office, 1987):

- The contracting parties should work to lower trade barriers in general, and to eliminate the use of import quotas in particular.
- Trade restrictions should be applied on a nondiscriminatory basis to all member countries. This is the so-called most favored nation principle.
- Once a tariff concession is made, it cannot be rescinded without compensation to other affected trade partners. Other forms of protection cannot be used to circumvent the effects of the concession.
- Trade disputes should be settled by consultation, whenever possible, using as a guide a set of codified and mutually accepted rules for the conduct of trade.

The major provisions of the GATT are summarized in table 1.

Despite these important principles, the GATT contains many exceptions and has limited enforcement power. It therefore provides a relatively weak legal framework for regulating the conduct of international trade. A stronger system was proposed in the Havana Charter of 1948 that would have involved an organization called the International Trade Organization (ITO), similar in status to the present International Monetary Fund. Many countries, including the United States, found the broad powers of the proposed ITO unacceptable. Congress failed to ratify the Havana Charter when it was submitted for approval in 1949, and the GATT, despite its limitations, became the legal framework governing the conduct of international trade.

The GATT seeks to promote greater transparency in the use of trade restrictions by favoring tariffs over quotas. The protection that tariffs provide by raising domestic prices is clearly visible in the size of the tariff. But the protection provided by nontariff barriers, such as import quotas, licensing, or state trading, is less transparent. The GATT also seeks to prevent discriminatory treatment of suppliers through the most favored nation principle and to provide a forum through which trade disputes can be resolved and barriers to trade progressively reduced. To further the last of these aims, a series of negotiating rounds has been held. The ongoing Uruguay Round, launched in Punta del Este in September 1986, is the eighth in the series.

The effectiveness of the GATT relies on the good faith of the contracting parties. Numerous exceptions to the basic principles were included in the original treaty to enable countries to sign without having to change their existing domestic policies or to limit their future policy options. Additional exceptions have been added, making it fairly easy for countries to find ways to circumvent the spirit, if not the letter, of the treaty. Few countries have failed to take advantage of one or another of these exceptions in conducting

TABLE 1. Major Provisions of the General Agreement on Tariffs and Trade

Provision	Description
Most favored nation (MFN) treatment	Specifies that tariffs should be applied equally to all members. This principle of nondiscrimination is called the MFN principle.
Binding tariff schedules	Legally binds members to tariff concessions by stating that tariffs should not be increased above the rates in each country's tariff schedule.
National treatment	Prohibits the use of nontariff barriers to offset tariff reductions. Internal taxes must apply equally to domestic and imported goods, and regulations must treat imported goods "no less favorably" than similar domestic goods.
Customs regulations	Curbs customs procedures that limit imports, such as rules of transit, customs valuation, fees and formalities, and marks of origin. States that laws and regulations regarding trade should be formulated and applied in a transparent manner, which requires public disclosure and the uniform and impartial administration of trade laws.
Antidumping and countervailing duties	Defines dumping and states that both dumping and injury to domestic producers must be proved in order for antidumping duties to be imposed. These duties should not exceed the dumping margin. Specifies similar rules for countervailing duties to offset foreign government subsidies.
Quantitative restrictions	Calls for the elimination of quantitative restrictions on trade, subject to the qualification that these may be used on a generally nondiscriminatory basis for the following purposes: (1) to safeguard the balance of payments, (2) provide temporary escape clause relief for domestic industries, and (3) allow developing countries to further development goals. This provision also limits the use of currency controls as a substitute for quantitative restrictions.

(Continued)

TABLE 1. *Continued*

Provision	Description
Subsidies	Discourages the use of subsidies in general and calls for the elimination of export subsidies for nonprimary products. Export subsidies for primary products should not cause a country to achieve more than an "equitable share" of world trade.
State-owned enterprises	Requires countries to choose among potential buyers and sellers according to normal business considerations, especially in terms of price, quality, and procurement.
Developing countries	Grants exemptions to most of the requirements of the GATT, but only under rigorous criteria. In reality, developing countries have justified the use of nontariff barriers and export subsidies as safeguards for balance-of-payments problems. The GATT acknowledges that developing countries are a special case and are eligible for special and differential treatment under the treaty, encourages developed countries to lower trade barriers unilaterally against products from developing countries, and encourages stabilization and improvement of market conditions for primary products.
Escape clause and other exceptions	Allows countries to withdraw concessions or use other measures to protect domestic producers from injury resulting from imports. Other justifications for restrictions—for example, national security reasons—are identified.
Consultation and dispute settlement	Provides for consultations among countries on trade disputes and for the establishment of panels of experts to review cases. The recommendations of the panels are not binding on the countries concerned.
Other procedural issues	Specifies principles for the establishment of free trade areas, rules for modifying tariff schedules, and criteria for the accession of new members. Periodic tariff negotiations are suggested.

Source: Adapted from U.S. Congress, Congressional Budget Office, 1987, *The GATT Negotiations and U.S. Trade Policy* (Washington, D.C., Government Printing Office), June.

their trade policies. Furthermore, the enforcement mechanisms of the GATT are weak. At the request of the contracting parties, panels can be formed to review alleged inconsistencies between the policies or actions of particular countries and the provisions of the treaty. Even if the allegations are ruled to be valid, there are no mechanisms for enforcement other than moral suasion, or the payment of compensation to injured parties, or the imposition of countertariffs by the injured party. In general, it is assumed that countries will make good faith efforts to resolve their differences for the wider good. Unfortunately, few countries have been prepared to behave with such magnanimity. The weakness of the rules and enforcement mechanisms in the GATT was not an accident; it was consistent with the wishes of the contracting parties.

Despite its limitations, the GATT has proved a useful vehicle for the reduction of tariffs, particularly on industrial goods. As can be seen from table 2, average U.S. tariffs have been reduced considerably since the first round of negotiations in Geneva in 1947. The GATT has faced increasingly difficult problems in recent years, however, for a number of reasons. One is the proliferation of nontariff barriers to trade, which are often used to circumvent limits on tariffs. Another is the inability of the treaty to deal with new developments in the global economy, such as the growing trade in services and the protection of intellectual property rights. The third is its failure to tackle the problems of agriculture. These are the main issues being addressed in the Uruguay Round. The outcome of these negotiations will determine whether the GATT remains a meaningful force in the international trade arena.

The United States, which has been one of the major supporters of the GATT, has become increasingly disenchanted with the treaty's inability to deal with pressing problems in international trade. Faced with a persistent trade deficit and perceptions that other countries are not "playing fair" in world markets, the U.S. commitment to multilateralism has weakened in recent years. There has been growing conflict between free market and protectionist sentiments in Congress, and between unilateralism and multilateralism. These conflicts are not new; they have been persistent themes throughout the history of U.S. trade policy.

THE EVOLUTION OF U.S. TRADE POLICY

Since the colonies declared their independence, the struggle over trade policy has revolved around three issues: (1) whether U.S. markets should be open to international competition or domestic industries should be protected, (2) the extent to which domestic trade policies should be constrained by international obligations and subject to international disciplines, and (3) the balance of power between the executive and legislative branches of government in implementing U.S. trade policy. All these issues have important implications for the conduct of trade policy and its relationship to agricultural policy.

TABLE 2. Reductions in U.S. Tariffs Under the General Agreement on Tariffs and Trade
(percent)

GATT conference	Proportion of dutiable imports subject to reduction	Average cut in reduced tariffs	Average cut in all duties	Remaining duties as proportion of pre-GATT tariffs
First Round Geneva, 1947	53.6	35.0	21.1	78.9
Second Round Annency, 1949	5.6	35.1	1.9	77.4
Third Round Torquay, 1950–1951	11.7	26.0	3.0	75.0
Fourth Round Geneva, 1955–1956	16.0	15.6	3.5	73.2
Dillon Round Geneva, 1961–1962	20.0	12.0	2.4	71.4
Kennedy Round Geneva, 1964–1967	79.2	45.5	36.0	45.7
Tokyo Round Geneva, 1974–1979	n.a.	n.a.	29.6	31.7

Note: n.a. means not available.
Source: Adapted from Lavergne, R. P., "The Political Economy of U.S. Tariffs" (Doctoral dissertation, University of Toronto, 1981), reproduced in U.S. Congress, Congressional Budget Office, 1987, *The GATT Negotiations and U.S. Trade Policy* (Washington, D.C., Government Printing Office), June.

Trade policy was transferred from the states to the federal government after the ratification of the U.S. Constitution in 1788. This policy shift created a common market among the thirteen states of the new United States of America and transferred the authority to levy tariffs to Congress. The first tariff act of 1789, which imposed tariffs averaging 5 percent, was primarily a means of raising revenue for the financially weak republic. However, it also identified the protection of domestic manufactures as an additional reason for the imposition of tariffs. The doubling of duties in the act of 1816 was, again, largely to raise revenue, this time to pay for the costs of the war with Britain.

Following the cessation of hostilities with Britain, the debate on protection was increasingly dominated by the growing rivalry between the North and the South. The North wanted protection from low-cost imports to aid the development of its manufacturing industry. The South, which had little industry, wanted access to low-priced imported manufactures. There was a bitter North-South struggle over the Tariff Act of 1828, which increased average duties on dutiable imports to roughly 50 percent. The new tariff became known as the "Tariff of Abominations" and precipitated an unsuccessful legal challenge by the state of South Carolina against the federal government. Continued southern objections to high tariffs led to conciliatory reductions by President Andrew Jackson in the Tariff Act of 1833 and further reductions in the 1840s.

As the likelihood of civil war increased, the Morrill Tariff of 1861 was passed to protect the vital iron and steel industry in the North. When war broke out, the need for additional federal revenues again resulted in sharp increases in tariffs. These high tariffs were generally maintained after the end of the war as the United States rapidly became a major industrial power. In response to pressures for change, Congress authorized the establishment of a Tariff Commission in 1882, but failed to enact its proposals for a significant reduction in tariffs. Indeed, with the McKinley Tariff of 1890, import duties reached a new postwar peak. Significantly, however, this legislation authorized the president to negotiate mutual reductions of duties on individual items with other countries. Prior to this, such reciprocal reductions had to be approved by Congress. In general, however, trade legislation was increasingly protectionist until the eve of World War I, both because of pressures for protection from domestic industries and the need for government revenue.

Following World War I, U.S. trade policy continued its protectionist stance. Declining farm prices prompted the passage of the Emergency Tariff of 1921, which imposed heavy import duties on agricultural products. This was followed in 1922 by the Fordney-McCumber Act, which allowed the president to raise or lower duties to "equalize production costs." With the onset of the Great Depression, Congress increased tariffs to one of the highest levels in U.S. history with the passage of the Hawley-Smoot Act in 1930. However, it rapidly became clear that the economic problems of the early 1930s could not be solved just by increasing import duties. Other nations, faced with their own economic difficulties, simply responded to U.S. actions by raising their own tariffs (McCalla, 1969). Thus, the volume of world trade contracted dramatically. Finally, the pendulum began to swing back toward liberal trade with the passage of the Reciprocal Trade Agreements Act of 1934 (RTAA).

Under the RTAA, the president was empowered to enter into reciprocal trade agreements with other countries through executive proclamation and without ratification by the Senate. The act included a most favored nation

(MFN) clause—meaning that a tariff reduction negotiated bilaterally could be extended to other countries with which the United States had bilateral agreements—and allowed the president to reduce duties by as much as 50 percent of their existing level. In earlier legislation, presidential authority had been limited by the findings of the Tariff Commission on the economic impact of tariff reductions. Under the new legislation, this limitation was eliminated. However, many of the later bilateral trade agreements concluded under the legislation contained an escape clause, through which part or all of the agreement could be made inoperative at the discretion of the president, if imports caused serious injury to domestic producers. The RTAA gave the president substantial power over trade policy that was continued through extensions of the legislation in 1937, 1940, and 1943. In 1945, presidential authority reached its zenith in a renewal of the act that allowed reductions of up to 75 percent in existing tariffs. The effect of the RTAA was substantial. During the period 1934–1945, thirty bilateral trade agreements were concluded, compared to only three in the ninety preceding years (Tontz, 1977). Average U.S. tariffs fell significantly because of these bilateral agreements and their MFN provisions.

In 1948, the RTAA was up for renewal. It was only extended for one year, which reflected the beginning of a gradual reduction of presidential authority over trade policy. A "peril point" clause was inserted, which required the Tariff Commission to establish a minimum rate of duty below which, in the commission's judgment, tariffs could not be lowered without damaging domestic industry. If the president reduced the rate of duty below this point, he was required to explain why to Congress. The 1948 extension was repealed in 1949, and the peril point provision was eliminated. When the RTAA was extended again in 1951, the peril point was reinstated. The use of the escape clause was made mandatory in the case of serious injury to domestic industries. The renewal of the act in 1955 and 1958 continued the trend toward limiting presidential negotiating authority by providing for tariff cuts of 15–20 percent, but constraining reductions through the peril point clause.

The limitations of the RTAA became apparent in the Dillon Round of GATT negotiations in 1961–1962. In 1962 President John F. Kennedy sought new negotiating authority, which was granted through the Trade Expansion Act of 1962 (TEA). The TEA did not include a peril point provision, but required the International Trade Commission (the successor to the Tariff Commission) to advise the president on the probable economic impacts of changes in U.S. trade restrictions. It also introduced a program of adjustment assistance for firms and workers whose economic viability or jobs were affected by increases in imports as a result of trade liberalization.

Presidential negotiating authority under the TEA lapsed at the end of the Kennedy Round of GATT negotiations in 1967. New legislation was not passed until the 1974 Trade Act. This act empowered the president to make further tariff reductions and to negotiate agreements to reduce or eliminate nontariff barriers. Such agreements could be vetoed by Congress during a ninety-day notice period. The act also liberalized the adjustment assistance provisions and permitted retaliation against other countries engaging in unfair or illegal practices affecting U.S. exports. However, these retaliatory measures could be initiated at the discretion of the president only after public hearings and could be vetoed by Congress. The 1974 act provided the authority under which the Tokyo Round of GATT negotiations was concluded. It included a "fast track" provision whereby the implementing legislation for the agreement reached in the Tokyo Round could not be amended by Congress, but only accepted or rejected in its entirety.

CURRENT U.S. TRADE LEGISLATION

The law governing U.S. trade policy at present is the Omnibus Trade and Competitiveness Act of 1988 (OTCA). This substantial piece of legislation, which runs to more than 500 pages, has ten titles: (I) Trade, customs, and tariff laws; (II) Export enhancement; (III) International financial policy; (IV) Agricultural trade; (V) Foreign corrupt practices amendments, investment, and technology; (VI) Education and training for American competitiveness; (VII) Buy American Act of 1988; (VIII) Small business; (IX) Patents; and (X) Ocean and air transportation. As stated in its preamble, the purposes of the act are to: (1) authorize the negotiation of reciprocal trade agreements, (2) strengthen U.S. trade laws, (3) improve the development and management of U.S. trade strategy, and (4) improve standards of living throughout the world (U.S. Congress, House of Representatives, 1988). The act contains a large number of provisions, but the focus in this chapter is on those that could directly or indirectly affect U.S. agriculture and agricultural policy.

The OTCA provides presidential authority until June 1, 1993, to negotiate trade agreements with other countries on a bilateral or multilateral basis. It permits tariff reductions of up to 50 percent over a period of up to ten years and allows for reduction or elimination of nontariff barriers. For nontariff barrier or bilateral agreements, the president is required to consult with the relevant congressional committees (including the House Ways and Means Committee and the Senate Finance Committee). As in the Trade Act of 1974, fast track procedures apply automatically to bills implementing trade agreements introduced before June 1, 1991 (so long as the consultation requirement is met) and at congressional discretion to bills introduced after then. The 1988 act continues the provision of the 1974 act that the International Trade Commission should advise the president on the potential economic impact on domestic industries of changes in U.S. tariff and nontariff barriers.

One of the most contentious parts of the OTCA is the so-called super 301 section and related provisions contained in Title I(C). Section 301 amends the provision in the 1974 act that gives the president the discretion to retaliate against other nations engaging in unfair trade practices and makes such action mandatory. Any person may file a petition with the U.S. Trade Representative requesting action under section 301. In addition, the Trade Representative is required to submit an annual report to Congress on U.S. trade liberalization priorities, identifying specific practices and countries, and to enter into negotiations with these countries, seeking either to eliminate the practices or to obtain compensation for their effects on the United States.[1] If no agreement is reached with the country concerned, retaliatory action—for example, the imposition of countervailing duties—can be taken under section 301.

The agricultural trade title of the bill (Title IV) states the following U.S. policy objectives:

1. To provide agricultural commodities and products for export at competitive prices, with full assurance of quality and reliability of supply;
2. To support "the principle of free trade and the promotion of fair trade";
3. To eliminate or substantially reduce constraints on fair and open trade;
4. To counter unfair foreign trade practices with export promotion programs and, if necessary, restrictions on U.S. imports of agricultural commodities in order to "encourage fair and open trade";
5. To provide for increased representation of U.S. agricultural trade policy interests in the formulation of national fiscal and monetary policy affecting trade (U.S. Congress, House of Representatives, 1988, section 4102).

The major means for achieving these objectives are improving the availability and collection of information on foreign trade practices; the activation of a marketing loan for 1990 crops of wheat, feed grains, and soybeans if there has been no significant progress in the GATT negotiations; the continuation of the Export Enhancement Program through fiscal year 1992 at a level of $2 billion; and funding for targeted export assistance through fiscal year 1990 at a level of $325 million.

AGRICULTURE AND THE GATT

Until the current round of negotiations, agriculture had largely been left out of the GATT process. To a large extent this was because of conflict between U.S. agricultural policies and the provisions of the GATT. The problem began with the enactment of import restrictions in 1948 under the provisions of section 22 of the Agricultural Adjustment Act of 1933, as amended. The act authorized the president to restrict imports (through the use of import quotas or fees) that threatened U.S. agricultural price support programs. It provided that any trade or other international agreement would not apply if it were inconsistent with section 22. The application of quotas originally was restricted to products that were subject to domestic production or marketing

[1]Brazil, India, and Japan were identified by the U.S. Trade Representative in June 1989 as "unfair traders." This has resulted in substantial criticism in the GATT and a common refusal to cooperate with the United States by the three countries.

limitations, such as grains. A subsequent amendment in 1951 made the use of quotas mandatory whenever any price support program was threatened. Further, national security reasons were cited for the imposition of quotas on dairy products during the Korean conflict in 1951. In 1955 the contracting parties granted the United States a waiver of its GATT obligations on the use of section 22 import quotas in order to deal with the legal inconsistency between U.S. farm policies and the GATT. The long-term negative implications of weakening the application of the GATT to agricultural trade seemed insignificant to the United States in the early 1950s, when agricultural trade was of relatively minor importance to the sector and short-term domestic farm policy imperatives took clear precedence over trade issues.

After the section 22 waiver was granted, it was tacitly accepted by the contracting parties that agriculture was a "special case." GATT members have been generally unwilling to expose their domestic farm programs to international discipline. Import quotas, export subsidies, and a variety of other measures have come to be permitted for agriculture within the GATT, even though they are outside the spirit of the treaty. Only with the growing importance of exports for U.S. agriculture and the increasingly visible effects of the protective agricultural policies of the European Community on international markets has the attitude toward the exclusion of agriculture from the GATT changed in the United States and other countries.

One reflection of this is the dramatic increase in the number of complaints brought before the GATT, many of which have been submitted to GATT panels. Over the past two years, a record number of panels have been created to deal with agricultural disputes. In 1987 four agricultural panels were established and in 1988 the number was twelve. Six of the 1988 panels were established at the request of the United States.

The increased dispute settlement activity in the GATT has led to some tangible results. For example, in 1988 Japan agreed to market-opening measures for twelve categories of agricultural products, including beef and citrus (Organization for Economic Cooperation and Development, 1989). There has been some limited success in previous negotiating rounds. The most notable development was the zero duty binding obtained from the European Community on soybeans and certain other products during the Dillon Round of negotiations. There have been some lost opportunities, too. During the Kennedy Round, the EC proposed that the problem of agricultural support be dealt with by placing a limit on the difference between domestic support prices and international prices for agricultural products (the so-called *montant de soutien*). At the time, the United States saw this as a means by which the Community could legitimize its protective agricultural policies, particularly its use of variable levies. The United States therefore refused to negotiate on this principle. With hindsight, the United States might have been wise to accept the offer, since at least then there would have been some limitation on the level of protection in the Community, which has been relatively unconstrained in the years since the Kennedy Round.

CURRENT U.S. TRADE POLICY AND THE GATT

To those who are unfamiliar with the U.S. political system, its structure and conduct are often bewildering. In other democratic societies, a political party, once elected, usually controls the executive and legislative functions and may exercise strong central control over the formation and implementation of policy. But a multipolar system characterizes government in the United States, making it difficult to identify "the" U.S. policy on, say, trade. There may be an official view associated with the executive branch, but views may differ among cabinet departments. Frequently, for example, the perspective of the State Department on U.S. trade policy differs from that of the Commerce Department. And the view on agricultural trade policy in these departments may differ from that in the Department of Agriculture. All three may have somewhat different policy perspectives from the Office of the Trade Representative, which actually negotiates with other countries on trade matters.

Added to the heterogeneity of attitudes and emphasis in the executive branch is the heterogeneity of Congress. In general, the House of Representatives tends to be more protectionist than the Senate, because its members are more subject to the pressures of employers or employees in their districts who are affected by import competition. However, attitudes in both the House and the Senate are influenced by local conditions, party affiliations, and the effectiveness of lobbyists, as well as personal beliefs and opinions. And there are many noneconomic but politically important factors, such as foreign policies toward particular countries (for example, communist countries), national security considerations, and even xenophobia. For these reasons, U.S. trade policies and legislation often seem to incorporate a curious mixture of objectives and measures that are not always consistent with one another.

Current U.S. trade legislation and its implementation reflect this tradition. If there is any consistency in the OTCA, it is in the continuing tendency for Congress to place additional limits on presidential authority in the area of trade. For instance, the act increases the reporting requirements to Congress by the executive branch; it mandates a review of the trade practices of other countries and negotiations to correct unfair practices; and it mandates action against countries that fail to take corrective action. In fact, the act specifies a number of steps to be taken against Japan, relating to services and supercomputers.

The OTCA continues the tradition of recent trade legislation in providing the means to conclude a trade-liberalizing agreement under the GATT, but also contains threats of unilateral protectionist measures—for example, the expansion of agricultural export subsidies—if there is a lack of progress in the multilateral negotiations. Indeed, the legislation reflects the "carrot and stick" approach to trade reform. It shows that Congress has not given up entirely on the multilateral process as a means to achieve reform, but that it is not convinced that a multilateral approach alone will succeed. Other weapons, such as bilateral negotiations and unilateral retaliation, are in the arsenal of current U.S. trade policy. These could become important if real progress is not made on agriculture and other pressing issues in the Uruguay Round.

AGRICULTURE IN THE CURRENT GATT NEGOTIATIONS

Agriculture is a crucial issue in the Uruguay Round of GATT negotiations and hence a key determinant of the future of international trading relations. The original Punta del Este Declaration of September 1986 launching the round stated that "there is an urgent need to bring more discipline and predictability to world agricultural trade by correcting and preventing restrictions and distortions . . . so as to reduce the uncertainty, imbalances and instability in world agricultural markets" (General Agreement on Tariffs and Trade, 1986, p. 1). To achieve this discipline and predictability, it was proposed that the negotiations should "aim to achieve greater liberalization of trade in agriculture and bring all measures affecting import access and export competition under strengthened and more operationally effective GATT rules and disciplines" (General Agreement on Tariffs and Trade, 1986, p. 6).

To the surprise of the other participants, the United States made a bold opening gambit on agriculture by proposing the elimination of all agricultural subsidies by the year 2000. This was countered by an EC proposal that called for short-term measures to improve conditions in international markets but contained no long-term commitment to reform. The Cairns Group of exporters (Argentina, Australia, Brazil, Canada, Chile, Colombia, Hungary, Indonesia, Malaysia, New Zealand, Philippines, Thailand, and Uruguay) made an intermediate proposal in which short-term measures would be part of a longer-term, significant reduction in protection.

The vast difference between the U.S. and EC proposals on agriculture generated some heated public statements but little real progress. At the midpoint of the negotiations, in December 1988, a meeting was held in Montreal to decide their future direction, but attempts to reach an agreement failed. Agriculture was one of the key areas in which it was not possible to reach a compromise. The Reagan administration was in its final weeks, and the Bush administration had to take office before the direction for the remainder of the GATT negotiations could be set. Negotiations between the major players continued after the Montreal meeting, and a compromise was finally announced at a meeting of the GATT's Trade Negotiations Committee in Geneva in April 1989.

Called the Mid-Term Review Agreement, the compromise contains both short- and long-term elements (table 3). In the short term, it calls for a freeze on trade restrictions and on support prices. It also requires a reduction in the level of protection in 1990. In the long term, it calls for a progressive reduction in the level of trade-distorting support and the strengthening of

TABLE 3. Summary of the GATT's Mid-Term Review Agreement on Agriculture

SHORT-TERM ELEMENTS

1. Until the completion of the negotiations, countries agree to bind existing levels of support and to keep tariff and nontariff barriers at 1987–1988 levels. No increase in support prices or levels of support is allowed.

2. The level of "support and protection" for 1990 is to be reduced based on an aggregate measure of support or on specific policy measures, the details of which are to be presented by the contracting parties by October 1989.

LONG-TERM ELEMENTS

1. There should be a substantial and progressive reduction in agricultural support and protection over an agreed period of time. This will be achieved through negotiations on the reform of specific policies and measures or through a reduction in an aggregate measure of support, or some combination of the two approaches.

2. The GATT rules and disciplines should be strengthened.

3. The specific measures to be addressed in the negotiations are tariffs, nontariff barriers, internal supports (including income and price supports) that directly or indirectly affect trade, export subsidies, and export prohibitions and restrictions. Special and differential treatment will be given to developing countries in the reform process and ways will be found to take into account the possible negative effects of the reform process on net food importing developing nations. Nontrade concerns, such as food security, will also be taken into account. A work program for the harmonization of sanitary and phytosanitary regulations is set out.

4. By December 1989, participants will submit proposals on how reforms are to be achieved—for instance, through the use of an aggregate measure of support; changes in rules and disciplines; special and differential treatment for developing countries; sanitary and phytosanitary regulations; tariffication, decoupled income support, and other ways to adapt support and protection; and means to compensate food importing developing countries.

5. By the end of 1990, agreement will be reached on the long-term reform program and its period for implementation. This agreement will be followed by notification of national plans for achieving reform.

6. The implementation of the first set of commitments under the reform program will be implemented in 1991 and will be monitored through multilateral surveillance.

Source: Adapted from General Agreement on Tariffs and Trade, 1989, "Mid-Term Review: Final Agreement at Geneva, *Focus* (GATT newsletter), no. 61, May.

GATT rules to deal with agriculture. The exact way in which the reform of domestic programs is to be achieved is left open. Countries could negotiate a reduction in an aggregate measure of support (AMS)—for example, the Producer Subsidy Equivalent made popular by the Organization for Economic Cooperation and Development—or on individual policies or policy instruments. A combination of both approaches could be used. New methods could be adopted in reforming existing programs, such as the conversion of nontariff barriers to their tariff equivalent (tariffication) or the conversion of existing forms of support to decoupled payments. The participants have until

December 1989 to advance proposals on these and other issues and until the end of 1990 to agree to a long-term reform program.

The Mid-Term Review Agreement on agriculture is substantive. It contains a set of clearly defined objectives, some alternative approaches to achieving these objectives, and a timetable. It does not set out the specifics of the final reform program, since these have yet to be negotiated. However, if the plan is successful, it could have significant implications for future U.S. agricultural policy.

IMPLICATIONS OF A GATT AGREEMENT FOR U.S. AGRICULTURAL POLICY

The GATT negotiations are unlikely to be concluded before the end of 1990, and the contents of a final agreement may not be sufficiently clear to have a major impact on the 1990 farm bill. However, the Mid-Term Review Agreement does contain short-term measures that have implications for current U.S. policies. These include the freeze on current levels of support and on barriers to trade and the requirement that the level of support be reduced in 1990 (table 3). The Mid-Term Agreement is unclear as to whether the reduction in support will apply to all commodities individually or to the aggregate level of support. Nor does it specify the magnitude of the reduction or how it will be calculated. The major candidates for reduction are the target prices for grains and cotton and the support prices for dairy products and sugar.

The United States has implemented annual reductions of roughly 3 percent in target prices for grains and cotton that will continue in 1990. Dairy support prices have also been reduced, with reductions being linked to the size of government purchases. It is probable that market conditions will trigger a further cut in support prices in 1990, but this could be preempted by legislative action. There has been no reduction in the support price for sugar under the current legislation, and a reduction in 1990 is unlikely. The future of the sugar program is uncertain because of a GATT panel ruling (accepted by the United States) that the use of import quotas is inconsistent with the provisions of the treaty. The program is supposed to operate at no cost to U.S. taxpayers. This has been achieved by reducing import quotas to keep domestic prices above the support price. In order to bring the program into conformity with the GATT, this system will have to change. But it will prove difficult to redesign the program to meet the requirements of the GATT and to satisfy the domestic sugar lobby. Both the dairy and the sugar programs will severely test whether the United States is willing to meet its obligations under the GATT or whether domestic politics will continue to dominate agricultural policies.

A long-term GATT agreement on agriculture could have substantial implications for U.S. agricultural policy, again assuming that Congress would be willing to subject domestic farm programs to international discipline. One of the major failings of the GATT has been its inability to impose constraints on the domestic agricultural policies of the major trading nations. This weakness was created by the exclusion of agriculture from the GATT as a result of the 1955 U.S. waiver for section 22 quotas. A meaningful GATT agreement on agriculture would imply that all countries would have to accept greater international discipline over domestic farm programs than at present. Countries could still pursue national agricultural policy objectives but would have to ensure that the achievement of these objectives did not negatively affect the trade or domestic markets of other contracting parties. Domestic policies would have to be formed within the constraints imposed by international obligations under the agreement. Compliance with these constraints would be a new experience for most countries and would certainly change agricultural policy formation in the United States. The nature of the constraints that would be placed on U.S. agricultural programs would be determined by the specifics of the final agreement, but some possible implications can be identified.

Effects on Import Quotas and Other Nontariff Barriers

The first implication of a GATT agreement is that programs involving the use of import quotas and other nontariff barriers (NTBs) would have to be changed. Quotas are already outside the spirit of the treaty and are tolerated only because of waivers and exemptions. It is highly likely that an agricultural agreement will target these for elimination. A prohibition on the use of quotas or their phased removal would have significant implications for some U.S. commodity programs, particularly dairy and sugar, since it would require the removal of the dairy quotas under section 22 and changes in the sugar program. If the agreement specified a particular reduction in the level of domestic support (the difference between domestic and world prices) or an increase in market access (the level of imports), gradual increases in quota levels over the life of the GATT agreement might be acceptable. In order to accommodate such an increase at reasonable cost, support prices for milk and sugar would have to be reduced or domestic production controls instituted. If the agreement placed restrictions on the level of trade-distorting support expenditures (as is likely), failure to reduce support prices or otherwise change the programs could put the U.S. government in the position of buying up "surpluses" of domestic or imported product to support domestic prices.

An alternative approach to the quota/NTB problem being pressed by U.S. negotiators is tariffication. This tack proved successful in bilateral negotiations over beef and citrus quotas with Japan, and the success seems to have prompted the United States to advocate the use of the concept in the Uruguay Round. With tariffication, quotas and other NTBs would be eliminated and replaced by an equivalent tariff (calculated as the domestic price minus the international price). This tariff would be bound (fixed) and then reduced gradually by an agreed percentage over a period of time specified in the GATT agreement. The tariffication of U.S. agricultural quotas would necessitate a radical change in some existing agricultural programs.

It would probably be very difficult to apply tariffs to the dairy program. With import quotas, prices in the domestic dairy market are not only above world market prices, they are also insulated from changes in world prices. Such insulation would not be possible with a tariff. With variable domestic prices it would be impossible to maintain a fixed support price for dairy products. The Commodity Credit Corporation (CCC) could find itself trying to support world market prices if the domestic support price were above the world price plus the tariff. It could still stabilize domestic prices in the short term by buying and selling dairy products, but it would have to do so under a trigger price that could adjust rapidly to changes in world market prices. The alternative would be to enter into a price stabilization or market-sharing arrangement with dairy exporters, a course unlikely to be politically attractive to the United States and one that may be technically infeasible.

Tariffication would raise two additional issues for dairy programs in the United States. First, it would make it more difficult to maintain higher and more stable prices for fluid milk than for manufacturing milk and to differentiate milk prices across milk sheds through federal and state milk marketing orders. It is hard to reconcile freer international trade with a system that maintains artificial regional price differentials by limiting the interregional movement of milk. Even if such limitations could be maintained—because most of the trade internationally (except possibly for trade with Canada) is likely to be in manufactured dairy products rather than fluid milk—it would prove difficult to "get prices right" in a situation where domestic prices in general were no longer insulated from world market prices.

A second issue is that tariffication could be implemented through either a specific (fixed) charge or an ad valorem (percentage) charge. A specific tariff provides more protection to a domestic industry when international prices are declining and less when they are increasing. An ad valorem tariff produces the opposite effect. Since world dairy prices are projected to rise with trade liberalization (Blandford, 1988), an ad valorem tariff would tend to favor the domestic dairy industry. But other things being equal, an ad valorem tariff leads to higher variability in domestic prices than a specific tariff (Bale and Lutz, 1979). A sudden, or even gradual, transition to tariffs could create additional problems for the dairy industry and dairy consumers who have been used to fifty years of relatively stable and predictable prices. Since the world prices of most U.S. agricultural exports are also expected to increase with liberalization (Blandford, 1988), the United States would presumably prefer other countries to use specific rather than ad valorem tariffs for those products it would export, such as grains to the European Community and beef to Japan.

The difficulties created by tariffication for U.S. agricultural programs are considerable, but no more so than for programs of other countries. Tariffication would require substantial changes in the price support system of the European Community, since it would mean replacing the variable levy with a fixed ad valorem or specific tariff. Thus, the current system of maintaining fixed support prices through support purchases would be undermined. Tariffication has the major advantage of making trade barriers more transparent and creates a closer linkage between domestic and international prices. However, the radical implications for existing support policies may mean that tariffication will not be easily accepted as a major instrument for reforming agricultural policies in the Uruguay Round.

Effects on Export Subsidies

The second implication of a GATT agreement involves reform of export subsidies. These have historically been part of U.S. (and other) agricultural programs, although in recent years they have primarily been used to provide surplus commodities to developing countries through Public Law 480 (the Food for Peace program) and other programs. The widespread use of direct export subsidies is a relatively recent phenomenon, being a product of the aggressive shift in farm policy with the buildup of commodity stocks and the drop in agricultural exports in the early 1980s. The Export Enhancement Program (EEP) has become an important method for using CCC surplus stocks to subsidize agricultural exports. The Food Security Act of 1985 established the Targeted Export Assistance Program to help the United States increase its export market share, partly by countering the effects of the subsidies of other countries in particular markets. Export credit programs were also expanded. Marketing loans for cotton and rice provide a direct payment equal to the difference between the price support loan rate on CCC loans and the world price at which these loans could be repaid. Thus, marketing loans provide an additional method for subsidizing rice and cotton exports. These measures were introduced in an attempt to recapture market share lost during the early 1980s, when the high value of the dollar and high loan rates priced the United States out of world markets. A major argument in favor of export subsidies is that they are needed to counter subsidized exports from other countries, particularly member nations of the European Community.

A GATT agreement will very likely call for the phasing out of direct export subsidies, such as those under the EEP. The elimination of these subsidies in exchange for parallel reductions in the export restitutions of the European Community would probably be viewed as a positive development by the United States. However, the exporting firms that have benefited from subsidy programs would not necessarily be glad to see them terminated. The status of indirect export subsidies—for example, U.S. deficiency payments—is less clear. Direct income supports may be allowed under the GATT if they are largely decoupled from production. The conversion of target price payments to decoupled payments might be one way to achieve acceptable policy changes.

Effects on Other Farm Program Provisions

A final agreement in the Uruguay Round to reduce, substantially and progressively, the overall level of support and trade-distorting subsidies as set out in the Mid-Term Review Agreement would raise for the United States and other countries a major question about the form of future agricultural support. There is considerable uncertainty on this point, since it is not yet clear in the GATT negotiations whether the target for reform is the level of support and government expenditures on agriculture per se or the forms of support and expenditure that create the greatest trade distortions. Even if it is eventually agreed to be the latter, there are many unresolved issues, such as what measures are to be included in the category of trade-distorting support, whether some supports are more distorting than others and therefore have a higher priority for reform, and what (if anything) constitutes non-trade-distorting support. Most forms of support in agriculture have some effect on production or consumption and, ultimately, on trade. If, on the other hand, all government expenditures are the target for reduction, then not only would all direct income support payments (such as deficiency payments) to producers have to be reduced, but the status of conservation payments, subsidized crop insurance, disaster relief, and numerous other types of programs would also be open to question.

At this time, it seems likely that most payments directly linked to production and input and marketing subsidies would be subject to greater discipline under a GATT agreement. Direct payments would probably be exempt if they could be shown to have only a limited impact on production and trade. Current U.S. programs for grains could be adapted fairly easily to meet this condition by fixing the payment base and program yields and capping the per-unit deficiency payment (see, for example, the discussion of Producer Entitlement Guarantees under the subsection "Intermediate Alternatives, Two-Price System," in the chapter by Phipps, Rossmiller, and Meyers in this volume). Whether production-restraining programs such as acreage reduction, paid diversion, or set-aside would be permitted is unclear. Production restraints are not specifically mentioned in the Mid-Term Review Agreement, although to the extent that they influence production and consumption they do distort trade. There would, however, be little need for such restraints if the United States moved to essentially free market prices for grains. Loan rates would either become inoperative (most of the time) because of the likely increase in world prices with liberalization, or they would be difficult to sustain because of restrictions on the use of export subsidies or on the level of government expenditures (Blandford, 1988). Government purchases and sales of commodities under a successor to the loan rate program could be permissible under a GATT agreement, providing such activities could be demonstrated to be for price stabilization purposes rather than income support. The credit facility provided by CCC loans could be continued, although any subsidy element would have to be counted in the overall level of subsidies subject to limitation under the GATT.

This discussion of some of the major changes that might be required in U.S. agricultural programs under a meaningful GATT agreement should demonstrate that the implications of such an agreement could be substantial. The United States entered into negotiations on agriculture in the Uruguay

Round because it saw a need to reduce the damaging effects of the support programs of other countries on its overseas markets; but trade negotiations, like trade itself, are a two-way street. Significant changes in U.S. agricultural programs will be required by other countries in exchange for the reforms they make. Should the GATT be successful in instituting such reforms, it is likely to change substantially agricultural policymaking in the United States.

IMPLICATIONS OF FAILURE IN THE GATT NEGOTIATIONS

Few economists are eager to contemplate the effects of a failure in the current GATT negotiations. Indeed, the political and economic implications of a complete failure are so unattractive that some sort of agreement on agriculture seems almost certain to emerge from Geneva. The essential issue is whether this agreement will make a major difference in the present situation in world agricultural trade or merely lead to cosmetic changes. If it is not seen as constituting significant progress, then it is likely to accelerate a shift to a more protectionist stance in U.S. trade policy.

The provisions of the OTCA suggest that in the event of an unsatisfactory outcome in the negotiations, the United States would move rapidly to take unilateral action to counter what it perceives to be the unfair trade practices of other countries. The OTCA contains provisions for increased use of agricultural subsidies, and its provisions for retaliation could lead to the imposition of tariffs or other restrictions on imports from nations that are judged to use unfair practices. The decision to use subsidies and sanctions, however, would likely depend on a number of factors. Included among these would be conditions in world and domestic markets, the availability of sufficient financial resources to meet the costs of additional subsidies, the likelihood of retaliation by other countries, and, above all, the political situation in the United States.

Bilateral agreements might be explored as an alternative to unilateral action or to the multilateral approach. It is difficult to imagine, however, a successful bilateral agreement between the European Community and the United States if the failure to reach a GATT agreement on agriculture were due to continuing EC-U.S. differences. Bilateralism has been somewhat successful in opening markets in the newly industrializing countries of Asia, although at some political cost, and in the recent trade agreement between the United States and Canada. The Canada-U.S. Free Trade Agreement, however, made only a modest contribution to the liberalization of North American agricultural trade, primarily by reducing tariffs (Government of Canada, Department of External Affairs, 1987).

Significant reductions in domestic agricultural subsidies are inherently difficult to achieve with a bilateral approach because present distortions in international agricultural markets are multilateral in origin. Bilateral liberalization, like unilateral liberalization, is a costly and inefficient way of dealing with global trade distortions and is difficult to engineer without providing a free ride to third parties.[2]

As far as agriculture is concerned, the failure of the GATT process would mean that few international constraints would be placed on domestic policies. The initial U.S. response would probably be to increase subsidies. The costs of agricultural programs would continue to rise unless expenditures were constrained—for example, through mandated reductions in the federal deficit. In the longer term, cost considerations as well as depressed prices could promote the adoption of more stringent production controls (mandatory supply controls for grains, production quotas for milk) or even some type of two-price plan through which producers are paid (and consumers pay) a higher price for domestically consumed product and the world price for product sold on international markets.

With no solution to the problem of agricultural protection, world market prices (under normal weather conditions) would continue to be depressed and would likely be quite volatile. An increasing number of countries would aim for self-sufficiency and use world markets as a means of disposing of domestic surpluses or making up for temporary shortfalls. U.S. agricultural programs would have to try to adjust to this instability or to accept the inevitable downsizing of agriculture that would result from less-outward-oriented agricultural policies. In conclusion, failure to conclude an agreement on agriculture in the GATT would leave the United States with greater flexibility in the choice of agricultural policies, but with little hope that these policies could substantially improve the outlook for U.S. agriculture.

REFERENCES

Bale, M. D., and Lutz, E. 1979. "The Effects of Trade Intervention on International Price Instability," *American Journal of Agricultural Economics* vol. 61, pp. 512–516.

Blandford, D. 1988. "The Implications of Freer Agricultural Trade: A Review of the Empirical Evidence." Paper prepared for the International Agricultural Trade Research Consortium Symposium, Annapolis, Md., August.

General Agreement on Tariffs and Trade. 1969. *Text of the General Agreement. Basic Instruments and Selected Documents, Volume IV,* Geneva.

———. 1986. *Ministerial Declaration on the Uruguay Round,* Geneva, September 20.

Government of Canada, Department of External Affairs. 1987. *The Canada-U.S. Free Trade Agreement,* Ottawa.

[2]The implications of unilateral versus multilateral liberalization are discussed in the chapter by Blandford, de Gorter, Dixit, and Magiera in this volume. The free-rider problem is evident in the Japanese beef agreement. It is unclear whether the primary beneficiary of the conversion of Japanese import quotas to tariffs will be the United States or competing exporters such as Australia and Canada.

McCalla, A. F. 1969. "Protectionism in International Agricultural Trade, 1850–1968," *Agricultural History* vol. 43, pp. 329–343.

Organization for Economic Cooperation and Development. 1989. *Agricultural Policies, Markets and Trade: Monitoring and Outlook 1989,* Paris.

Tontz, R. L. 1977. *Foreign Agricultural Trade Policy of the United States, 1776–1976.* Report No. ERS-662 (Washington, D.C., U.S. Department of Agriculture, Economic Research Service).

U.S. Congress, Congressional Budget Office. 1987. *The GATT Negotiations and U.S. Trade Policy* (Washington, D.C., Government Printing Office), June.

U.S. Congress, House of Representatives. 1988. *Omnibus Trade and Competitiveness Act of 1988.* Report 100–576 (Washington, D.C., Government Printing Office).

PART 4

PRACTICAL OPTIONS FOR A 1990 FARM BILL

The following chapter draws heavily on the combined expertise and insight of a task force convened to develop the policy options discussed and analyzed herein. The members of the task force contributed their time and analytical skill as individuals, not necessarily as representatives of their institutions. The task force met three times over the summer of 1989 to prepare, circulate, discuss, and debate a number of background briefs. John Schnittker graciously agreed to write the chapter, and the personnel from the Center for Agricultural and Rural Development/Food and Agricultural Policy Research Institute at Iowa State University ran the models of the scenarios. The final result, therefore, represents the combined efforts of fine minds keenly attuned to the agricultural policy process. As such, it should not be construed as representing the views of any particular organization. Nor should it be read as advocating any particular policy or set of policies.

The members of the task force were—

Vernon McMinimy, Task Force Chairman; Consultant, Rolling Meadow, Ill.

Daniel G. Amstutz, former Undersecretary of Agriculture and former Ambassador and Chief Agricultural Negotiator

Richard J. Goodman, Vice President, Continental Grain Co.

Stanley R. Johnson, Professor of Economics, Iowa State University and Director of the Center for Agricultural and Rural Development

Ross C. Korves, Economist, American Farm Bureau Federation

Howard F. Lyman, Legislative Assistant, National Farmers Union

William H. Meyers, Professor of Economics, Iowa State University, and Associate Director of the Center for Agricultural and Rural Development

B. H. Robinson, Associate Administrator, Economic Research Service, U.S. Department of Agriculture

Fred H. Sanderson, Senior Fellow, National Center for Food and Agricultural Policy, Resources for the Future

John A. Schnittker, President, Schnittker Associates

Patrick W. Westhoff, Research Associate, Center for Agricultural and Rural Development, Iowa State University

Jeffrey A. Zinn, Specialist in Natural Resource Policy, Congressional Research Service

The input of all participants is gratefully acknowledged.

KRISTEN ALLEN
Volume Editor

PRACTICAL OPTIONS FOR A 1990 FARM BILL

JOHN A. SCHNITTKER

Developments in the U.S. farm sector and in world agricultural trade since 1985 have led to a firm consensus that U.S. farm policy in the 1990s can continue on the course set in recent years with only limited amendments. But budget austerity, negotiations under the General Agreement on Tariffs and Trade (GATT), economic instability, or efforts to brand U.S. agriculture as nonsustainable because of its heavy reliance on chemicals may shatter that consensus. As the debate began on the 1990 farm bill, however, the range of practical choices on key commodity, trade, and environmental issues was relatively narrow. Commodity program options and their linkages to the budget, GATT, and the environment are described briefly in this chapter, and references are made to other chapters in this book that deal in detail with relevant issues.

Analytical and political activity in the U.S. farm and food sectors picks up speed every four or five years as the farm bill debate is renewed. This ritual dates back at least to the Agricultural Act of 1949, although the number of years between its occurrence has varied greatly, depending on the length of the extension of the preceding farm bill and the need for changes in the legislation arising from economic and political circumstances.

Participation in the analytical and political activity has broadened significantly over time. Analysis of farm bills as they developed twenty to thirty years ago was limited largely to the U.S. Department of Agriculture (USDA) and the Office of Management and Budget (OMB). In recent years, university, foundation, private, and governmental analysts have provided an extensive list of ideas and amendments for consideration. The chapters in this volume represent a small but select sampling of the ideas in play for the 1990 farm bill.

Political interest until twenty years ago went little beyond the House and Senate committees on agriculture and the farm and commodity organizations. Today, everyone is in on the act, including environmentalists, conservationists, consumer representatives, universities, research centers, and trade specialists, in addition to the congressional and farm interests (see the chapter by Allen in this volume). Foreign attention to U.S. farm programs has increased over time and will reach a new level in 1990, as domestic farm policies are debated in the GATT as well as in Congress. Officials from other countries lobby freely to influence reforms of U.S. farm policies. This proliferation of interest and activity promises that 1990 will be an extraordinary year in farm and food circles. It has also changed the nature of the farm bill debate and increasingly led to different results than if the matter had been left strictly to farm interests and their political allies.

The federal budget deficit will also exert a major influence on this debate and its outcome. In fact, budgetary considerations since 1985 have forced a yearly debate on farm programs to design and adopt provisions that would limit planned spending to the Gramm-Rudman-Hollings guidelines[1] established in 1987 and to make limited mid-course corrections in policy. The farm financial crisis of 1983–1987 and emergency legislation in 1988 and 1989 to offset the results of serious droughts also contributed to the impression that farm bill debates are becoming continuous.

Renewing the farm bill while completing the Uruguay Round of trade negotiations in 1990 adds both a sense of drama and an air of confusion to the debate. The Bush administration sees the GATT negotiations as the centerpiece of agricultural policy action in 1990 and proposes that farm subsidies be phased down and out on a worldwide basis. Officials minimize the importance of the 1990 congressional debate on farm issues, arguing that whatever is passed may have to be revised dramatically in 1991 to adjust U.S. agricultural policy to GATT agreements on subsidy reduction and market access.

Congressional leaders doubt that significant agricultural policy reforms will be offered by other countries in Geneva to set the pace for U.S. reforms. Rather, they look to congressional action driven largely by domestic considerations in 1990 as the key event setting the course of U.S. farm policy. Congress expects any amendments to the 1990 farm bill made necessary by the outcome of the Uruguay Round to be as routine as the annual adjustments needed to conform to budget policy.

This difference of opinion, and the rhetoric that has accompanied it, was illustrated by the following sharp statement from Congressman E. (Kika) de la Garza (D-Tex.), Chairman of the House Committee on Agriculture, entitled "In Response to Certain Press Reports" (presumably about making farm policy in Geneva), on September 27, 1989: "It is the responsibility of Congress to legislate and to decide how long legislation we pass shall have effect. The Congress will decide whether or not to accept or reject any agreement made by the Administration in Geneva or any other place."

This dispute will cast a long shadow on farm bill and trade policy debates for several years, as the executive and legislative branches of government compete for power over U.S. agricultural policy.

POWER AND INFLUENCE

Time has changed the locus of political power that sets the terms of the farm bill. The broader base of interest has diluted the once dominant personal roles of agriculture committee chairmen, the Secretary of Agriculture, and a

[1] The Balanced Budget and Emergency Deficit Control Reaffirmation Act of 1987.

handful of farm leaders. The greatest change from early farm bill debates, however, is in the dominance of Congress through House subcommittees and the bipartisan leaders of the Senate committee, and the accompanying decline in the influence of the executive branch. This shift was partly the result of strong leadership and greatly increased staffing in Congress, including establishment of the Congressional Budget Office in the 1970s. Agricultural policy proposals by the executive branch far outside the mainstream of political feasibility in 1981 and 1985 also contributed to the ascendancy of Congress and commodity organizations, especially in connection with the last two farm bills.

The executive branch has the opportunity in 1990 and 1991 to exert a greater influence on agricultural and food policy than has been typical in recent years. This could develop if the administration adopts a cooperative and transitional approach in its farm bill negotiations with Congress and in its Uruguay Round strategy. Speaker of the House Thomas Foley (D-Wash.), Senator Patrick Leahy (D-Vt.), Secretary of Agriculture Clayton Yeutter, and other key actors in this drama are well-known conciliators. And President Bush certainly seems to prefer working with Congress, rather than confronting it, on a variety of issues. Such a strategy aimed at limited agricultural and environmental objectives in 1990 would contrast sharply with 1981 and 1985, when the Reagan administration ceded leadership to Congress by proposing unrealistic agricultural reforms, but later supported the 1981 and 1985 farm bills, although it had little sympathy for either. It would, however, be consistent with the two-stage farm bill revision that includes changes arising from the Uruguay Round, in which the executive branch plays the active role while Congress is only a marginal participant. Finding common ground on agroenvironmental questions, however, may be more difficult than on conventional farm bill issues.

Executive branch influence in 1990 and 1991 will be greatest, of course, if through strong international and internal pressures other nations make major concessions on agricultural protection in the Uruguay Round. In that case, U.S. agricultural reforms that go well beyond the 1990 bill would be set by the administration after nominal consultations with Congress. Once agreed on in Geneva, these accords will almost inevitably be approved by Congress under fast-track, "all-or-nothing" procedures. They would face rejection only if agricultural and other trade agreements went far beyond the congressional consensus or were considered to be "unbalanced."

POLICIES FOR A PROSPEROUS AGRICULTURE

The dramatic financial recovery of the U.S. farm sector under the Food Security Act of 1985 has established farm income benchmarks in 1987 and 1988 that will be difficult to maintain (see the chapter by Thompson in this volume). Record expenditures for farm programs in 1986 and 1987, a favorable economic climate, liquidation by late 1989 of $17.5 billion in commodity assets that were acquired before 1986, and higher farm prices as a result of two severe droughts contributed significantly to the farm recovery.

If the 1990 farm program debate coincides with big crops, declining prices, and the prospect of increased spending, Congress will face the possibility of voting in an election year on a farm bill that does not sustain aggregate income at recent levels. The alternative would be to go to the bank again, with a farm bill designed to cost, say, $14 billion to $16 billion or more annually in its first few years. This would break the implicit budget compact, which appears to allocate only $10 billion to $12 billion a year or less to farm programs in the early 1990s. This is based on 1989–1990 spending levels, which were reduced sharply by the droughts (see the chapter by O'Brien, Meyers, and Morton in this volume).

Budget costs as estimated for the farm bill debate might well be kept within the targets just cited, however, either with an extension of 1990 support levels to 1995, or with slow and delayed reductions in supports more consistent with progress in the trade talks. But this might heighten the controversy over future farm income levels, as is illustrated later in this chapter.

"STAY THE COURSE" IN 1990

"Stay the course" became a popular slogan in 1986 in response to public criticism, after it became clear that farm programs would cost nearly $26 billion in fiscal year 1986 and nearly that much in fiscal year 1987, instead of $18 billion as planned. Since then, "stay the course" has evolved into an all-purpose aphorism in support of keeping the framework if not the detail of existing farm policies (see Thompson in this volume). It developed partly as a defensive strategy designed to derail two alternative approaches that seemed for a time to have a chance to replace the existing farm policy framework in the 1990s.

One of these approaches was the proposed Family Farm and Full Production Act of 1985, commonly known as the Boschwitz-Boren bill, after Senator Rudy Boschwitz (R-Minn.) and Senator David Boren (D-Okla.). It also became known as the "decoupling" approach for its separation of federal farm payments from production. Boschwitz-Boren would have taken deficiency payments back to their original function in academic literature and legislative lore—that is, as income support devices, not rewards for producing crops. Boschwitz-Boren also incorporated an idea that had appeared in President Reagan's first (1981) and subsequent farm policy proposals to Congress—to reduce federal farm payments and then end them within a few years (see the chapter by Phipps, Rossmiller, and Meyers in this volume).

This approach was adopted by the Reagan administration in its proposals to reduce and end trade-distorting farm subsidies by the year 2000 under the Uruguay Round of trade negotiations. Boschwitz-Boren would also have ended U.S. programs to limit crop acreage as a condition of eligibility for income payments. This provision was vulnerable to opposition by farm groups, whose members receive most of the benefits of existing programs. It could also have been very costly, especially in an era of excess production capacity and huge grain surpluses.

Boschwitz-Boren as an idea peaked with the agribusiness sector and agricultural economists around 1986–1987, when the United States was piling up crop surpluses and breaking the budget. It never developed a strong farm constituency, however. Decoupling, reducing, and ending farm subsidies were not popular ideas among farmers during the financial crisis of 1983–1987. Elements of the decoupling approach will have an important place in the 1990 debate, however, and in the GATT negotiations. Limited decoupling is practical under the existing farm program framework: it represents one route to greater flexibility on the farm and to income support as distinguished from production payments. It is adaptable to targeting benefits to certain farmers as well (see the chapter by Sumner in this volume). Budgetary and environmental considerations also make payment reduction an inevitable issue in 1990 and later, even though prospective declines in farm income may limit the possibility of reducing payments to farmers.

The Family Farm Reform Act of 1985 and successor bills were a second and less realistic threat to the Food Security Act of 1985. Called the Harkin-Gephardt bill, for Senator Thomas Harkin (D-Iowa) and Representative Richard Gephardt (D-Mo.), the bill included techniques designed to reduce budget costs through limited production, more effective marketing, and support of farm prices at higher levels than in the Food Security Act. These were not revolutionary ideas by any means, since they had been used for wheat, cotton, and rice from 1938 to the 1960s and are still used for tobacco and peanuts.

Harkin-Gephardt developed only a narrow farm constituency. In fact, it generated strong opposition and ran counter to a twenty-year history of talk and sporadic action to reduce rather than increase federal intervention in decision making on the farm. It also ran counter to a number of studies showing that supply control programs and high price supports involve high consumer costs (see the chapter by Hertel in this volume).

Just as the 1985 Food Security Act bowed to Boschwitz-Boren by reducing target prices systematically beginning in 1987 and 1988, after an initial pause at the high 1985 levels, it included provisions that could have led to mandatory wheat acreage controls and much higher wheat price supports in 1987 (as in Harkin-Gephardt). Lack of farmer support in a referendum and an unfriendly administration doomed this effort in 1986, however. And the broad Harkin-Gephardt framework, never a serious threat to existing law, was forgotten by 1989.

Elements of the mandatory supply control approach persist, though, in actions to fine-tune existing Acreage Reduction Programs to limit stocks and costs. Harkin-Gephardt ideas may also reappear in the 1990s for dairy, sugar, or both. Developments in the GATT talks may require the United States to open its market to imports, while political necessity at home may require maintenance of price supports and market prices well above world levels. Techniques related to Harkin-Gephardt provisions may have a small role in the transition to increased sugar and dairy product imports (see the chapter by Blandford in this volume). But approval for them, over the strong opposition of the executive branch, seems doubtful unless the administration adopts a very pragmatic attitude toward program mechanics in pursuit of important economic objectives after the Uruguay Round.

AMENDING THE FOOD SECURITY ACT OF 1985

"Stay the course" implies a framework and a direction, not simply keeping existing farm laws as they are. Budget and trade policy pressures, changed circumstances in world markets, experience with the Farmer-Owned Reserve program, the unnecessary rigidity of the 1985 act with respect to cropping decisions, and inequitable target prices for key commodities provide further basis for change. Ideas represented by key code terms that have developed for the 1990 debate—including decoupling, targeting, restructuring of payment systems, limiting of export subsidies, flexibility, and competitiveness—all represent practical modifications of the Food Security Act.

Concern about declining water quality and questionable food safety, and linkage of these problems to excessive use of farm chemicals, provide a rationale for environmentally inspired amendments building largely on the 1985 Conservation Reserve Program and conservation compliance framework. At least a dozen ideas for making marginal but positive progress toward environmental objectives by tapping into commodity programs are under discussion, in addition to separate agroenvironmental programs (see Reichelderfer's chapter in this volume).

The Budget as a Limitation in 1990

There is general agreement that average spending targets for loans, deficiency payments, the Conservation Reserve Program, and other farm programs should not exceed $10 billion to $12 billion per year in the early 1990s. This figure may well be lower during the 1990 debate, but it is not likely to be higher. The president's 1991 budget or other administration statements may propose substantially lower average spending for the next two to three fiscal years to meet Gramm-Rudman-Hollings targets and to show reductions from 1988, 1989, and 1990 levels. Budget strategies, however, often hinge on Machiavellian choices and can scarcely be predicted. What is often overlooked is that recent low farm program spending levels were possible largely because the 1988 and 1989 droughts raised farm prices and reduced government payments. Huge commodity surpluses that had been built from 1984 to 1987 were turned into cash via Payment-in-Kind certificates, replacing out-of-Treasury expenditures.

Whatever their level, farm program spending limitations in the early 1990s will be extremely elastic in practice. Price supports and deficiency payments are entitlements that take precedence over budget targets. Larger crops, lower exports, and lower prices in 1991 or 1992 than assumed in connection with budget estimates for the farm bill may lead to actual spending well above targets. Stocks and reserves policies to be adopted in 1990 may also add to budget costs, as will environmental programs.

The Uruguay Round and the Farm Bill

Trade provisions in the 1985 Food Security Act also require some changes. Retaliatory measures as illustrated by the Export Enhancement Program, Targeted Export Assistance, and the marketing loan are built into the 1985 act. Similar authority in the 1988 Omnibus Trade and Competitiveness Act designed to put pressure on other countries will also have an important influence on the 1990 farm bill debate. This is especially so in view of the low expectations held by many members of Congress for the Uruguay Round and the general belief in Congress that the United States will have to "get tough" after limited settlements in the negotiation. This attitude is softened somewhat by strong rhetorical support by members of Congress for administration efforts to make balanced progress in worldwide reduction of farm subsidies.

Negotiating guidelines in the Uruguay Round, together with U.S. budget restrictions, effectively prevent legislated increases in U.S. farm supports (levels of protection) in the 1990 farm bill for the crop years 1991–1995. Strictly speaking, the April 1989 accord now governing the Uruguay Round requires subsidy cuts in 1990 as down payments on longer-term reforms. The European Community has indicated that continuation of its recent support reductions meets that goal. The United States has announced a number of cuts for 1990. Congress may not subscribe to U.S. farm subsidy cuts for 1991, however, in view of the tradition of freezing most farm supports for one to two years before authorizing any necessary reductions in the later years of a four- to five-year bill. Rudimentary targeting of payments via slightly reduced per-farm limits could be adopted, however, even if overall payments were not reduced.

Opposition by Congress and the executive branch to what officials call "unilateral disarmament" on farm subsidies may limit or delay U.S. payment and price support reductions. Concern on this score may lead to legislation in 1990 making future reductions in U.S. supports and payments and any opening of U.S. import markets contingent on reciprocal concessions by others, agreed and bound in GATT, beginning about 1991 or 1992. Uruguay Round rules providing credit for subsidy reductions made since the beginning of the negotiations in 1986 were designed to maintain the momentum of the EC and U.S. reforms, but may not have that effect in the United States.

Authority under which the U.S. Trade Representative may offer and Congress would later approve reductions in U.S. farm support or subsidy levels beginning in 1992 or 1993 also provides a rationale for postponing the start of trade-based reforms to 1992 or later. The recent tendency to adjust farm program details from year to year to conform to budgetary necessity and low congressional expectations for the Uruguay Round have the same effect. A delay strategy is also useful to farm groups who want to preserve existing payment levels or other benefits. Delayed reforms may be especially applicable to the most politically difficult commodities such as sugar and dairy products, where even moderate reforms may be possible only under the cover of a trade agreement. That strategy applies equally to grains and cotton for at least one to two years.

Finally, Congress has often garnered enough votes to pass a controversial farm bill by holding the line on subsidy cuts for a year or two. Leaders can promise urban members some budget savings on farm programs in a few years, while holding out hope to farm groups that economic conditions will change so that future budget cuts against farmers can be avoided. This strategy will depend on an overall budget policy that can absorb the extra costs of holding farm supports at 1990 levels for one to two years to await the close of the Uruguay Round.

THE MOMENTUM OF CONSERVATION AND ENVIRONMENTAL ISSUES

Momentum generated by the popularity and success of conservation programs in the 1985 Food Security Act, information developed in recent years on declining water quality, and a burgeoning interest in the alternative agriculture movement also contribute to the climate for change in 1990. The farm income, crop production, and trade expansion agenda of the farm and agribusiness participants are very different from the soil-saving, water quality, human health agenda of the newer environmental and conservation-oriented participants in the farm bill debate.

Four factors strengthen the hand of the latter groups for 1990 and beyond:

1. Water quality and food safety have a much larger and more committed actual and potential constituency in Congress than do farm issues. Urban members of Congress have usually voted for farm bills over the years. In 1990, however, some votes may seem to pit farm aid against clean water and safe food. Farm and agribusiness interests may need to compromise on such votes, or lose them.

2. The ingenuity of farmers and scientists has created some interesting alternatives to present levels of farm chemical use and holds out the promise of more. Congress may wish to join many state governments to give these techniques a federal boost. Banning a few farm chemicals via revised Environmental Protection Agency procedures and reducing the use of others via farm bill provisions involving the "purchase of environmental services" may yield rather large gains in water quality and food safety in five to ten years (see Reichelderfer's chapter in this volume).

3. Farmers will be fairly responsive to the environmental agenda, just as they were to soil conservation in past years. They may already be leading

the way to safer chemical use through more careful and reduced usage. A barrage of anecdotal evidence appeared in farm publications and other journals in 1989 in support of this phenomenon.

4. There is still some excess farmland available on which to grow crops to replace any production lost from reduced chemical applications. At least 15 million to 20 million acres could be put back to crops quickly. This significantly undermines, for a few years, the economic argument that reduced chemical use on the farm will lead inevitably to higher food costs or reduced exports. Continued crop failures on the order of 1988 and 1989, however, would bring idled land back into production quickly.

Budget limitations may prove to be a greater constraint on adoption of the agroenvironmental agenda than farm-based concerns over lost production, reduced income, or declining exports. Three basic possibilities arise in connection with the competition of farm and environmental programs for federal funds:

1. The cost of environmental measures must be absorbed in the anticipated $10 billion to $12 billion per year maximum farm program budget target for 1991–1992. Payments to farmers may have to be squeezed to provide money to begin or expand environmental programs. Political strains would be extreme in this situation if environmental groups decided to play hardball to increase spending on the agroenvironmental agenda.

2. Alternatively, environmental considerations could provide a rationale for setting higher farm bill budget targets than now implied by the national budget compact. This may be an unlikely outcome, but it should not be ruled out that Congress might add $0.5 billion to $1.0 billion per year by 1992 or 1993 to foster the agroenvironmental agenda, if farm program spending is low enough.

3. If exports increase while crop yields rise only slowly, some increased spending in pursuit of agroenvironmental objectives may be possible with limited political strains. An overall budget policy that is extremely restrictive could offset declining farm program costs, however.

SPECIFIC COMMODITY POLICY OPTIONS FOR 1990

Three similar options, with subtle differences, constitute the baseline and two alternative scenarios for 1990. They are described briefly below, and in more detail in the appendix to this chapter.

The three generic options described and evaluated here are all based on existing farm law, including some features that date from 1938. There is only the remotest possibility that fundamental farm program design will be scrapped, whether the 1990 bill is a five-year or a one-year program. In late 1989, there seems to be little chance that the Bush administration will draft a comprehensive new approach to agricultural support and stabilization to present to Congress.

Other scenarios could be developed, and other approaches will certainly be proposed in 1990. Those described here, however, represent a persistent theme that will still be recognizable at the end of 1990 no matter what budget, political, or economic developments may arise during the year. Another drought disaster in the United States by mid-year, prospects for record crops by late summer, or an early deadlock and failure in the Uruguay Round would all lead to different results, but largely within the framework described below.

The basic options include—

1. *The Baseline.* Extend the 1985 Food Security Act, possibly with some changes in operating provisions as described for scenario 1 below; freeze 1990 target price and other support levels through 1995.
2. *Scenario 1.* Adopt a two-year standstill on income and price support levels (for 1991 and 1992 crops), assuming that unusually stringent budget guidelines do not force immediate continuation of target price reductions. Follow this with 2 percent per year reductions in income and price support levels beginning in 1993, either as required by budget policy or contingent on reasonable success and reciprocity in the Uruguay Round. A number of important but not fundamental amendments could be made in grain reserves, planting flexibility, acreage reduction, payment limits, the Export Enhancement Program, and other program features, effective in 1991 (see the appendix to this chapter). Some of these amendments would be driven by environmental consideration, among other forces.
3. *Scenario 2.* A budget-driven reform package. Extend the farm support reductions of the late 1980s into the early 1990s without interruption at a 3 percent annual rate. The operating amendments cited above for scenario 1 would also be included. This option may be the strongest environmental measure possible in 1990, since it would reduce the profitability of heavy use of farm chemicals through its effect on overall farm returns.

Details of the baseline option and the two commodity program options are given in the appendix to this chapter. Economic estimates for the three options are described in the final section. These estimates are presented as illustrations only, since they may be out of date when economic and commodity data are revised early in 1990.

ENVIRONMENTAL AND CONSERVATION SCENARIOS FOR 1990

Environmental and conservation provisions of the farm bill will have a separate set of objectives but will be closely related to agricultural objectives and to the budget. Dr. Garth Youngberg, of the Institute for Alternative

Agriculture in Bethesda, Maryland, has advanced the idea of making sustainability the overriding theme and test of the commodity program provisions of the 1990 farm bill.[2] If that were done, the race would be on to brand many current programs as nonsustainable because of their environmental effects and to identify practices that would be consistent with long-term sustainability (see Reichelderfer's comments on integration of environmental and agricultural programs in the farm bill debate in her chapter in this volume).

The two basic approaches considered here are linked to the two commodity scenarios described above, but could also go with the baseline if it becomes a realistic option. Few specific program features can be identified in view of the proliferation of ideas and the limited stage of development of legislative proposals in this area. The general outline of agroenvironmental programs is fairly visible, however.

Scenario 1: Limited Expansion of Low Input Sustainable Agriculture (LISA) Programs, 1991–1995, Beyond the 1990 Level of Activity

- Relatively slow implementation from 1991 to 1995 of voluntary, limited chemical input regimes, based principally on incentives and education, should be assumed.
- Farm chemical use per acre could decline slowly partly as a result of forces already in motion and partly as a result of new incentives offered to farmers under a bill (S. 970, 101st Congress) proposed by Senator Wyche Fowler (D-Ga.).
- Low input crop production incentive programs in S. 970 are assumed to be operated on a demonstration scale, so as to be essentially budget neutral. Increased acreages (due to smaller acreage reductions) of program crops on demonstration farms would be the principal incentive to reduce chemical use or to encourage other environment-enhancing practices on cropland.
- Funds for research and education might reach five to ten times the $5 million per-year level of 1986–1990. Any substantial funds authorized for program operations as distinguished from research and education would have to be offset by specific savings in commodity programs resulting from LISA programs.
- Denial of deficiency payments for farm chemical use not conforming to safe water guidelines would be delayed until the 1995 farm bill, or at least until 1992 or 1993. Better data on water quality will then be available, and the voluntary, incentive-based programs described here would have had an initial test. This is practical and consistent with the environmental agenda, in view of the annual farm bill debate now held in connection with the budget.

---

[2]Testimony before a subcommittee of the House of Representatives, July 19, 1989.

• The Conservation Reserve Program (CRP) should be limited to 40 million acres by the end of 1995 (adding about 5 million acres to the expected 1990 area). This would direct new CRP contracts principally to conservation and environmental objectives such as wildlife habitat, wetlands protection, forestry, and water quality. Little new attention would be given to soil conservation or reduction of excess crop production capacity via long-term contracts.

• A bill (S. 1063, 101st Congress) introduced by Senator Richard Lugar (R-Ind.) could, along with S. 970, become the vehicle for tree planting, well testing, wetlands conservation, and a wide range of new or expanded programs. Plans for keeping present CRP land in conserving uses might also be established in the 1990 legislation.

Scenario 2: Rapid Expansion of LISA Programs, 1991–1995

• Much more rapid implementation of low input farm production systems, despite a very limited information base, should be assumed.

• This approach would provide an annual budget of $1 billion for environmental and conservation practices described above (in practice, spending might begin at $250 million to $500 million in 1991 and rise to $1.5 billion after a few years), over the initial 1991–1992 commodity program and CRP budget target.

• The LISA demonstration programs described above would move at a relatively fast pace toward national coverage by 1995. They would include cash payments for reduced chemical use to supplement acreage incentives and could have tangible effects on crop production.

• "Water conservation compliance" would be adopted in 1991, to become effective by about 1995. This would limit farm program benefits, subject to appropriate chemical use and other cultural practices.

• Limit the CRP to 50 million acres by the end of 1995, concentrating on the special objectives cited above.

• Contracting of more land to meet traditional soil conservation and production adjustment objectives could be done with the higher budget and acreage targets. Little productive land devoted to wheat and other export crops would be added to the CRP in the 1990s, however.

• Land already contracted would be retained in conserving uses beyond the contract period via incentives.

• Provide for strongly funded research and educational activities, including a nationwide demonstration program under the LISA banner, well testing, tree planting, and handling and application of farm chemicals (S. 779, 101st Congress, the Senator David Pryor (D-Ark.) bill).

PROGRAM RESULTS

Projected economic impacts, especially annual budget expenditures and aggregate farm income, will be critical factors in determining acceptable provisions of the 1990 farm bill. Budget targets will be set by the budget committees of Congress, based to a degree on the president's 1991 budget and on baseline estimates by the Congressional Budget Office (CBO) early in 1990. Farm income estimates will be the result of program assumptions and estimates by the USDA and CBO of farm prices, exports, production costs, and the like. Estimates for 1991 and 1992 will be the most influential in determining farm bill provisions and the budget, since more extended projections are notoriously questionable.

Overall budget policy, and CBO estimates for specific bills early in 1990 and again at mid-year, will be the decisive determinants of budgetary limits on farm bill provisions. Private estimates made late in 1989 and in 1990 provide both a preview of 1990 estimates and a concurrent critique of the official budgetary line. They, too, can influence the course of the 1990 farm bill. (See the chapter by O'Brien, Meyers, and Morton in this volume for comparisons of the assumptions of the Food and Agricultural Policy Research Institute and the CBO.)

Budget Costs of Farm Programs

Estimates by the Food and Agricultural Policy Research Institute (FAPRI) at Iowa State University and the University of Missouri in October 1989 provide a ray of hope that budget costs in the early 1990s can be held at or near the guidelines cited earlier in this chapter. This prospect appears to be valid whether principal farm program features now in force are extended (baseline) or income support levels are reduced significantly during the early 1990s. Given the yield and export demand assumptions that follow, manageable budget costs do not depend on drought at home or abroad. Faster yield increases and slower export expansion produce substantially higher budget estimates, however.

Expenditures for both the Commodity Credit Corporation's price support and related programs and the CRP under the baseline option (continuation of 1990 programs through 1995) would be around $13.3 billion per year in fiscal years 1991 and 1992 (the first seasons under the 1990 farm bill) and would fall to $7.7 billion in fiscal year 1996 (table 1). Expenditures of the Commodity Credit Corporation (CCC) alone (without the CRP) would be around $2 billion per year lower.

The costs for scenarios 1 and 2 would be roughly the same as or slightly below baseline levels in the early years, but would be significantly lower in 1995 and 1996.[3] Estimates of farm program costs for the 1990 fiscal year should not change drastically by early 1990. Cost projections when bills for

[3]FAPRI estimates do not explicitly take into account possible program amendments involving greater planting flexibility, realignment of target prices, and some other minor changes described in scenario 1. Details of the FAPRI estimates will be published in Westhoff, Patrick, Deborah Stephens, Michael Helmar, Brian Buhr, and William H. Meyers, 1990, "An Evaluation of Policy Scenarios for the 1990 Farm Bill," CARD Staff Report 90-SR43 (Ames, Center for Agricultural and Rural Development, Iowa State University).

TABLE 1. Cost of Farm Programs, Fiscal Years 1986–1989, and Projections to 1996
(billions of dollars)

			Projected cost		
Fiscal year	Actual cost	Fiscal Year	Baseline option	Scenario 1	Scenario 2
1986	25.8	1991	13.3	13.5	13.1
1987	22.4	1992	13.2	13.5	12.2
1988	13.3	1993	10.2	10.2	8.3
1989	14.0[a]	1994	8.7	7.8	6.7
1990	11.4[b]	1995	8.8	7.0	6.0
		1996	7.7	5.7	5.6

Note: "Farm programs" include the Conservation Reserve Program and Commodity Credit Corporation programs.

[a]This CBO estimate is as of November 1989.

[b]This CBO projection is as of November 1989.

actual floor debate are "costed" by the CBO in comparison with baseline estimates early in 1990 may well be similar to those shown here. Estimates may change significantly, however, in the mid-1990 budget review if crop conditions are far different from those incorporated in the original baseline assumptions.

If baseline spending estimates by mid-1990 are well above the OMB-CBO targets for farm programs, Congress would need to consider real cuts in spending by reducing target prices and other support levels below baseline levels, along the lines suggested in scenarios 1 and 2. The possibilities of a presidential veto and a tough stand by environmental interests in and out of Congress could force severe cuts of this type in support levels.

If the budget allocation to farm programs for fiscal years 1992, 1993, and 1994 is extremely low (say $8 billion to $9 billion per year, under the influence of the 1990 spending level and drastic overall budget cuts necessitated by Gramm-Rudman-Hollings considerations), price and income support levels may need to be reduced immediately, possibly far more rapidly than in scenario 2.

Farm Income

Recovery of crop production in the early 1990s, followed by lower livestock and crop prices than in 1989–1990 and higher production costs, would erode net farm income, even if 1990 program provisions were to apply (baseline assumptions). Note in table 2 that net farm income estimates hold up well through 1992, before sliding sharply in the baseline and alternative scenarios. This will be the critical time frame in which farm income estimates may significantly influence the farm bill. With three and five years of declining

TABLE 2. Net Farm Income, 1986–1989, and Projections to 1996, Baseline Assumptions Compared with Slower Production Cost Increases
(billions of dollars)

			Projected income with—	
Year	Actual income	Year	Baseline[a]	Slow production cost increases[a]
1986	38	1990	47	47
1987	47	1991	44	45
1988	46	1992	42	44
1989	49	1993	38	42
		1994	37	42
		1995	35	43
		1996	36	49

[a]Baseline production costs increase 3–5 percent per year while the alternative rate of cost increase is 2 percent per year (slow).

deficiency payments under scenarios 1 and 2, net farm income would fall well below baseline levels by 1995–1996.

Farm income projections are highly sensitive to the rate of increase in farm production costs. Projections are also heavily dependent on a lack of any increase in livestock receipts during the 1991–1996 period, as a result of lower grain prices than in 1989. If production costs were to increase only 2 percent per year for 1990–1996, however, net farm income would be $4 billion higher than the baseline in 1993 and $13 billion higher in 1996. That outcome would roughly preserve 1987–1989 net farm income levels to 1996 in nominal terms, even though real aggregate income would decline. This is considered by some analysts to be more realistic, under the economic conditions expected, than a sharp drop in net farm income in the 1990s.

The lessons for policy choices in 1990 are (1) that concern about serious declines in net farm income under baseline assumptions may reinforce Uruguay Round considerations, helping to derail other options, and (2) that any major farm income declines appear to be two to three years ahead. That would permit Congress to discount them, since farm budget and income issues will be revisited yearly in the early 1990s.

Acreages

The three options examined in the econometric analysis all generate large crop acreages in the early 1990s, similar to those of the immediate post-drought years (1989 and 1990) (see table 3). This is the result of small acreage reduction percentages, which can be continued to 1995 with little change, according to FAPRI yield and demand assumptions.

Sustained yields for grains only 2–3 percent above the levels assumed, or several years of spectacular yields early in the 1990s, would either force the

TABLE 3. Baseline Acreage Planted to Crops for 1987 and 1989, and Projections for 1991 and 1995
(millions of acres)

	Actual		Projected	
	1987	1989	1991	1995
Corn	66	72	74	75
Wheat	66	77	78	79
Soybeans	60	60	57	61
Fifteen major crops	252	266	266	272
Acreage Reduction Program	61	29	19	18
Conservation Reserve Program	16	30	40	40

USDA to reduce the area in crops (increase acreage reduction percentages) or face sharply increased stocks, low prices, and increased budget costs by 1993 or 1994.

Crop Yields

Crop acreage, production, stock accumulation, and excess capacity estimates are, of course, a direct function of yield and demand assumptions. The FAPRI model projects slightly lower yields for scenarios 1 and 2, primarily in response to lower target prices. The scenario 2 corn yield, for example, is 125 bushels per acre in 1995. Yield assumptions are also similar to those used by the CBO (see the chapter by O'Brien, Meyers, and Morton for comparisons).

Baseline yields are illustrated in table 4 for the principal grains and for cotton. Yields listed for 1991 are slightly below to slightly above the highest yield ever realized in the late 1980s for each crop.

Program costs are highly sensitive to crop yields. In any year, costs would be increased by about $1 billion if yield increases as described below were to occur without offsetting increases in demand:

- For corn, about five bushels per acre above the normal yield assumed in the model. National average yields in the area of 125 bushels per acre or more are clearly possible at any time.

TABLE 4. Baseline Crop Yield Assumptions, 1987–1995

	1987	1991	1993	1995
Corn (bushels per acre)	119.4	120.8	124.3	127.2
Soybeans (bushels per acre)	3.7	34.7	35.1	35.8
Wheat (bushels per acre)	37.7	39.1	39.5	40.4
Cotton (pounds per acre)	706	661	681	701

TABLE 5. Baseline Exports for 1987–1989, and Projections to 1995

	Average 1987–1989	1991	1993	1995
Corn (millions of bushels)	1,950	2,030	2,240	2,525
Wheat (millions of bushels)	1,450	1,425	1,560	1,690
Soybeans (millions of bushels)	640	620	650	690
Cotton (millions of bales)	7.0	7.5	7.8	8.4

- A combination of wheat and soybean yield increases around two to three bushels per acre over the yields listed above. These are realistic possibilities, especially for wheat.

If events in the 1990s show that FAPRI yields listed above are too low, on average, the resulting increased production would require the Secretary of Agriculture to reduce the area in particular crops once stocks had risen to a desired level. This would be done under acreage reduction authority.

If yields were to rise less rapidly than shown above because of poor weather or reduced chemical use, annual Acreage Reduction Programs would be very small, or might not be operated at all during the early 1990s.

Commodity Exports

Program cost estimates in any year are extremely sensitive to export projections. The export projections that were used to make the cost and income projections are given in table 5. Projected exports in 1995 range from 29 percent above the 1987–1989 average for corn to 17, 8, and 20 percent above the 1987–1989 averages for wheat, soybeans, and cotton, respectively.

Commodity Surpluses

Large acreages, rising yields, and domestic plus export demand combine to maintain wheat, corn, and cotton stocks on the low side of the normal reserve consensus in the FAPRI estimates (table 6). Soybean stocks, on the other hand, rise well above the normal reserve level of around 250 million bushels, although not to the record level of 1987.

TABLE 6. Baseline Year Ending Stocks for 1986–1987, and Projections to 1995

	1986–1987	1992–1993	1994–1995
Corn (millions of bushels)	4,882	1,536	1,529
Wheat (millions of bushels)	1,821	862	784
Soybeans (millions of bushels)	436	310	337
Cotton (millions of bales)	5.03	4.88	4.80

Maintenance of relatively constant stocks levels will require good luck and careful fine-tuning, using Acreage Reduction Programs, export aids (the Export Enhancement Program), and other program tools. There should be no illusions about holding wheat stocks, for example, in a narrow range near 800 million bushels for five years. It may be practical, however, to keep wheat stocks in a range of 600 million to 1,200 million bushels most of the time, based on a policy target in the range of 800 million to 1,000 million bushels of stocks.

APPENDIX

Program crops include feed grains, wheat, cotton, and rice. Along with soybeans, they usually account for the bulk of agricultural exports, farm program costs, and political interest in farm legislation. The options described here also include dairy products, sugar, peanuts, and tobacco.

BASELINE OPTION

Continue all essential 1990 program features, especially price and income support levels, for five years. Technical improvements involving planting flexibility, realignment of target prices among eligible crops, targeting of payments to smaller operations, and improved reserve operations that would not have major budget or income effects could be incorporated in the baseline option. Many of these are listed below with scenario 1. Similarly, the scenario 1 agroenvironmental programs cited in the text could be included in the baseline without material effect on commodities or costs in the 1991–1995 period.

SCENARIO 1

Scenario 1 constitutes a two-year standstill on support levels followed by automatic price and income support reductions.

Grains, Cotton, Rice, and Soybeans

• Freeze target price and other support levels for 1991 and 1992 crops at 1990 levels; reduce target prices 2 percent per year in 1993, 1994, and 1995 contingent on an acceptable outcome of the Uruguay Round and/or on budget policy, whether or not other nations are reducing their farm subsidies and opening their markets systematically. Target prices for grains could be realigned slowly beginning in 1992 or 1993 as a part of this package without any substantial impact through 1995.

• Provide farmers the option of planting other crops on some of their "program" acres, in pursuit of greater flexibility and a more market-directed output mix. This could be done with a version of "triple base" plans under which 10–20 percent of the crop base is not eligible for deficiency payments and could be devoted to nonprogram crops. It could also be done via a return to the "normal crop acreage plan" in effect during the late 1970s, or by using "flexibility" language similar to that of the 1988 drought bill (see Phipps, Rossmiller, and Meyers, in this volume). These approaches also involve marginal decoupling, so they might draw broad support from environmentalists and agribusiness interests (see the chapters by Reichelderfer and by Blandford et al. in this volume).

• Keep price support loan levels in a narrow range near current levels by continuing the present price support formula based on recent average market prices. Some price support levels might increase slightly as a result of 1988 and 1989 prices, but they would be offset by lower maximum deficiency payments. This would not affect production levels and would not be considered an increase in the (GATT) level of protection. No great harm would be done if Congress were to return loan rates for grains to 1988 or 1989 levels for a year or two to limit budget exposure, since average market prices under conditions assumed here would be well above loan levels throughout the period (see Barkema, Drabenstott, and Tweeten, in this volume for a discussion of competitiveness).

• Build or deplete grain stocks to around the levels suggested in present law, using Acreage Reduction Programs to add to stocks or to limit stock increases as needed (0.8 billion to 1.0 billion bushels of wheat and 1.6 billion to 2.0 billion bushels of corn).

• Continue and amend the Farmer-Owned Reserve and food security wheat reserve, including new provisions to permit reserve grains to be marketed at reasonable prices in the range of 80–100 percent of target price levels. Shorter reserve contracts and phased release options would also increase farmers' marketing options. Commodity certificates would not be used or would be used in small amounts to limit CCC acquisitions.

• Use the Export Enhancement Program for several years to ensure export tonnages for grains and cotton as shown later in this section, increasing slightly from levels established in 1987–1989, pending completion of the Uruguay Round. After that, a coordinated cut in export subsidies by the United States and European Community may limit use of the Export Enhancement Program, and would possibly end it by 1996.

• Limit sales or deliveries of surplus commodities from CCC stocks to periods when prices are 15 percent or more above recent average market prices, at least 50 percent above the current loan levels, and not far below prevailing target prices. These provisions would preclude use of the marketing loan, would keep U.S. commodities competitive abroad, and would limit expenditures of deficiency payments.

Dairy

• Freeze price supports at $10.10 per 100 pounds of milk through 1991. Beginning January 1, 1992, reduce dairy supports $0.50 per 100 pounds per year, contingent on expected annual CCC purchases of dairy products equivalent to 5 billion pounds of milk or more. Assume that any trade concessions made on dairy products in the Uruguay Round would not lead to materially increased import levels until 1994 or 1995.

Sugar

• Freeze the 1986–1990 price support level of 18 cents per pound of raw sugar for the 1991 and 1992 seasons. Provide for 3 percent per year reductions in the loan and market stabilization prices, beginning in 1993, contingent on an acceptable outcome (to the United States) of the Uruguay Round. Increased import levels during the mid-1990s would depend on the trend in U.S. sugar and sweetener production during the 1990s, as guaranteed prices decline in the United States (and presumably in other countries).

Tobacco and Peanuts

• The programs for tobacco and peanuts would not be changed in the 1990 farm bill under scenarios 1 and 2. Any reforms would be made as concessions in the Uruguay Round; amendments to domestic programs could be made in 1991 or 1992, effective in 1992 or 1993. Meaningful U.S. trade concessions or farm policy reforms for tobacco and peanuts would tend to follow the pattern of dairy products and sugar described above.

SCENARIO 2

Under this scenario, target price reductions begun in 1987 and 1988 would be continued in 1991 and subsequent years at 3 percent per year. Make room for increased dairy and sugar product imports at a slightly more rapid rate than in scenario 1.

Grains, Cotton, Rice, and Soybeans

• Reduce target price levels for grains, rice, and cotton and the soybean loan level 3 percent per year beginning with 1991 crops. This could be the result of budget pressures and/or rapid reductions in levels of support in the EC, Japan, and other countries.

• Adopt other "improving" provisions as listed under scenario 1. These amendments would provide planting and marketing flexibility, ensure adequate stocks to meet domestic and export needs, maintain the U.S. competitive position, and limit budget costs of the farm programs.

Dairy

• Freeze supports indefinitely at $10.10 per hundredweight. Introduce an unspecified program limiting milk marketing to the level of commercial use plus school lunch and other program needs for dairy products, thereby preventing accumulation of excess stocks. Increase dairy product import quotas by 1 percent of U.S. dairy product demand per year beginning in 1992, contingent on reciprocity in the Uruguay Round.[4]

Sugar

- Freeze the loan and market stabilization levels for five years at the high 1990 levels. Limit U.S. sugar and corn-based sweetener marketings from year to year to levels that would accommodate about 1.5 million, 1.75 million, 2.0 million, and 2.25 million tons of raw sugar imports in the 1992 through 1995 marketing years, without adding to stocks or government costs.

[4]This change in dairy policy would maintain consumer costs higher than scenario 1 and would support farm income while reducing government costs.

PART 5

EPILOGUE

U.S. AGRICULTURE BEYOND THE FARM BILL

GEORGE E. ROSSMILLER, LYNN DAFT, and KRISTEN ALLEN

The prologue to this volume asked, What does U.S. agriculture want from the agricultural programs? Most of the individual chapters focused on the changing goals for and approaches to the major agricultural programs. They concentrated on issues likely to be debated in the reauthorization of farm legislation upon the expiration of the Food Security Act of 1985, specifically the programs in the commodity and trade titles and the conservation and environmental provisions. However, many important issues covered by the act were not addressed, including credit; agricultural research, extension, and teaching; food stamps and related provisions; marketing; processing, inspection, and labeling; special studies and commissions; and animal welfare. Rather, the purpose of the project behind this volume was to focus on the more contentious sections of a new bill, and the decision was made early on not to dilute its main thrust by trying to consider in depth all issues important to agriculture.

It is appropriate at this point to put the results of this project in perspective and the 1990 farm legislation in the context of the broader set of concerns that will affect the food and agricultural system in the years ahead. This epilogue, therefore, looks at the long-term forces that will affect U.S. agriculture in the coming decades. These include macroeconomic policies; U.S. foreign policies; economic policies of other countries and the resolution of developing countries' external debt; global warming; environmental policies and regulations; food safety at home and abroad; the promises, challenges, and threats arising from the uses of new technologies—including biotechnology—in food production, processing, and marketing; concerns about animal welfare and animal rights; and the economic plight of rural communities. Also relevant are concerns about poverty and homelessness and the attendant problems of health and nutrition among America's poor. Our purpose here is not to be comprehensive or definitive, but rather to illustrate the breadth of the public policy agenda that will influence the fortunes of the food and agriculture system in the future and the (likely decreasing) role of farm policy legislation in that agenda.

EMERGING CONCERNS

Agriculture is no longer just farming and, in fact, has not been for some time. No matter how much some people might yearn for simpler times, the fact is that the agricultural sector is an integral part of the national and global economies. Recognizing this means accepting the notion that little can be done to the traditional U.S. commodity price and income support programs to change significantly agriculture's fortunes. Therefore, if we accept the premise that the goal of U.S. agricultural policy should be to promote a healthy, dynamic, responsive sector that is part of the world economy as we approach the twenty-first century, rather than to perpetuate an anachronistic set of programs that make the sector dependent upon the kindness of strangers, we must take a more comprehensive approach to designing agricultural policy than has so far been done.

Protecting the Natural Resource Base and Maintaining the Integrity of the Food Supply

The four years since the passage of the Food Security Act of 1985 have seen increasing scrutiny of modern agricultural techniques and their effects on the physical environment. Although some environmental concerns have long been part of agricultural policy, the Food Security Act elevated them to greater prominence with the Conservation Reserve Program, the sodbuster and swampbuster programs, and conservation compliance provisions. These programs are built on the assumption that agriculture does have some detrimental effects on the environment and attempt in various ways to reduce the damage. Over the past four years, however, there has been a shift in the focus of those groups concerned with environmental degradation and agriculture, away from emphasis on soil loss per se to the effects of agricultural practices on water quality more generally, including soil and chemical contamination of groundwater and surface water.

In the not-so-distant future, agriculture stands to be profoundly affected by what has been called the "third wave" of environmental regulation (Carrick, 1989). In writing on California's Proposition 65, Carrick points out that an increasingly skeptical public is sending a strong message that it wants closer monitoring and regulation of chemical use in the environment. For U.S. agriculture the upshot could be that many of the practices that have contributed to the gains in productivity over the past forty years—in particular chemical use—will be much more closely controlled or even prohibited in the future.

A great deal of attention is now centered on what is termed "alternative" or "sustainable" agriculture. Despite all the attention, there appears to be little consistency in defining the terms or in systematically laying out the goals of a sustainable agricultural system and the means for achieving them. Lockeretz (1988) provides a useful review of a number of general concepts pertaining to sustainable agriculture. Key among them are biological diversity; ecological soundness in choices of crops and livestock; crop rotations; composting and mulching; and a reduction in the use of purchased inputs in general and synthetic chemicals for crops and concentrated rations for livestock in particular. While this view is broader than that championed by many of the current proponents of sustainable agriculture, who seem to focus primarily on pesticide use, it is perhaps still too narrow. In a search for agricultural sustainability, we need to consider also the health and welfare of farm families, the economic sustainability of the sector, and the continuing integration of agriculture into the general economy. As we move into the twenty-first century, a truly sustainable agricultural system, in addition to not contributing to the irreversible degradation and contamination of the environment, should recognize the needs for farm operators to make a profit, for the sector to export some portion of its production or face a significant downsizing, and for the industry to adapt to changing production and processing technologies, to new ways of marketing products, and to changing consumers needs, tastes, and preferences. The areas of consumers' health interests—both food safety and diet/nutrition/health linkages—as well as changing demographics and consumption and food purchasing behavior patterns are especially important. New trends will require the U.S. food and agricultural sector to be much more responsive, and less doggedly defensive, than it has been in the past. A real concern for a truly sustainable agricultural system is perhaps the most important thing that policymakers can take into the debate on the 1990 bill.

Food safety has become the "hot" issue of the late 1980s. It began slowly, with a small, dedicated faction arguing for improvements in seafood, poultry, and meat inspection services. In early 1989, however, a new aspect was catapulted to the fore with the Natural Resources Defense Council's exposé of Alar use on and residue in apples. The public took the council's report seriously and began questioning the safety of much of the U.S. food supply. The belated protests and pronouncements from the U.S. Department of Agriculture (USDA) and other organizations and individuals in the sector that the U.S. food supply was the safest in the world fell largely on deaf ears.

The reaction to the Alar incident indicates most pointedly what little credibility federal agencies—in this case the USDA, the Food and Drug Administration, and the Environmental Protection Agency—currently have with the American public. Because more incidents of this nature will likely occur in the future, there is a real need now to establish a credible source that can evaluate and disseminate information about the integrity of the U.S. food supply.

Tying together the concerns about food safety and environmental quality, and the underlying concern over total food supplies, should perhaps lead us to consider the broader issue of a safe and secure food supply. Focusing on any one of its component parts seems to generate what could be termed overreaction from some quarters. A rational, long-term approach to policy should recognize that trade-offs are inevitable, but also that substitutions can be viable options.

Chemicals used in the production, storage, processing, and marketing of agricultural products have enabled a large supply of high-quality food to be made available to the American public at relatively low cost. While it is almost certainly possible to produce these foods using fewer chemicals, it is also likely that to do so at the current level of production would require more intensive inputs of management and labor. Such a change in the input mix could increase the cost of agricultural products, possibly by a very significant amount. In addition, some of the characteristics of the products may change. Producers are now rewarded for meeting (and penalized for not meeting) USDA attribute standards for many agricultural products. For example, smaller, less colorful, less picture-perfect fruits and vegetables produced with fewer chemical inputs may be no less nutritious than their chemically treated counterparts, yet producers may be precluded from marketing them, or receive lower prices for them, if they do not conform to the current standards. Both a reevaluation of the attribute standards and consumer education may be needed to effect a move away from purely cosmetic use of chemicals. Such a move could alleviate some of the concerns about chemical residues in foods without jeopardizing the quality or the size of the food supply. On the other hand, grain contaminated with aflatoxin (a known carcinogen) is visually indistinguishable from uncontaminated grain, but in fact poses a serious health risk. Restricting the use of chemicals, and thus requiring alternative measures to be used to prevent the development of the aflatoxin fungus, may be extremely expensive and not necessarily as effective in preventing the health hazard posed by the contaminated grain.

Another trade-off is that between land and chemical inputs. Traditionally, U.S. agricultural policy has used land retirement (both short and long term) to meet various goals—environmental and commodity supply management in particular. As long as weather and economic conditions are favorable and commodity stocks adequate, there is little chance that land retirement, especially short term, will have a negative impact on total agricultural commodity supplies. In addition, it may prove to be beneficial in terms of such environmental goals as reducing soil loss and providing or protecting wildlife habitat. A common criticism of land retirement, however, is that it encourages overuse of chemical inputs on the land remaining in cultivation to make up for lost production. Economic theory suggests that rational producers will not overuse any inputs unless high prices provide an incentive for extra production. Therefore, keeping commodity prices at reasonable levels and maintaining the freeze on yield bases can limit the overuse of chemical inputs. But a series of unfavorable years in terms of weather conditions, depletion or tying up of stocks, or policies that artificially raise prices could significantly change the equation in a fairly short period. Without any flexibility to respond to these events by moving land into production, farmers would almost certainly turn to more intensive use of the land available, which would likely mean more intensive chemical use. Thoughtful balancing and integration of pricing, stockholding, land use, and environmental policies are essential in maintaining a safe and secure food supply.

As these few examples illustrate, there is not a single, common problem in the debate over chemical use in agriculture and the safety, security, and sustainability of the food and agriculture system. It is, therefore, highly unlikely that a single, simple solution will be found.

Concern About New Technologies

One enduring characteristic of the agricultural sector is the relentless march of technological change and the sector's ability to take advantage of productivity-enhancing technologies. Already U.S. agriculture has been through the hand and animal power era, the machine era, and the chemical era. It now appears to be on the threshold of the information technology and biotechnology eras. The interface of the chemical and biotechnological eras is, however, a turbulent time for both technologies, marked largely by uncertainty, ignorance, and fear. New technologies and processes are becoming available that could offer alternatives to current levels, forms, and timing of chemical use in agriculture. Yet many of the new technologies—especially those based on biological or genetic engineering, such as bovine somatotropin and herbicide-resistant crops—are being viewed with suspicion by various groups. In the past, technology has been viewed as the savior, the way out of the dilemma of rising expectations and limited resources. Now, fears about misuse of technologies and the potential for releasing something monstrous into the environment, as well as institutional rigidities that preclude developers of new technologies from fully capturing their benefits, may limit how much technology can come to the rescue in the future. Certainly there is a need to proceed cautiously, but we must proceed nevertheless.

While it is not possible to know the full range of biotechnology research presently under way (since much of it is being done in proprietary laboratories), anecdotal evidence indicates that a significant number of products will be turned out over the next few years. Many, if not most, will be of concern to environmentalists and food safety activists. The debate will continue, regulatory policy will be developed and changed, and accepted practices and uses of the new technology will be defined even as it is invented and implemented. To the extent that agricultural productivity continues to be increased through further refinement of chemical technology and the innovation of new biotechnology and information technology, pressures will be placed on traditional agricultural policy as well as on environmental and food safety policies.

Concerns About Rural Communities

Agriculture functions as part of a larger rural economy that is increasingly more heterogeneous and diversified. It is also an economy that is confronted with serious problems that extend beyond the scope of farm policy.

Farming no longer dominates the rural economy. To the contrary, agriculture represents a small and shrinking portion. The population living on full-time commercial farms (those with gross sales of $100,000 or more) now accounts for less than 0.5 percent of the rural population. Farm households are themselves substantially dependent upon earnings from off-farm employment for their income. In 1988 the aggregate income of farm families from off-farm employment exceeded income from farming by 13 percent. It is fair to say that agriculture has now become as dependent on the rural economy as the rural economy is dependent on agriculture.

At the risk of oversimplification, the public policy needs of rural areas can be separated into three distinctly different types:

- Needs associated with adjustment to changing economic circumstances;
- Needs associated with low income;
- Needs associated with the realization of development potential.

Adjustment Needs. Rural areas in some parts of the country, particularly those regions that are sparsely populated and heavily dependent on industries based on natural resources (including agriculture), are confronted by major problems of adjustment. The loss of employment opportunities as these industries experience economic downturns ultimately leads to a loss in population, erosion of the tax base, and a reduced capacity to provide needed community services and facilities. Rural areas confronted by such economic adjustments account for much of the recent, sizable outmigration from rural areas in total. Compared to a net immigration of around 350,000 people per year during the 1970s, migration out of rural areas has accelerated to a rate of 632,000 people per year in 1985–1986. This represents a change in migration patterns of major consequence. As a result of this reversal, almost half (1,160) of all nonmetropolitan counties lost population during 1983–1985.

Many of the smaller communities in the areas affected by this population loss face severe adjustment problems. As their populations shrink, so does their capacity to provide basic public and private services. While reversal of these population trends often is not feasible, changes in the form and organization of public services can make an important contribution to helping communities cope with the effects of this adjustment.

Low Income Needs. There are other rural areas where the most pressing problem is rooted in the poverty of the region. Although this is more concentrated in the South and in Appalachia, it occurs in isolated pockets throughout the nation. As a result, the overall incidence of poverty in rural areas is considerably higher than it is in urban America. In fact, in 1983 it was more than 40 percent higher.

Areas suffering from high concentrations of poverty represent the most desperate form of rural need. Frequently these are areas that are poor because they lack a resource base sufficient to support the local population. The problem is often cyclical. Low income results in a workforce that is poorly equipped to compete for jobs, either locally or in other labor markets. Lack of jobs results in low incomes. Thus, the process repeats itself and the stagnation persists.

Development Potential. There are other rural areas that confront a still different set of needs—needs that can be characterized as developmental. These are areas that, because of a variety of circumstances—good access to markets and transportation linkages, favorable climate, scenic beauty, and so on—have developmental potential but lack one or more of the preconditions required to translate this potential into reality. For example, the local labor force might lack needed skills, the community might lack needed infrastructure, or the governmental structure might be ill-prepared to cope with the demands of a developing economy.

Assignment to these three categories is not mutually exclusive, of course. Some rural areas fall into more than one category; some fall into none of them. But a large share of rural areas are affected by one or more of these types of need.

Policy Approaches. Attempts have periodically been made to deal with the needs of rural areas through national policy, dating back at least to the Country Life Commission of 1908. Numerous programs and agencies have been created to deal with various dimensions of the issue. There was a flurry of activity during the New Deal and again during the days of the War on Poverty. While some of these programs have effectively addressed particular functional needs or the needs of specific regions, they have largely been episodic, temporary, and disconnected from other programs. To some extent, it seems to have been implicitly assumed that the economic needs of rural areas were being effectively addressed through farm programs. But that has not been true for many years. There has been no umbrella policy framework to tie the pieces together into a consistent whole or to ensure continuity of effort. The appropriate division of responsibility among local, state, and federal governments still is not clear. Many past federal programs, particularly those that sought to alter trends in economic growth through investment in infrastructure or that tried to impose standard, national solutions on highly variable local needs, ended in failure.

Policy approaches to the problems of the rural community have not always been realistic. Many of the economic trends that have given rise to these problems stem from fundamental changes in the structure of major industries, changes that frequently are far beyond the capacity of a few modestly funded government development programs to alter. While growth in employment and population is possible under certain circumstances for some of these areas, for many others it is highly unlikely. Established interests understandably have a bias toward protecting investments that are attached to particular places. But policies that emphasize the development or preservation of rural places without regard for these underlying forces have too often created unrealistic expectations.

Instead of focusing on the development of places, a strong case can be made for policies that emphasize people and the investments required to make them healthy, productive citizens, better able to cope with the economic adjustments that characterize so many rural areas. Some programs have this orientation, but many do not. With the benefits of good health and a good education, individuals in the workforce can adapt to changing economic opportunities. They can even change occupations and residences if necessary. Programs that emphasize places, in contrast, are far less flexible. By their nature, they are tied to a particular location and often to a particular industry.

Political support for a coherent rural policy at the national level is weak and diffuse since, in reality, the problem cannot be characterized as a single issue but rather comprises a disparate set of many loosely connected issues, each with its own small constituency. Furthermore, since the various functional needs—health, transportation, education, community facilities, employment, and income security—cut across the functional makeup of the federal government, there is fragmentation both in support for the programs and in program administration. Although the USDA is the federal institution with the firmest attachment to rural America, it is an attachment based largely on the administration of programs related to agriculture and forestry. It is poorly equipped to deal with the other dimensions of the issue, especially those relating to human resources. No other cabinet agency, however, has the breadth of responsibility required to deal with such a broad range of needs. Furthermore, the larger, more powerful urban interests tend to dominate the agenda of the human resource agencies, making it unlikely that rural constituencies could compete very effectively for attention and resources.

Thus, the economic and social conditions of rural areas are of vital importance to the large number of people who live outside the nation's cities. The issue is not agricultural, although most of the federal institutions closest to the problem are those with program responsibilities in agriculture or forestry. While rural communities are not totally lacking in public policy attention, the level and composition of assistance as well as the unity of direction, particularly at the federal level, leaves much to be desired.

In charting the course for future federal assistance to rural areas, the focus needs to be on rural people, not just the places where they live. Particular attention needs to be given to such topics as education and training, health care, income security, housing, sanitation, job counseling, and employment information. Maximum attention should be devoted to keeping the assistance flexible so that it can be adapted to widely varying local situations.

The federal government requires a better mechanism for giving the condition and needs of rural areas greater visibility and for harnessing the many programs and policies that have potential to provide needed assistance. While the USDA should continue to play a lead role, the direct involvement of the Executive Office is also required. Despite the need for additional federal support and cohesion, the principal direction of effort must come from state and local governments, whether operating individually or collectively.

Concerns Arising from the Changing Global Environment

Given the tremendous increase in U.S. agriculture's dependence on international markets during the 1970s and the pain suffered by the sector through the mid-1980s from the loss of export markets, trade and international economic issues will continue to play a prominent role in agriculture's and the nation's policy agendas. Important dimensions of the issues surrounding trade have been detailed in earlier chapters. Several additional points and linkages bear highlighting.

First, it is impossible to know the specific nature of the agreement that might come out of the agricultural negotiations of the General Agreement on Tariffs and Trade (GATT), or, indeed, whether any agreement at all is likely. It is possible that an agreement will be reached that will establish principles for the use of domestic policy mechanisms in order to minimize trade-distorting effects, as well as the direction of policy reform toward trade liberalization. With the principles and direction firmly established, future rounds of trade negotiations under the GATT could be expected to make further, substantial progress toward agricultural trade liberalization. But it is also possible that the Uruguay Round may not produce a significant agricultural agreement. If progress in the round is less than deemed acceptable to both the U.S. administration and Congress, quite a different scenario could emerge.

The Omnibus Trade and Competitiveness Act of 1988 (OTCA) contains provisions in section 301 of Title I(C) that could be triggered if the president certifies dissatisfaction with the course and progress of the present negotiations. Under section 301, when the United States determines that other nations are engaging in unfair trade practices, it is *required* that the president or the U.S. Trade Representative take retaliatory action against those nations. This is in contrast to the 1974 Trade Act, in which the decision to retaliate was left to the discretion of the president. Also under section 301(4), the United States can use export promotion programs and, if necessary, restrict U.S. imports of agricultural commodities in order to "encourage fair and open trade" and counter unfair foreign trade practices (for a more detailed discussion of the OTCA, see the chapter by Blandford in this volume). If any of these provisions are triggered, the OTCA will have a greater impact on the functioning of the agricultural sector than any of the likely trade provisions in the 1990 farm legislation.

Second, the developing countries are the largest potential markets for U.S. agricultural products. A major constraint to the realization of that potential is sluggish economic growth in the developing economies, exacerbated by the onerous external debt burdens of many of these countries. How the United States, other developed countries, and the international institutions, especially the International Monetary Fund and the World Bank, respond to rationalization of the debt and the creation of a climate for strong, positive growth in the developing world will have a major effect on the ability of U.S. agriculture to prosper in the international market.

Third, the conduct of U.S. monetary and fiscal policies will continue to influence the fortunes of U.S. agriculture, including how well U.S. agricultural products perform in the international market. The effect of macroeconomic policies on the value of the dollar in foreign exchange markets is felt directly by U.S. producers through the foreign price of U.S. agricultural exports, thus affecting U.S. competitiveness. Macroeconomic policies also affect the interest rate, which influences the cost of holding stocks and debt service costs. Debt service costs, in turn, affect the foreign exchange availabilities for agricultural imports by indebted developing countries. Some policymakers argue that macroeconomic policies have a greater impact on U.S. agriculture than does farm legislation.

Fourth, events and policy changes on the horizon in other parts of the world could have a profound effect on U.S. agriculture. The 1992 market harmonization in the European Community and reforms being contemplated or already under way in eastern Europe, the USSR, and the Peoples Republic of China could, if successful, have major impacts on the international market for agricultural products over the next several years. How fast and how sweeping those reforms turn out to be will bear close scrutiny and analysis. In addition, agricultural policy reforms in the industrialized countries have been taking place, and indications are that they will continue, regardless of the specific outcome of the GATT negotiations on agriculture. A few examples serve to illustrate that agricultural policy, worldwide, is not as static as some might believe. The European Community has instituted supply control mechanisms that are now binding on milk and cereals. The Japanese have liberalized their beef and citrus markets and, even with the current political setback, are likely to continue to push for rationalization of their rice policies domestically, the effects of which will eventually be felt in the international market. New Zealand has unilaterally withdrawn government intervention in agricultural and other markets. Australia and Canada have also taken steps to liberalize their agricultural policies. These countries are likely to continue their agricultural policy reforms at their own pace and in their own self-interest. The U.S.-Canada Free Trade Agreement was successfully negotiated and took effect at the beginning of 1989. While full implementation will come slowly, the agreement might nevertheless serve as a model for future bilateral arrangements between the United States and other countries, as well as for other possible trading blocs in different regions of the world. This could be either a positive or a negative development for U.S. agriculture, depending upon how those blocs relate to the rest of the world.

Finally, many have argued that the United States is concentrating too much on bulk agricultural commodity exports when it should be emphasizing promotion of high-value and value-added exports. Proponents often cite statistics on the additional jobs and economic activity that would be generated by exporting wheat in the form of flour or bakery products, or corn and soybeans in the form of beef steaks, pork chops, or chicken TV dinners. They also argue that U.S. producers should be more aggressive in tapping export markets, especially in developed countries, for high-value products such as nuts, fresh fruits, and other specialty crops.

Detractors point out the escalating tariff schedules in most countries for processed products versus bulk commodities, the health and sanitary restrictions on fresh products that often amount to nontariff barriers, and the disservice the United States would do to developing countries by concentrating on exports of processed products. Moreover, some argue that U.S. production agriculture has benefited greatly from intensive investment in capital and technologies and thus can compete very well in bulk commodity exports, while the U.S. agricultural processing industry does not have that competitive edge.

The truth, as usual, probably lies somewhere between these two extremes. The United States can be, and is, competitive in some high-value and value-added products. The debate is likely to continue, as will the search for policies that can facilitate a stronger move into such markets. A new development that will bear watching is the multinational mergers and purchases among food-processing firms in the global industry and the effect of this activity on U.S. export potentials.

Sleepers

The preceding sections have covered factors that are already being discussed, if not necessarily incorporated into the general debate on agricultural policy. There are, however, many issues that have not yet fully (or even partially) infiltrated the general policy arena but that could potentially have a significant impact on U.S. agriculture. A few of these "sleepers" are examined in this section to illustrate how vulnerable the agricultural sector is to a very large set of seemingly remote factors and events.

U.S. agriculture has been fortunate in having an excellent physical endowment and a strong infrastructure—including transportation, information, education, and research and extension systems—that have contributed much to its past and present competitiveness. Their importance to a truly sustainable agricultural system should not be downplayed in the push to raise prices and incomes while simultaneously reducing spending. In the long run, investment in infrastructure will be as important to the health of the sector as are the price, stock, and production manipulations that sometimes give U.S. agricultural commodities a short-term advantage in international markets.

An issue that is partly a subset of infrastructure and partly an emerging environmental concern is subsidized federal water supply contracts. It is particularly important to agriculture in the western United States, where many forty-year contracts are now coming due for renewal and will thus, over the next few years, attract considerable attention from agriculturalists, environmentalists, and policymakers (see Lancaster, 1989a, 1989b). As long as the United States and the world continue to produce surplus agricultural commodities, these contracts will continue to draw fire from environmentalists and possibly from urban interests, who view them as yet another example of the special treatment for agriculture running to excess at the considerable expense of other segments of society. (For a comprehensive study of western water and the issues involved, see Wahl, 1989.)

An issue that some consider frivolous and others dead serious is animal welfare, escalating all the way to animal rights. The effect of bovine somatotropin use on the health of dairy cows, the effect of contemporary broiler housing on the health of chickens, and the raising of veal calves for market in dark and physically confining pens, as well as the use of animals in medical research, have become the concern of activists who view other animals as equal to humans in the right to a decent existence. While this may seem far-fetched to some, note that Sweden and Denmark have traded their international competitiveness in broilers for chicken welfare. Obviously, the issue is real. How the United States deals with the trade-offs will be important to its ability to compete in both domestic and international markets, now and in the future.

Finally, there are two other latent issues that could become important in the next few years. Already, quiet murmurings have been directed at the question of transportation of agricultural products and what will be allowed to be backhauled in containers used to transport food products. Also, the investigation of the activities of members and traders at the Chicago Board of Trade and the Chicago Mercantile Exchange, begun in 1989, and Tokyo's interest in establishing a commodity futures market, both raise questions about who will control agricultural and futures markets and how they will be regulated to facilitate the efficient functioning of global agricultural markets.

Any or all of these issues could, with little or no warning, ignite to become *the* agricultural issue(s) of 1990, just as Alar and chemical residues in food became the issues in 1989 largely on the strength of a single report that was given prominence by extensive and sensationalized popular media coverage. Or they could die on the vine, for any number of reasons, despite their potential importance to the future performance of the U.S. food and agricultural sector.

CONCLUSIONS

The focus in most farm bill debates tends to be much narrower than the wide range of factors that affect agriculture. Therefore, it is likely that most of the forces discussed in this chapter will be overlooked as the 1990 agricultural legislation is written. It is, however, important to keep them in mind, and to begin to make U.S. agriculture and agricultural policy responsive to the change that these forces will bring.

The prologue and some of the chapters in this book looked at the broadening concerns and goals of U.S. agricultural policies and how many of the goals are difficult to achieve within the current framework of agricultural programs. For example, targeting program payments to "needy" farmers, stabilizing farm income, protecting environmental quality, and providing economic stimulation to rural areas are all laudable objectives, but they are often beyond the scope and ability of the traditional agricultural programs.

This epilogue, we hope, has raised a small warning flag: traditional agricultural policy cannot do everything for the agricultural sector. The sector is now, and increasingly will be, both buffeted and benefited by events and policies outside its scope of influence. Responsiveness may therefore be its most important goal in the new decade.

REFERENCES

Carrick, Roger Lane. 1989. "California's Proposition 65: Environmental Law's Third Wave," *Environmental Hazards,* premier issue, pp. 26–30.

Lancaster, John. 1989a. "White House Backs Environmentalists on Water," *Washington Post,* November 29, p. A4.

———. 1989b. "Interior to Renew Western Water Pacts," *Washington Post,* November 30, p. A3.

Lockeretz, William. 1988. "Open Questions in Sustainable Agriculture," reprint from the *American Journal of Alternative Agriculture* vol. 3, no. 4 (Fall), pp. 174–181.

Wahl, Richard W. 1989. *Markets for Federal Water: Subsidies, Property Rights, and the Bureau of Reclamation* (Washington, D.C., Resources for the Future).

ABOUT THE AUTHORS

Kristen Allen is a policy associate at the National Center for Food and Agricultural Policy, where she focuses on U.S. agricultural and trade issues. She edited, with Katie Macmillan, *U.S.-Canadian Agricultural Trade Challenges* (Resources for the Future, Washington, D.C., 1988). Prior to joining the National Center, she was an analyst for the Congressional Budget Office. Her concentration there was agricultural policy, with emphasis on 1985 agricultural legislation for U.S. crop and dairy subsectors. A native of Australia, Ms. Allen received her M.S. in agricultural economics from Michigan State University.

Alan Barkema is a senior economist with the Federal Reserve Bank of Kansas City, where he is a member of the Regional Economics Group in the Economic Research Department. He has written and spoken widely on the effects of macroeconomic policy on the farm economy, the determination of farmland values, and other topics in agricultural policy and finance. He holds an M.S. from Cornell University and a B.S., an M.S., and a Ph.D. from Iowa State University.

David Blandford is a professor at Cornell University, where he teaches and researches issues in agricultural policy, international trade, and market analysis. During 1982 and 1983 he worked as an economist in Paris for the Organization for Economic Cooperation and Development. He received his Ph.D. in agricultural economics from the University of Manchester. His most recent project is with the Pacific Economic Cooperation Conference, studying agricultural trade issues in the Pacific Basin countries.

Lynn Daft is vice president of Abel, Daft and Earley, an agricultural economic consulting firm that examines the economic implications of changes in commodity, food assistance, and international agricultural trade policy. Previously he was vice president of Schnittker Associates, associate director of the White House Domestic Policy Staff from 1977 to 1981, and the Agriculture Team leader for the Carter-Mondale Transition Planning Group. He earned his Ph.D. in agricultural economics from Michigan State University.

Praveen Dixit is an agricultural economist in the Agricultural and Trade Policy Branch of the Economic Research Service, U.S. Department of Agriculture. He is on a USDA task force that analyzes policy reform proposals under consideration in the current negotiations of the General Agreement on Tariffs and Trade. Formerly he was a research assistant at Purdue University, where he received his Ph.D. in 1984. A native of Nepal, he worked at the Agricultural Projects Services Center in Kathmandu for two years after earning his M.A. in economics at the Delhi School of Economics in India.

Mark Drabenstott is an assistant vice president and economist with the Federal Reserve Bank of Kansas City, where he heads the Regional Economics Group in the Economic Research Department. He is the author of numerous articles on such subjects as farm policy, rural development, farm exports, and agricultural finance. He currently chairs the Federal Reserve System Committee on Agriculture and Rural Development. He received his Ph.D. in agricultural economics from Iowa State University.

Harry de Gorter is an assistant professor at Cornell University, where he specializes in GATT and agricultural policy reform issues. He has published journal articles on agricultural policy in the United States and the European Community and the political economy of agricultural policy. Previously he was an assistant professor at the University of Guelph in Canada and from 1983 to 1986 studied agricultural trade liberalization for the government of Canada. He received his Ph.D. in government policy and trade from the University of California, Berkeley, in 1983.

Thomas W. Hertel is an associate professor at Purdue University. His research and teaching activities emphasize economywide analysis of farm and food policy issues. He works closely with the Economic Research Service of the U.S. Department of Agriculture and has served as a consultant to the Ford Foundation, the Organization for Economic Cooperation and Development, and the World Bank. He has a Ph.D. in agricultural economics from Cornell University.

James P. Houck is a professor of agricultural and applied economics at the University of Minnesota, where he has worked since 1965. His main research interests are agricultural prices, trade, and trade policy. He has written several journal articles and a textbook in this area. He received his B.S. and M.S. in agricultural economics from The Pennsylvania State University and his Ph.D., also in agricultural economics, from the University of Minnesota. He was editor of the *American Journal of Agricultural Economics* from 1981 to 1983.

Joseph V. Kennedy received his law degree in 1983 and his M.S. in agricultural economics in 1986 from the University of Minnesota. He currently is an attorney and economic analyst in Washington, D.C., specializing in policy issues involving agriculture, the environment, and international affairs. Previously he was a budget analyst at the U.S. Office of Management and Budget.

Stephen Magiera is the leader of the Economic and Agricultural Policy Section, Economic Research Service, U.S. Department of Agriculture. He was an organizer for the International Agricultural Trade Research Symposium on "Bringing Agriculture into the GATT" and has been a consultant with the Food and Agriculture Organization on agriculture toward 2000, with the Organization for Economic Cooperation and Development on the Ministerial Trade Mandate, and with Winrock International on trade problems in developing countries. He has an M.S. in operations research from the University of California, Berkeley.

William H. Meyers is a professor of economics at Iowa State University, where he teaches and researches agricultural policy issues and conducts trade and price analysis. He is also the associate director of the Center for Agricultural and Rural Development, codirector of the Food and Agricultural Policy Research Institute, and executive director of the Midwest Agribusiness Trade Research and Information Center. He received his M.S. from the University of the Philippines and his Ph.D. in agricultural economics from the University of Minnesota.

Andrew S. Morton is a budget analyst at the Congressional Budget Office. His area of expertise is agriculture, and his major focus is the preparation of cost estimates for bills that affect the federal farm programs, especially the feed grains and soybean programs. He is coauthor of *Diversity in Crop Farming: Its Meaning for Income Support Policy, Crop Price-Support Programs: Policy Options for Contemporary Agriculture, and The Outlook for Farm Commodity Program Spending, Fiscal Years 1988–1993,* all CBO publications. He earned his Ph.D. in economics from Iowa State University.

Patrick O'Brien is the director of the Commodity Economics Division at the Economic Research Service, U.S. Department of Agriculture, where he administers and directs the focus of the agency's outlook and research in the commodity analysis area. He has had several different roles in the International Economics Division at ERS, as economist, regional analyst, commodity analyst, project leader, chief of the World Analysis Branch, and assistant director. He received his M.S. from The Johns Hopkins University in 1972.

Tim T. Phipps is an associate professor of agricultural economics at the University of West Virginia. From 1985 to 1989 he was a fellow at the National Center for Food and Agricultural Policy, where he focused on agriculture's effects on the environment. Prior to that he was an assistant professor in the Department of Agricultural and Resource Economics, University of Maryland, and a research agricultural economist in the Department of Agricultural Economics, University of California, Davis, where he earned his Ph.D.

Katherine Reichelderfer is a senior fellow with the National Center for Food and Agricultural Policy at Resources for the Future and leader of RFF's program on agriculture and the environment. Prior to joining RFF she was associate director of the Resources and Technology Division at the Economic Research Service of the U.S. Department of Agriculture, where she worked for twelve years. She received her Ph.D. in agricultural and resource economics from the University of Maryland in 1977.

George E. Rossmiller is the director of the National Center for Food and Agricultural Policy. In 1986 he joined the National Center as a senior fellow specializing in international trade. Previously he served as a senior economic adviser and the director of the planning and analysis staff, Office of the Administrator, Foreign Agricultural Service. He has a Ph.D. in agricultural economics from Michigan State University. Dr. Rossmiller joined the faculty there in 1965 and attained the rank of professor in 1977.

John A. Schnittker has filled many different roles within the realm of agriculture—from Kansas farmer; to staff economist in the Office of the Secretary, director of agricultural economics, and Undersecretary of Agriculture, all in the U.S. Department of Agriculture; to professor of economics at Kansas State University. He has headed Schnittker Associates, an economic research and policy group, since 1972 and publishes widely in the academic and popular press. He received his Ph.D. from Iowa State University.

Daniel A. Sumner is a professor in the Division of Economics and Business at North Carolina State University. From 1987 to 1989 he was a senior economist on the President's Council of Economic Advisors, where he was responsible for agricultural policy issues. He has a Ph.D. in economics from the University of Chicago. He is the author of numerous scholarly articles and recently edited the book *Agricultural Stability and Farm Programs* (Westview Press, Boulder, Colo., 1988). From 1986 to 1987 he was a Resident Fellow at the National Center.

Robert L. Thompson has been the dean of agriculture at Purdue University since 1987 and professor of agricultural economics there since 1974, where his principal areas of work are agricultural trade policy, U.S. agricultural policy, and world agricultural development. He currently chairs the advisory council of the National Center for Food and Agricultural Policy. He served as Assistant Secretary for Economics at the U.S. Department of Agriculture and was senior staff economist for food and agriculture on the President's Council of Economic Advisors from 1983 to 1985. He received his Ph.D. in agricultural economics from Purdue University.

Luther G. Tweeten is the Anderson Professor of Agricultural Marketing, Policy, and Trade in the Department of Agricultural Economics and Rural Sociology at Ohio State University, where he has worked since 1988. Previously he was on the faculty at Oklahoma State University. From 1982 to 1987 he was also on the core staff, Agricultural Policy Analysis Project, U.S. Agency for International Development. His research emphases are on public policy for agriculture and on regional, national, and international economic development and trade. He has a Ph.D. in economics from Iowa State University and has published more than 300 journal articles.

Jon Visser is a student at Westmont College in Santa Barbara, California. He served as an intern in domestic agricultural policy at the National Center for Food and Agricultural Policy as part of an American Studies exchange program. A political science major and economics minor, he will receive his B.A. in May 1990. His interest in agricultural policy stems from being raised on an almond and cotton farm near Bakersfield, California.

APPENDIX

MEMBERSHIP
FOOD AND AGRICULTURE COMMITTEE
NATIONAL PLANNING ASSOCIATION

*__Lynn Daft__ (FAC Chairman), Abel, Daft and Earley, Alexandria, Va.

*__R. J. Hildreth__ (FAC Executive Committee Chairman), Farm Foundation, Oakbrook, Ill.

Ron O. Aines, Northbrook, Ill.

Merle Anderson, Climax, Minn.

Sandra S. Batie, Virginia Polytechnic Institute, Blacksburg, Va.

*__Anne Chaffin Bostick,__ Nabisco Brands, Inc., East Hanover, N.J.

Robert Caldwell, North Carolina State Grange, Greensboro, N.C.

Raymond Casey, Ohio Farm Bureau Federation, Columbus, Ohio

Fred Clark, National Rural Electric Cooperative Association, Washington, D.C.

*__Mark R. Drabenstott,__ Federal Reserve Bank of Kansas City, Kansas City, Mo.

Kenneth R. Farrell, University of California, Oakland

*__Robert M. Frederick,__ National Grange, Washington, D.C.

Ray A. Goldberg, Harvard Business School, Boston, Mass.

J. B. Grant, National Association of State Departments of Agriculture, Washington, D.C.

Robert D. Havener, Winrock International Institute for Agricultural Development, Morrilton, Ark.

Dana Herron, Washington Association of Wheat Growers, Connell, Wash.

Maureen K. Hinkle, National Audubon Society, Washington, D.C.

*__Dale M. Hoover__ (FAC Membership Chairman), North Carolina State University, Raleigh, N.C.

Joseph A. Kinney, National Safe Workplace Institute, Chicago, Ill.

*__Dean Kleckner,__ American Farm Bureau Federation, Park Ridge, Ill.

Richard Krumme, *Successful Farming,* Des Moines, Iowa

Elmer W. Learn, University of California, Davis

***Charles K. Mann,** Harvard Institute for International Development, Cambridge, Mass.

Peter Marten, Canadian Federation of Agriculture, Ottawa, Ontario

***Edward E. Masters,** National Planning Association, Washington, D.C.

Edward R. Mauldin, First Colbert National Bank, Muscle Shoals, Ala.

***Vernon R. McMinimy,** Rolling Meadow, Ill.

***John Miller,** Sandy Spring, Md.

Owen J. Newlin, Pioneer Hi-Bred International, Inc., Des Moines, Iowa

Emiel Owens, University of Houston, Houston, Texas

Dean A. Peterson, Marietta, Ga.

Ralph Raikes, Ashland, Nebr.

John M. Reams, Bryan, Cave, McPheeters and McRoberts, Kansas City, Mo.

William E. Reed, Greensboro, N.C.

Don Reeves, Bread for the World, Washington, D.C.

Len Richardson, *California Farmer* Publishing Company, San Francisco, Calif.

Allen Rosenfeld, Public Voice for Food and Health Policy, Washington, D.C.

***George E. Rossmiller,** National Center for Food and Agricultural Policy, Resources for the Future, Washington, D.C.

Vernon W. Ruttan, University of Minnesota, St. Paul

Esther Ruud, Montana Cattlemen's Association, International, Malta, Mont.

Thomas Reese Saylor, Cell Systems Limited, Cambridge, U.K.

Lyle P. Schertz, *Choices,* Herndon, Va.

Leonard Schnell, Nationwide Mutual Insurance Company, Apple Creek, Ohio

Monte Sesker, *Wallaces Farmer,* West Des Moines, Iowa

Harry Snyder, Consumer Union, San Francisco, Calif.

F. Robert Starr, Redwood Falls, Minn.

***John Stencel,** Rocky Mountain Farmers Union, Denver, Colo.

Gregg Suhler, Food and Agricultural Policy Research Institute, University of Missouri–Columbia

Gene Swackhamer, Farm Credit Banks of Baltimore, Baltimore, Md.

Wayne E. Swegle, Winrock International, Morrilton, Ark.

R. Lee Taylor II, Holly Farms Corporation, Memphis, Tenn.

Robert L. Thompson, Purdue University, West Lafayette, Ind.

***Richard Wagner,** GROWMARK, Inc., Bloomington, Ill.

E. Morgan Williams, Scott, McLeod, and Pires, Washington, D.C.

Robert Willis, American Federation of Grain Millers, Minneapolis, Minn.

Ralph S. Yohe, Mt. Horeb, Wis.

HONORARY MEMBERS

Beverly Everett, New Sharon, Iowa

W. E. Hamilton, Wheaton, Ill.

A. C. Hoffman, Bull Shoals, Ark.

Theodore W. Schultz, University of Chicago, Chicago, Ill.

Lauren Soth, The Lodge of Ashworth, West Des Moines, Iowa

MEMBERS ON LEAVE OF ABSENCE

Barbara Huddleston, Via Delle Terme Di Caracalla, Rome, Italy

William H. Yaw, The Farm Clinic, Galesburg, Ill.

---

*Executive Committee